Girt Reiman
2514 Soma Ave.
Bellmore, N.Y. 11710

Girt Reiman

# Principles of Extracorporeal Shock Wave Lithotripsy

# Principles of Extracorporeal Shock Wave Lithotripsy


Edited by

Robert A. Riehle, Jr., M.D.

Assistant Professor of Surgery (Urology)
Cornell University Medical College
Assistant Attending Surgeon
Director, Lithotripsy Unit
The New York Hospital–Cornell Medical Center
New York, New York

Associate Editor

Robert C. Newman, M.D.

Assistant Professor of Surgery
Chief, Clinical Stone Service
Division of Urology
University of Florida College of Medicine
Gainesville, Florida

Churchill Livingstone
New York, Edinburgh, London, Melbourne 1987

**Library of Congress Cataloging-in-Publication Data**

Principles of extracorporeal shock wave lithotripsy.

Includes bibliographies and index.
1. Ultrasonic lithotripsy. I. Riehle, Robert A.
[DNLM: 1. Kidney Calculi—therapy. 2. Lithotripsy.
3. Ureteral Calculi—therapy. WJ 356 P957]
RD574.P75 1987 616.6′22064 87-671
ISBN 0-443-08513-7



Distributed in the United Kingdom by Churchill Livingstone, Robert Stevenson House, 1–3 Baxter's Place, Leith Walk, Edinburgh EH1 3AF, and by associated companies, branches, and representatives throughout the world.

Accurate indications, adverse reactions, and dosage schedules for drugs are provided in this book, but it is possible that they may change. The reader is urged to review the package information data of the manufacturers of the medications mentioned.

Copy Editor: *Margot Otway*
Production Designer: *Jocelyn Eckstein*
Production Supervisor: *Jane Grochowski*

Printed in the United States of America

First published in 1987

Dedicated to

Lee, Erik, Win, Mary Ann,
Mary, Tom, Doug, Zorica,
Liz, Gayle, Joan, Klaus,
Pat, Nancy, Phil, Scott,

and

The New York Hospital–Cornell ESWL Unit

# Contributors

**Henry C. Alder**
Manager, Technology Planning, Division of Clinical Services and Technology, American Hospital Association, Chicago, Illinois

**Herbert Brandl, M.D.**
Assistent der Urologischen Klinik und Poliklinik, Klinikum Grosshädern; Oberarzt der Urologischen Abteilung des Akademischen Lehrkrankenhauses Harlaching, Munich, Federal Republic of Germany

**Christian G. Chaussy, M.D.**
Chief, Department of Urology, Stadt Krankenhaus Harlaching, Munich, Federal Republic of Germany; Co-Director, UCLA Stone Center, UCLA Medical Center, Los Angeles, California

**Prof. Dr. med. Ferdinand Eisenberger**
Ärztlicher Direktor der Urologischen Klinik, Katharinenhospital, Stuttgart, Federal Republic of Germany

**Gerhard J. Fuchs, M.D.**
Visiting Professor of Surgery, Division of Urology, University of California, Los Angeles, UCLA School of Medicine; Co-Director, UCLA Stone Center, UCLA Medical Center, Los Angeles, California

**Patrick T. Hunter II, M.D.**
Clinical Assistant Professor of Surgery, Division of Urology, University of Florida College of Medicine, Gainesville, Florida; Medical Director, MML Mobile Lithotripsy Unit, Central Florida Affiliated Hospitals, Orlando, Florida

**Alan D. Jenkins, M.D.**
Assistant Professor of Urology, University of Virginia School of Medicine, Charlottesville, Virginia

**Priv. Doz. Dr. med. Dieter Jocham**
Oberarzt der Urologischen Klinik, Ludwig-Maximilians Universität, Klinikum Grosshädern, Munich, Federal Republic of Germany

**James E. Lingeman, M.D.**
Director of Research, Methodist Hospital Institute for Kidney Stone Disease, Methodist Hospital of Indiana; Associate Clinical Instructor of Urology, Indiana University School of Medicine, Indianapolis, Indiana

**Martin L. Madorsky, M.D.**
Medical Director, Kidney Stone Center of South Florida; Clinical Instructor of Urology, University of Miami School of Medicine, Miami, Florida

**J. David Mullins, AIA, B.Arch.**
J. David Mullins, AIA, Inc., Consulting Architects and Planners for Health Care, Peachtree City, Georgia

**Erik Näslund**
Research Associate, Division of Urology, The New York Hospital–Cornell Medical Center, New York, New York

**Robert C. Newman, M.D.**
Assistant Professor of Surgery, and Chief, Clinical Stone Service, Division of Urology, University of Florida College of Medicine, Gainesville, Florida

**Dr. med. Jens Rassweiler**
Physician, Urologische Klinik, Katharinenhospital, Stuttgart, Federal Republic of Germany

**Robert A. Riehle, Jr., M.D.**
Assistant Professor of Surgery (Urology), Cornell University Medical College; Assistant Attending Surgeon and Director, Lithotripsy Unit, The New York Hospital–Cornell Medical Center, New York, New York

**Thomas W. Schoborg, M.D.**
Assistant Clinical Professor of Surgery, Division of Urology, Medical College of Georgia School of Medicine, Augusta, Georgia; Director, Atlanta Stone Center, and Chief, Division of Urology, Georgia Baptist Medical Center, Atlanta, Georgia

**Keith N. Van Arsdalen, M.D.**
Assistant Professor of Surgery, Division of Urology, University of Pennsylvania School of Medicine; Director, Extracorporeal Shock Wave Lithotripter Center, Hospital of the University of Pennsylvania, Philadelphia, Pennsylvania

# Foreword

In the recent past, we have experienced a tremendous change in the management of symptomatic urolithiasis, which has led us to completely reassess our traditional treatment concepts. Beyond any doubt, the surgical skills to perform open surgical procedures for stone removal and/or correction of underlying anatomic abnormalities must continue to be a basic cornerstone in the urologist's therapeutic armamentarium. However, noninvasive treatment of urinary stones has become the treatment of choice in the vast majority of symptomatic stone cases. Open surgery has rapidly been superceded by percutaneous and extracorporeal lithotripsy, and has become the exception at those centers where all treatment modalities are available.

Public opinion has accepted the new treatment concepts very quickly. Government officials, journalists, and health care analysts have treated noninvasive stone disintegration as something of a medical miracle.

This book is an impressive collection of clinical experience concerning all aspects of ESWL. Each author clearly and effectively emphasizes that there is still more to modern stone treatment than pressing buttons. For this reason it should be mandatory for every contemporary urologist to study this book.

Six years ago we performed the first successful ESWL treatment in humans, after extensive experimental research. After 1983 the high success rate and low complication rate allowed the range of indications to be carefully expanded. Since then, more than 150,000 treatments have been performed worldwide.

In the initial phase of machine distribution, centralized referral centers accumulated a high degree of experience within a short period of time. General indications were developed and endorsed by these lithotripsy centers, and, in general, treatment guidelines are uniformly followed. Nevertheless it is interesting to observe the various modifications and tricks described by the authors in this book. These step-by-step comments and advice concerning patient selection and education, secondary procedures, and indications for endoscopy after ESWL were generated at university centers after careful analysis of prior and current clinical experience. They should serve as a guide for decision making for individuals at the very beginning of their learning curve for ESWL. New operators should not incorporate unproven and often hazardous variations on the treatment guidelines. Originality neither can nor may ever be sought at the cost of the patient's well-being and safety.

Obviously, stepwise evolution of the ''art'' of lithotripsy will continue to emerge from the initial indications which are backed by reproducible results. New styles and techniques, including ureteral stenting, tubless lithotripters, and laser-tripters will be developed and implemented.

A word must be said about outpatient ESWL. There is indeed the theoretical possibility that a certain percentage of stone patients may receive ESWL treatment as an outpatient procedure. However, as for any other procedure, the appropriate indications and exclusion criteria must be established on behalf of the patient in an unbiased way, and results must be compared to the safety, reliability, and reproducibility of inpatient therapy. Having had myself quite some experience with ESWL, starting in 1980 with carefully selected patients (stones $< 10$ mm, no medical-risk patients) and gradually enlarging the range of indications, I am truly impressed with the results reported for outpatient ESWL, even at the beginning of the learning curve. Certainly these results from outpatient facilities can contribute to a redefinition of the strategies for the use of ESWL. Nevertheless, precise analysis of treatment results and follow-up must continue. At least for me, there are still questions which need answers. Is the secret of the virtual absence of periprocedural complications explained by the coincidence of a 48-hour stenting period and the definition of a primary 48-hour follow-up period? The crucial question then is, what happens after stent removal?

We do not know for certain whether we have reached the end-point of development of ESWL and its clinical applications for the treatment of urinary stone disease. I suspect that this book, written by urologists for urologists, represents rather a starting point for future technologic advances than a final statement on ESWL. However, it is certain that this progress made by urologists, for the sake of urology, represents a dramatic change in the management of urinary stones. Invasive approaches to symptomatic calculi have been superceded by a noninvasive procedure. From now on, all other methods employed for the treatment of kidney and ureteral stones will have to be judged against the results of this new methodology.

*Christian G. Chaussy, M.D.*
Munich and Los Angeles
June, 1986

# Preface

A book about ESWL, or a computerized weekly newsletter? In the decade of zap mail, video text and data banks, word processors, and modem links, a new technology such as shock wave lithotripsy evolves so quickly that a book covering the clinical application of the technique is almost obsolete before its publication date. The incredible speed at which 60 Dornier lithotripters were installed and brought into operation in the U.S.—before the first journal article on the United States experience with ESWL appeared—contrasts with the 9-month lag in journal publications and the 1½ year interval between the signing of a book contract and the shipping of the bound book to the purchasers.

Nevertheless, a record of the first clinical experiences and developments is still needed—as a textual touchstone to educate lithotripter users and to serve as a foundation of fact for future developments, modifications, and treatment evolution. This book also records both the sound principles that have been developed for safe administration of shock waves to humans and the success achieved to date, which must serve as a record for second-generation technologies to equal or exceed.

At this moment, new-generation extracorporeal lithotripters are under investigation, but only preliminary results are available. Over the last 2 years, the rumors and press releases concerning the new methods of lithotripsy have outweighed the facts and results, and consequently discussion of these devices has been left to other authors and future publications.

The idea for this book evolved from the insight of a friend. Over a pasta dinner, Dr. John Sherman, a creative and accomplished plastic surgeon with more ideas than he himself can implement, expressed astonishment that a book on the United States experience with ESWL was not forthcoming. By the time brandy arrived and the coffee cups were cleared, he had selected my publisher, outlined the chapters, and projected the publication date. All I had to do was fill in the blanks.

The authors were selected with an intent to involve young endourologic surgeons who had direct and continuous experience with shock wave lithotripsy for clinical symptomatic stone disease. The authors for this book know their topics well because each was instrumental in either developing the lithotripter or introducing the technology to the U.S. They have lived through the experiences about which they write. The application of lithotripsy far preceded the literature documenting its success, so in lieu of a long, computer-researched bibliography, they write from their own successes and failures, as surgeons have done for centuries.

I am indebted to the entire crew at ESWL Nireinstein Therapie, Klinikum Grosshädern, Munich, F.R.G. for their patient translation and instruction about ESWL, their advice concerning Fasching customs, and their instruction as to which mustard to use on weisswurst at noon.

My sincere appreciation to Dornier GmbH for the development of a quality product that is yet to be equaled, and especially to Eckhard Polzer and Klaus Koelsch whose expertise, attention, concern, and friendship epitomize the precise yet personal attitude of Dornier Medical Systems, Inc.

A special note of thanks to Robert Newman, who joined me as associate editor and contributed selflessly to our final product. Also, special gratitude to Christian Chaussy who endorsed this project and contributed with interest and a personal sincerity that has made him friend and admired colleague to many.

Thanks also to E. Darracott Vaughan, M.D. and William Fair, M.D. who dealt with national, state, hospital, and governmental politicians, insurers, financiers, investigators, and consultants, and left me to the patients, the machine, and the manuscripts.

A special acknowledgment to Birdwell Finlayson, M.D., Ph.D. and J. S. Gravenstein, M.D. (University of Florida at Gainesville) for review of portions of the manuscript and suggestions concerning revisions.

Special thanks to Mary Lombardi, our first and only ESWL surgical assistant, whose personal warmth and knowledge served to educate and manage both patients and visiting physicians unfamiliar with the technology.

Certainly, the ultimate appreciation and admiration to Erik Näslund, whose computer program allowed the analysis of the New York Hospital–Cornell Medical Center ESWL experience and whose energy and inimitable style challenged and refreshed me when my enthusiasm thinned.

*Robert A. Riehle Jr., M.D.*
*New York, New York*

# Contents

# 1

# Historical Development of ESWL

Dieter Jocham

The first clinical application of extracorporeal shock wave lithotripsy (ESWL) was on February 2, 1980 in Munich. Extensive experimental investigations had preceded and had proven the safety and reproducibility of in vivo shock wave application. In Munich, the initially restrictive indications for ESWL were extended, and between 1980 and 1982, the internationally valid indications for the ESWL were established. These have not changed to date. Fortunately, it can be stated that the treatment strategies and therapeutic results of this early clinical phase were fully confirmed by the results provided by the numerous subsequent ESWL users.[1–8,9] This noninvasive method for removal of urinary stones has now gained wide and unusually rapid acceptance among urologists. Numerous lithotripters (November 1986, $n = 228$) have been installed in many countries worldwide. By the end of November 1986, with their help, more than 250,000 urinary stone carriers had been successfully treated.

Without question, the concept of ESWL has revolutionized the urologic treatment of urinary stone disease. Coupled with valuable auxiliary techniques such as percutaneous lithotripsy and ureterorenoscopy, ESWL has reduced the necessity of open operative stone removal to less than 10 percent of cases in which the stones would previously have been removed surgically. Although today ESWL is quite accepted in the strategy of urinary stone therapy, few are aware of the idea source for the medical application of shock waves and the difficulties of developing a clinically practical lithotripter. Therefore, a few aspects of the historical development of ESWL are outlined below.

## EARLY USE OF SHOCK WAVES

In 1969 the idea of urinary stone destruction aided by shock waves was conceived. At that time only a few routine applications of shock waves were known. In medicine the only application of shock waves was the so-called "Urat" apparatus, the Russian lithotripter for endoscopic destruction of bladder stones by means of direct contact between the shock wave source and the stone. Further developments were limited because of the shock waves properties generated in air. These properties are quite different from shock waves generated in fluid or solid media.

Physicists at Dornier Systems, Ltd. in Friedrichshafen, F.R.G., had already been doing experimental work since 1963 regarding questions of air and space travel and the shock wave phenomenon associated with high velocity. A broad infrastructure for research on shock wave effects had been established. The awareness of the effect of shock waves on human tissues arose accidentally. In 1966, a Dornier engineer felt a sensation similar to an electrical shock by touching a shock wave target at the exact moment of the shock wave action. This incident stimulated further investigations which showed that it was shock waves entering the body that elicited those sensations.

State-financed studies of the potential degradation of tissues resulting from shock waves were conducted in animals at Dornier Systems, Ltd. from 1969 to 1971. These studies showed, among other things, that shock waves generated in water were transmitted into and through the body without appreciable energy loss. Only the lung tissue demonstrated sensitivity to the shock waves because of a high acoustical impedance at the air–tissue interface. Brittle materials within the body (i.e., measuring probes) were easily destroyed by the shock waves. From this background of detailed observation, the idea of using shock waves for the treatment of urinary stones was born. Exact information about the origin of the idea is not available; however, it is known that the possibility of ESWL for the treatment of kidney stone disease was first discussed by three physicists, Armin Behrendt, Eberhard Häussler, and Günther Hoff. Not until 1972 did in vitro studies prove the destructibility of human kidney stones by shock waves transmitted in water.

It is seen as a great service from Professor E. Schmiedt, the director at the urologic clinic of the Ludwig-Maximilians University, Munich, that he, in contrast to many other urologists, recognized at that time the revolutionary potential of the application of shock waves for the disintegration of kidney stones. It is thanks to his efforts and that of his then co-worker, Prof. Eisenberger, that after the passage of only 1 more year, in 1974, the necessary medical/technical research project with government support was begun. That project was a collaborative effort between Dornier Systems, Ltd., the urologic clinic, and the Institute for Surgical Research located at Ludwig-Maximilians University in Munich. The studies were conducted at the Institute for Surgical Research (Brendel). Figure 1-1 shows the contract between the vari-

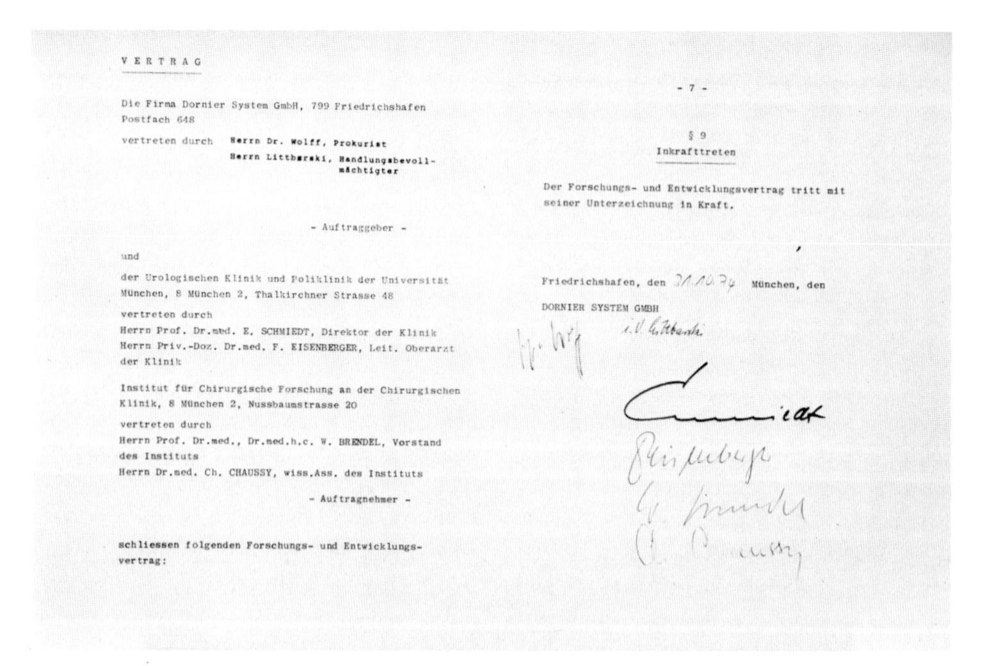

V E R T R A G

Die Firma Dornier System GmbH, 799 Friedrichshafen
Postfach 648

vertreten durch Herrn Dr. Wolff, Prokurist
Herrn Littbarski, Handlungsbevollmächtigter

\- Auftraggeber -

und

der Urologischen Klinik und Poliklinik der Universität München, 8 München 2, Thalkirchner Strasse 48

vertreten durch
Herrn Prof. Dr.med. E. SCHMIEDT, Direktor der Klinik
Herrn Priv.-Doz. Dr.med. F. EISENBERGER, Leit. Oberarzt der Klinik

Institut für Chirurgische Forschung an der Chirurgischen Klinik, 8 München 2, Nussbaumstrasse 20

vertreten durch
Herrn Prof. Dr.med., Dr.med.h.c. W. BRENDEL, Vorstand des Instituts
Herrn Dr.med. Ch. CHAUSSY, wiss.Ass. des Instituts

\- Auftragnehmer -

schliessen folgenden Forschungs- und Entwicklungsvertrag:

\- 7 -

§ 9
Inkrafttreten

Der Forschungs- und Entwicklungsvertrag tritt mit seiner Unterzeichnung in Kraft.

Friedrichshafen, den München, den

DORNIER SYSTEM GMBH

Fig. 1-1. Documentation of the cooperative agreements between technical and medical partners (1974).[4]

Fig. 1-2. Ellipsoid of the first experimental shock wave lithotripter with an electrode. At the second focal point of the ellipsoid is a human kidney stone in a plastic container.

ous institutions in 1974. Professor Chaussy, who at that time was on the staff at the Institute for Surgical Research, was assigned to the urologic clinic and coordinated an ever-broadening range of research activities over several years under the direction of Professor Schmiedt. This author has worked on this project since 1977.

In 1974, the technical principle of focusing shock waves generated in a water bath was discovered. The focusing was achieved with the help of a water-filled metal semi-ellipsoid, the geometry of which is still in use (Fig. 1-2).

## EXPERIMENTAL LITHOTRIPTER MODELS

The first experimental lithotripter had no water bath. Instead, the water-filled semi-ellipsoid was covered with a rubber membrane through which the shock wave was introduced into the body of an animal (Fig. 1-3). This system had no positioning mechanism, and the rubber membrane did not permit effective transmission of the shock wave into the body. This led to the construction of a new lithotripter model in which the research animal was placed in a water bath (Fig. 1-4). Since this lithotripter had no positioning mechanism, treatment was only sporadically successful. The integration of a positioning device necessitated the construction of a completely redesigned lithotripter. It seemed appropriate to use ultrasonography for target location. Therefore, a compound scanner was situated beside the animal's body and the image plane was arranged to coincide with the plane of the second focus of the shock wave wave front (Fig. 1-5). This technique was only partially successful for initial localization of the stone, and satisfactory control of stone disintegration was still not possible. From 1976 to 1977, much time was taken to optimize the ultrasonic localization technique. Finally, it was recognized that ultrasonic localization in the lithotripter water bath was simply not succeeding.

Further attempts were aimed at exploring the potential of an x-ray localization system. The requirement for a three-dimensional stone localization and the necessity of optimizing the image without distortion of the x-rays in the water bath presented many problems. In spite of this, a new lithotripter model (Fig. 1-6) was built in early 1978 with two integrated x-ray systems for localizing stones at the second focus of the shock waves. This permitted increasingly reproducible disintegration of stones in the animal model (Fig. 1-7). Chaussy and co-workers have reported extensively on the years of the experimental phase concerning the in vitro and in vivo animal experiments that successfully demonstrated application of ESWL in vivo.[1–8,10–20] Some of the main experimental knowledge and results of that time are outlined as follows:

1. Urinary stones of different chemical compositions were reproducibly disintegrated when placed in the focal area of the shock wave.

Fig. 1-3. Shock wave lithotripter with a rubber membrane over the ellipsoid (right).

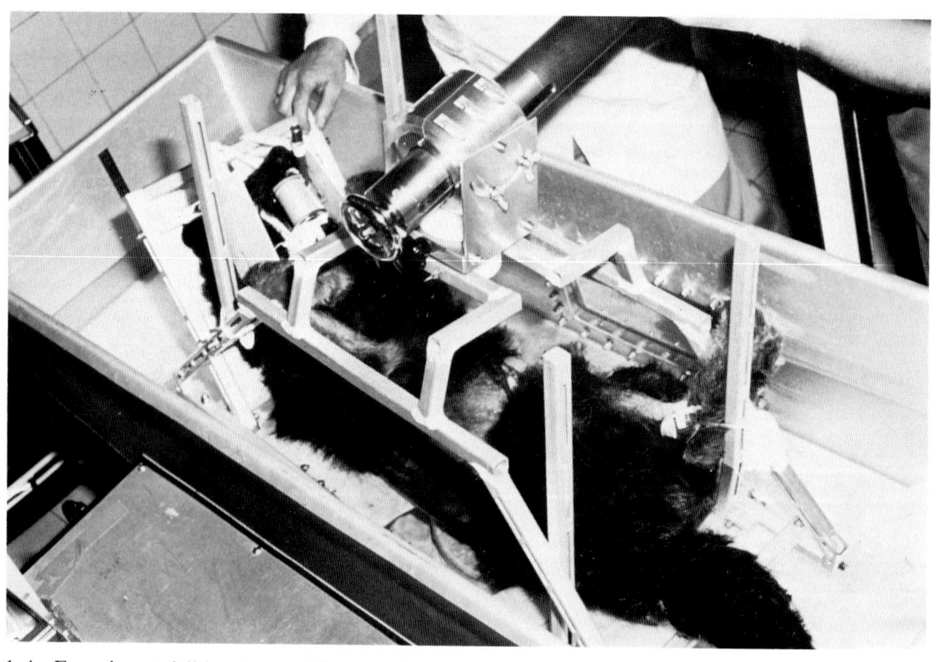

Fig. 1-4. Experimental lithotripter with water bath and movable sling over the ellipsoid. In the sling is an anesthetized animal with one of the kidneys implanted with a human kidney stone.

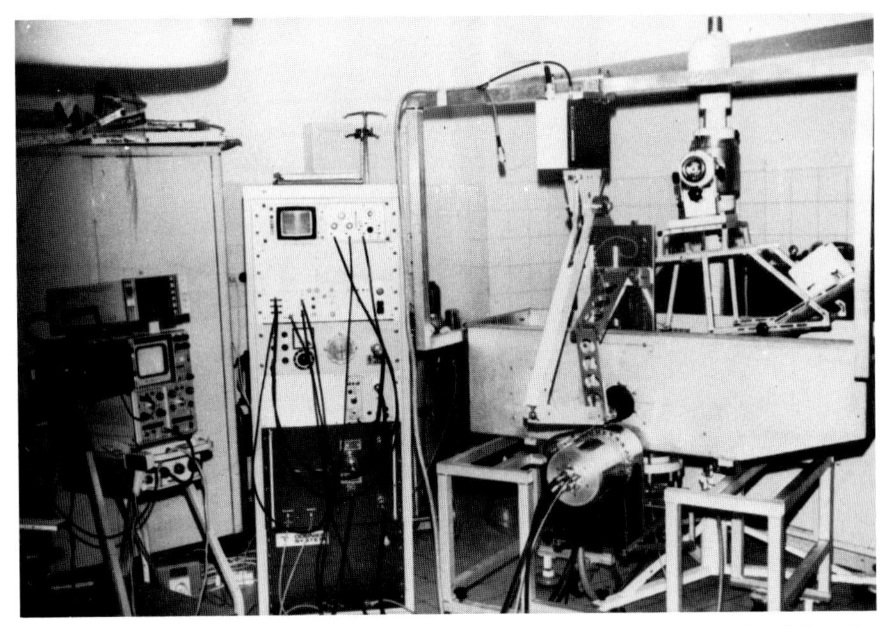

Fig. 1-5. Experimental shock wave lithotripter (right) with ultrasound location device (left and center).

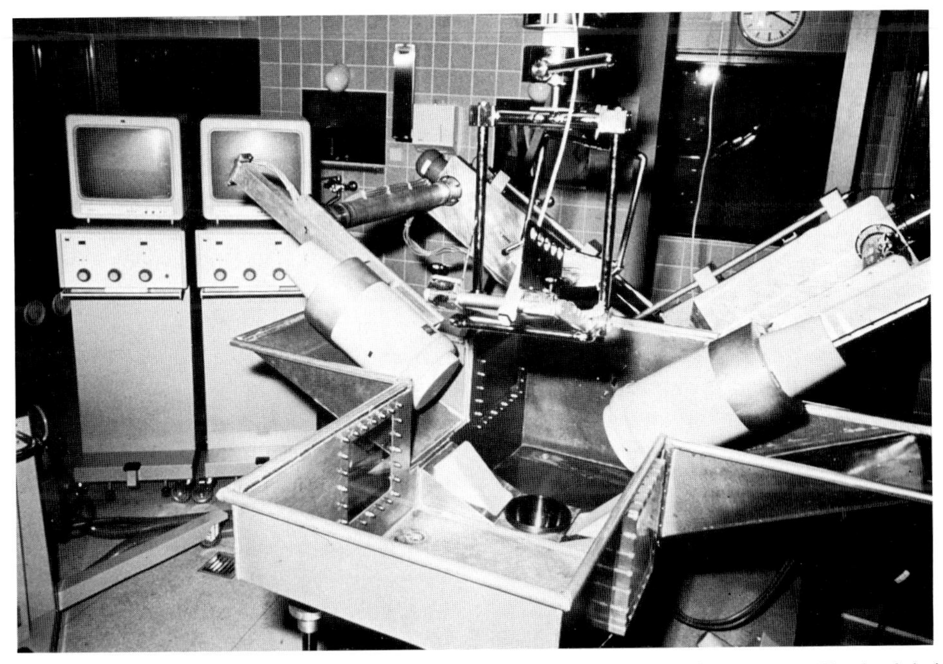

Fig. 1-6. Experimental shock wave lithotripter with integrated x-ray location system. To the left is the attached x-ray monitor.

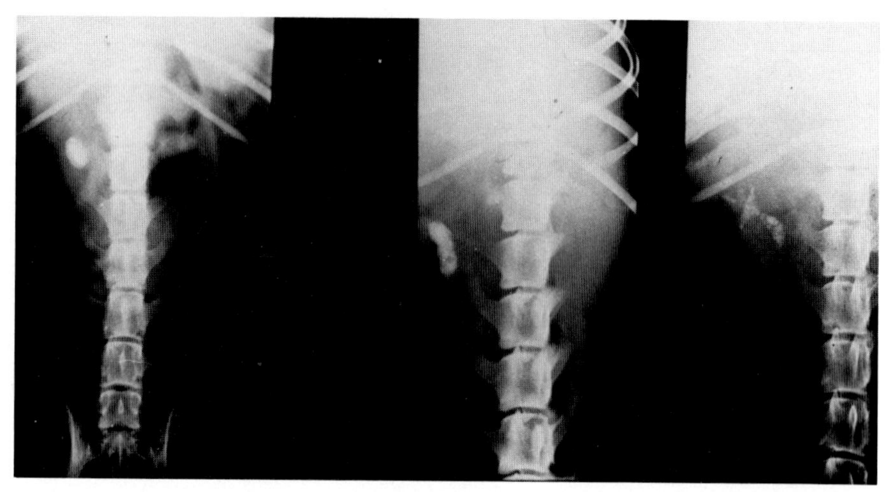

Fig. 1-7. Radiograph of a shock wave lithotripsy in a dog model. Left: the stone in the right kidney before shock wave lithotripsy. Center and right: stone disintegration showing reduced portions of the pelvic concretion.

2. Exposure of fresh blood samples demonstrated some hemolysis with repeated shock wave treatment. Treatment of isolated organs including liver, kidney, intestine, and bone failed to demonstrate any deleterious effect.

3. These results were confirmed with shock wave application in the rat. With the exception of the exposure of the thoracic region, which led to pulmonary alveolar damage, there was no tissue damage, either from specific exposure of individual organs or general exposure of the abdominal region.

4. Mixed lymphocyte cultures exposed to shock waves did not demonstrate cytolysis. There was no significant reduction in the proliferation rate induced by mitogen- or lymphocyte-derived different irradiated responder cells.

5. After development of an experimental stone model in dogs, in vivo urinary stone destruction was achieved by shock wave exposure. It was proved that kidney stones were destroyed with shock waves without any pathologic changes in either the kidney or the surrounding tissue.

6. After unsuccessful experiments with ultrasound for stone localization, it was shown that reproducible localization of kidney stones could only be obtained through the integration of a biplanar x-ray system.

On the basis of these and numerous other experimental findings and according to the judgment of independent experts, it seemed justified to pass the threshold to clinic application. Against this backdrop, the first clinically proven human lithotripter model with an integrated x-ray localization system (Fig. 1-8) was installed in a room of the Institute for Surgical Research of Ludwig-Maximilians University, Munich, Klinikum Grosshädern. Based on sound animal experimental data, the first human lithotripter model with an integrated x-ray localization system was built in September 1979 (Fig. 1-8).

The completion of more animal studies in this device and an improvement of the patient support device to overcome the buoyancy of anesthetized patients in the water bath permitted the first human application. This was successfully conducted in the spring of 1980 by Christian Chaussy and Dieter Jocham at the urologic clinic, Ludwig-Maximilians University, Munich, West Germany.

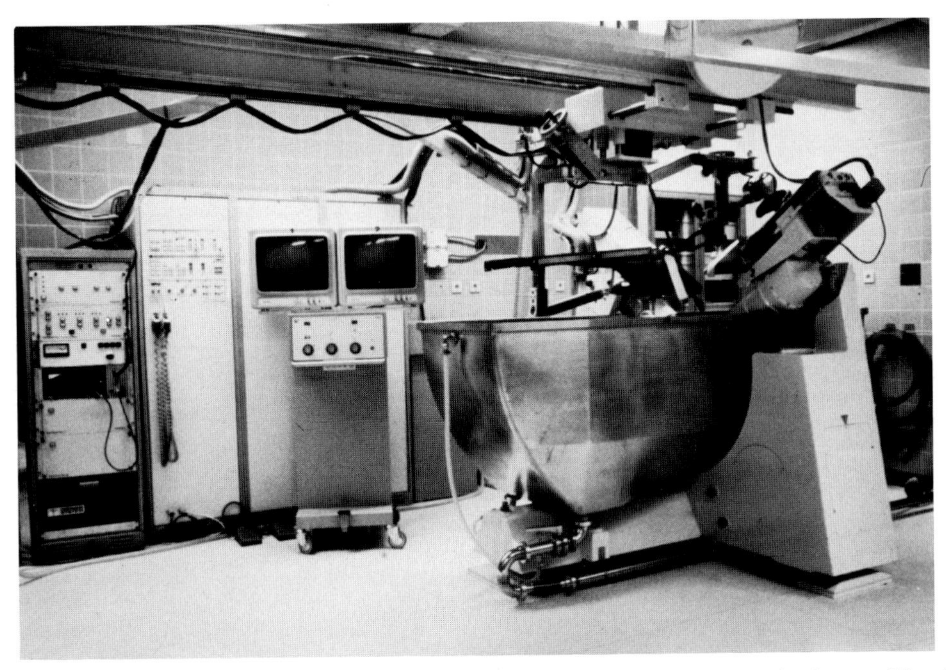

Fig. 1-8. Human model 1 (HM-1) as a prototype for the recently developed modern shock wave lithotripter. The first shock wave lithotripsy of a human was conducted with the HM-1 in February 1980.

## CURRENT USE OF ESWL

Clinical experience gained on the HM-1 model led to further improvements resulting in the construction of another clinical prototype, HM-2, and finally the building of the production model, HM-3, by Dornier Medical Systems, Inc. (Fig. 1-9). Even during this phase of clinical ESWL application performed exclusively at the Munich clinic, it was possible to prove, on a comprehensive scale, the safe and reproducible application of the procedure to treat the large spectrum of urolithiasis. Within a relatively short time, the initial compilation of contraindications (e.g., infected stones, size greater than a cherry, high risk patients, ureteral stones, obstruction, translucent stones) could be reduced to cases requiring open surgery due to an obstruction located distal to the stone and blood coagulation disorders that could not be corrected by therapeutic measures. A great number of visitors from all over the world who came to the Munich stone center had the chance of seeing for themselves the effectiveness and safety of the procedure. Last, but not least, one has to ascribe this to the circumstance that the ESWL had gained rapid worldwide acceptance and dissemination in an especially impressive way.

After 2½ years of using ESWL exclusively in the urologic clinic at Grosshädern, a second lithotripter was installed in the fall of 1983 in the urologic department of the Katharinenhospital in Stuttgart (director, Prof. F. Eisenberger). The installation of more ESWL units in Germany followed in rapid succession (Fig. 1-10, 1-11). The first lithotripter in the United States was installed in March 1984 in Indianapolis, Indiana. As of February 1986, 133 machines were in operation worldwide (Tables 1-1, 1-2). Future developmental possibilities of this device are as unpredictable as the changing pattern of kidney stone disease. We live in a time in which different formulations for shock wave generation and for stone localization are being discussed and implemented toward development of an improved type of machine. These develop-

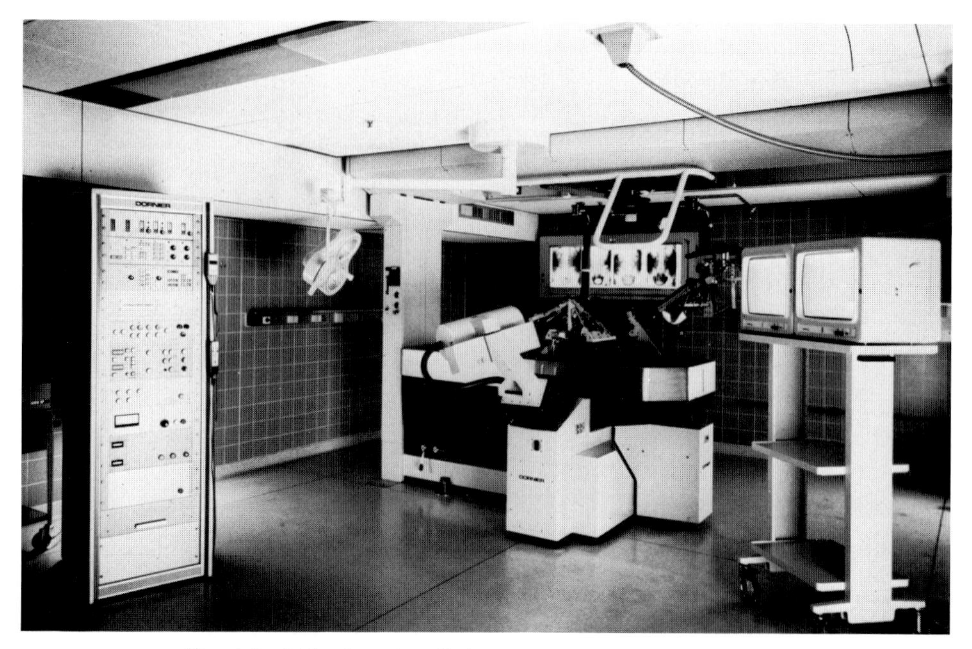

Fig. 1-9. Lithotripsy installation HM-3 model (Dornier type).

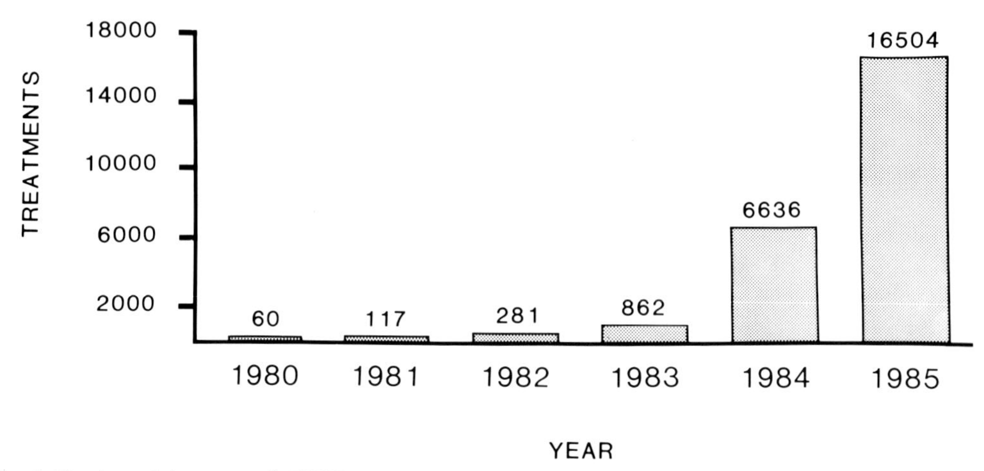


Fig. 1-10. Annual increase of ESWL treatments in the FRG since the first clinical application of this method in Munich in February 1980. This increase was made possible by the growing number of ESWL centers (20 by 1985) and the greater number of treatments performed in each center because of ever-increasing routine and better organizational forms.

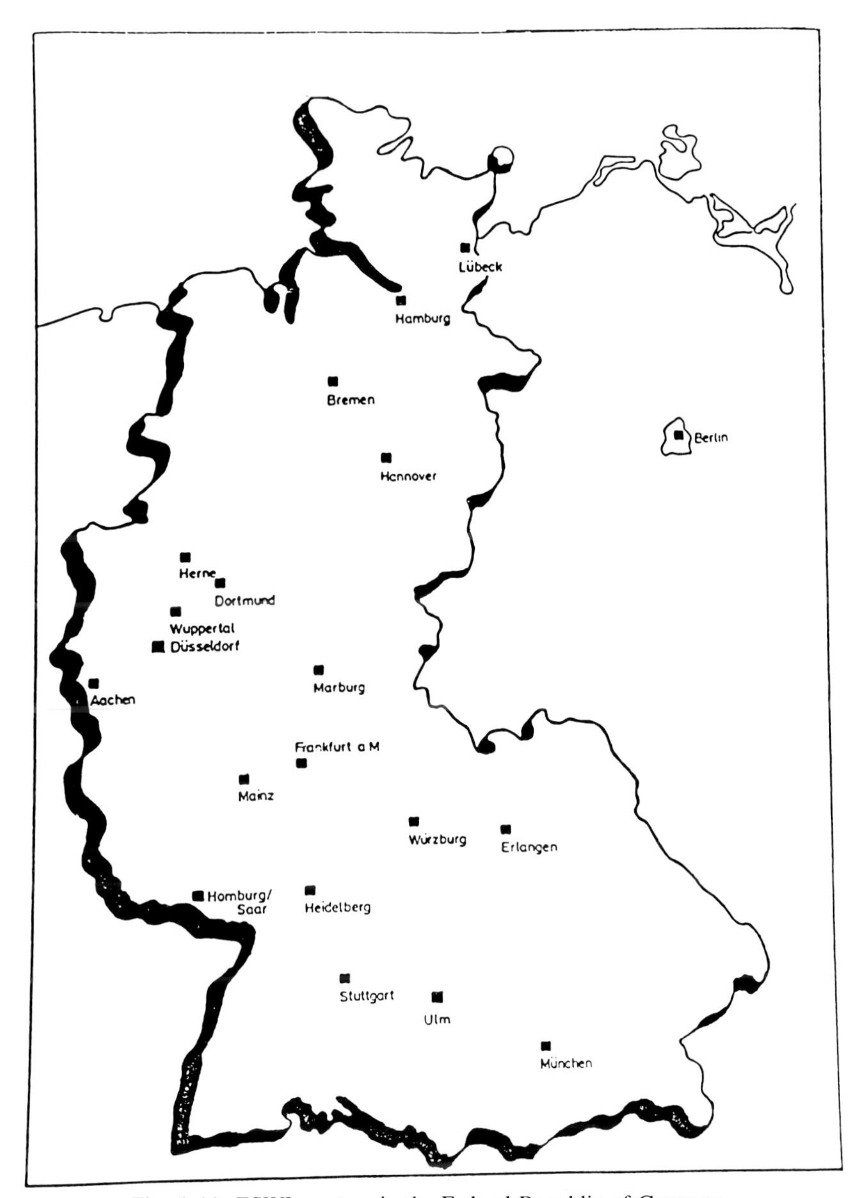


Fig. 1-11. ESWL centers in the Federal Republic of Germany.

**Table 1-1. Historical Background of ESWL**

| | |
|---|---|
| 1966 | Accidental discovery of shock wave effect on humans |
| 1969 | Emergence of the idea of stone disintegration |
| 1969–71 | Research programme, "interaction between shock waves and tissue in animals," sponsored by the Ministry of Defense |
| 1972 | First in vitro experiments for stone disintegration |
| 1972 | Letter from Professors Schmiedt and Eisenberger stating, "It would be revolutionary to disintegrate kidney stones without surgery." |
| Jan 1974 | Start of the ESWL research project sponsored by the Federal Ministry of Research and Development |
| 1974–78 | Various lab models of lithotripter with water bag and ultrasound system for stone location |
| 1978 | First water bath model with a two-axis x-ray system for animal experiments |
| 1978–79 | Animal experiments |
| 1979 | Lithotripter model HM-1 |
| Feb 1980 | Treatment of the first patient |
| 1980–82 | Treatment of 220 patients with HM-1 |
| May 1982 | Inauguration of the model HM-2 in Munich |
| Oct 1983 | Inauguration of the first series model HM-3 in Stuttgart |
| 1983–84 | IDE procedure in USA |
| Dec 1984 | Pre-market approval from the FDA |
| Nov 1985 | Approval from the Japanese Ministry of Public Welfare |
| Feb 1986 | 130 Dornier lithotripters in operation worldwide |
| Feb 1986 | More than 90,000 patients treated |

**Table 1-2. Worldwide Distribution of Dornier Lithotripters as of February 1986[a]**

| | |
|---|---|
| USA | 62 |
| Germany | 20 |
| Italy | 7 |
| Spain | 9 |
| Japan | 9 |
| Saudi-Arabia | 5 |
| Austria | 3 |
| Switzerland | 3 |
| France | 3 |
| Great Britain | 2 |
| Netherlands | 1 |
| Norway | 1 |
| Sweden | 1 |
| Egypt | 1 |
| Hong Kong | 1 |
| Singapore | 1 |
| Canada | 1 |
| Yugoslavia | 1 |
| Taiwan | 1 |
| Australia | 1 |
| Total | 133 |

[a] Other lithotripter systems are only in the initial clinical investigational phase.

ments are welcomed and reflect technical advances. However, even in the face of such a rapidly advancing technology, one should not forget the sources of ESWL—the accidental contact of a technician with a shock wave target, the innovative visions of nonmedical engineers, the willingness of doctors to adopt unconventional ideas, and finally, the close cooperation between technology and medicine. Last, but not least, a project of this magnitude can only be successfully conducted with adequate financial support which, in the case of the development of ESWL, was provided by the government ministry for research and technology of West Germany.

## ACKNOWLEDGMENT

The support of Dr. W. Hepp is gratefully acknowledged by the author.

## REFERENCES

1. Chaussy C, Schmiedt E, Brendel W: Instrumentelle Harnsteinentfernung. In Vahlensieck W. (hrsg): Urolithiasis 2. Springer, Berlin, 1979
2. Chaussy C, Forssmann B, Brendel W et al: Berührungsfreie Nierensteinzertrümmerung durch extrakorporal erzeugte fokussierte Stosswellen. In Chaussy C, Staehler G (eds.): Beiträge zur Urologie. Karger, Basel, 1980
3. Chaussy C, Brendel W, Schmiedt E: Lancet 1265, 1980
4. Chaussy C, Schmiedt E, Jocham D et al: Dtsch Ärztebl 18:881, 1981
5. Chaussy C (ed): Extracorporeal Shock Wave

Lithotripsy—New Aspects in the Treatment of Kidney Stone Disease. Karger, Basel, 1982

6. Chaussy C, Schmiedt E, Jocham D et al: Therapiemöglichkeiten des Harnsteinleidens mit extrakorporal erzeugten Stosswellen. In Pathogenese und Klinik der Harnsteine, Bd. 8: Fortschritte der Urologie und Nephrologie. Steinkopff, Darmstadt, 1982
7. Chaussy C, Schmiedt E, Jocham D et al: J Urol 127:417, 1982
8. Chaussy C, Schmiedt E, Jocham D: Nonsurgical treatment of renal calculi with shock waves. In Roth RA, Finlayson B. (eds): Stones—clinic management of urolithiasis. William & Wilkins, Baltimore, 1982
9. Riehle RA Jr., Fair WR, Vaughan DE, Jr.: Extracorporeal shock wave lithotripsy for upper urinary tract calculi. One year's experience at a single center. JAMA 255:2043, 1986
10. Chaussy C, Eisenberger F, Wanner K et al: Urol Res 4:175, 1976
11. Chaussy C, Eisenberger F, Wanner K et al: In vitro-Untersuchungen und erste in vivo-Untersuchungen mit fokussierten Stosswellen. Biophysikalische Verfahren zur Diagnose und Therapie von Steinleiden der Harnwege. Wissenschaftliche Berichte, Meersburg, 1976
12. Chaussy C, Eisenberger F, Wanner K, Forssmann B: Aktuel Urol 9:95, 1978
13. Chaussy C, Eisenberger F, Wanner K: Urologe (Ausg A) 16:35, 1977
14. Chaussy C, Wieland W, Jocham D et al: Verh Dtsch Ges Urol 30:333, 1978
15. Chaussy C, Schmiedt E, Forssmann B, Brendel W: Eur Surg Res 11:36, 1979
16. Eisenberger F, Chaussy C, Wanner K: Entwicklung eines steintragenden Hundemodells zur in vivo-Untersuchung der Wirkung fokussierter Stosswellen auf Nierensteine. Biophysikalische Verfahren zur Diagnose und Therapie von Steinleiden der Harnwege. Wissenschaftliche Berichte, Meersburg, 1976
17. Eisenberger E, Schmiedt E, Chaussy C et al: Dtsch Ärztebl 17:1145, 1977
18. Eisenberger F, Chaussy C, Wanner K. Aktvel Urol 8:3, 1977
19. Forssmann B, Hepp W, Chaussy C et al: Verh Dtsch Ges Pathol 11:4, 1976
20. Forssmann B, Hepp W, Chaussy C: Biomed Technol 22:164, 1977

# 2

# The Physics and Geometry Pertinent to ESWL

Patrick T. Hunter, II

In the past, the unfocused shock waves of electrohydraulic lithotripsy (EHL) have been used to treat urinary calculi.[1,2] Chaussy and his colleagues subsequently developed a method of using focused shock waves to fragment urinary calculi noninvasively in vivo.[3–6] This technique, extracorporeal shock wave lithotripsy (ESWL), is rapidly emerging as the treatment of choice for most urinary tract calculi.

During EHL and ESWL, shock waves are created by discharging electricity across two electrodes. The EHL probe consists of coaxial electrodes separated by insulation, whereas the ESWL shock plug uses two opposing electrodes surrounded by a Faraday-like cage which electrically isolates the electric discharge. In either case, the underwater discharge or spark vaporizes the water near the electrodes so rapidly that the surrounding water is pushed away with a sudden jolt that creates a shock wave. Some time later, the gas bubble collapses and a second wave occurs.

Near the electrode tips, the shock waves are very strong. However, as the shock waves travel radially away from the tips, the shock wave energy becomes weaker. Because the shock wave impact must be great enough to crack a human calculus, either the electrodes must be very near the calculus or the shock waves must be focused onto the calculus if the electrodes are distant.

With EHL, the probe is placed near the calculus endoscopically so that the shock wave impact will be great enough to cause calculus fragmentation. With the Dornier extracorporeal lithotripter, the target calculus is fluoroscopically positioned at F2, the second focus of a semi-ellipsoid reflector, while the shock plug electrodes are located at F1, the other focus of the ellipsoid (see Fig. 2-1). The shock waves are created at F1 and are reflected and geometrically concentrated at F2. With ESWL, the shock wave emerges from the mouth of the ellipsoid reflector in the shape of a concave arc. This arc becomes smaller, more concentrated, and more powerful as it travels toward F2. With the patient correctly positioned and with the calculus located at F2, the arc-shaped shock wave enters the body with very little attenuation of energy. The arc becomes so concentrated on the calculus that the hydraulic impulse of the shock wave is sufficient to crack it. It is the ellipsoid reflector that concentrates the shock wave and permits it to be created extracorporeally. It is the water bath that couples the energy created at F1 to its target at F2 and provides the correct electrode environment for shock wave creation.

Following the introduction of ESWL in the

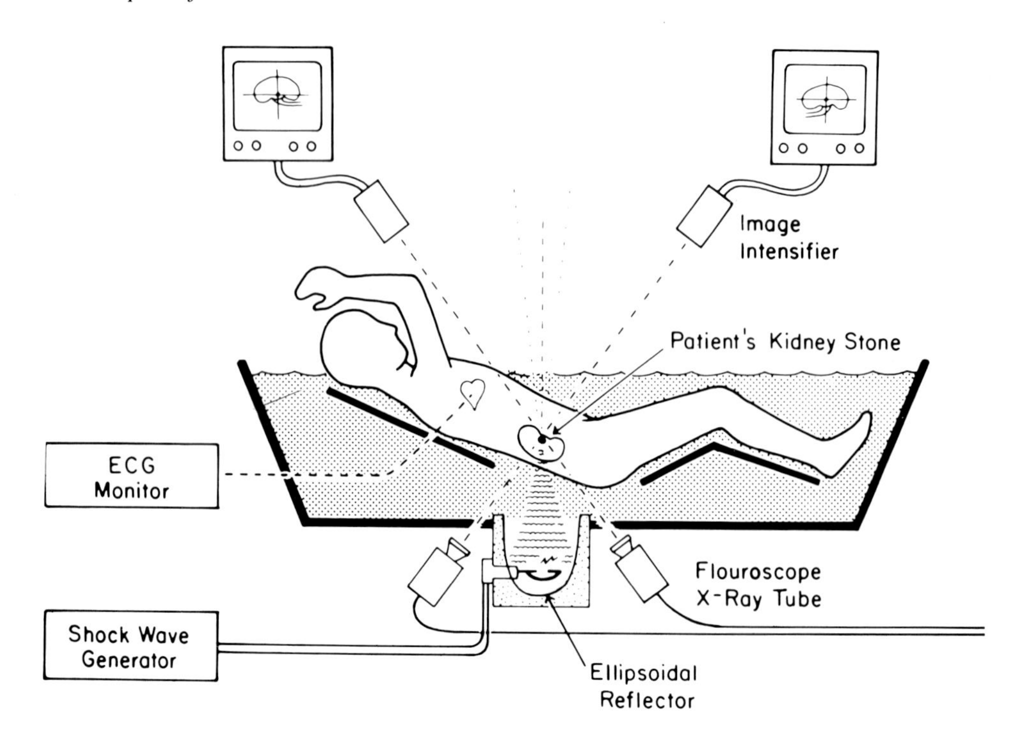


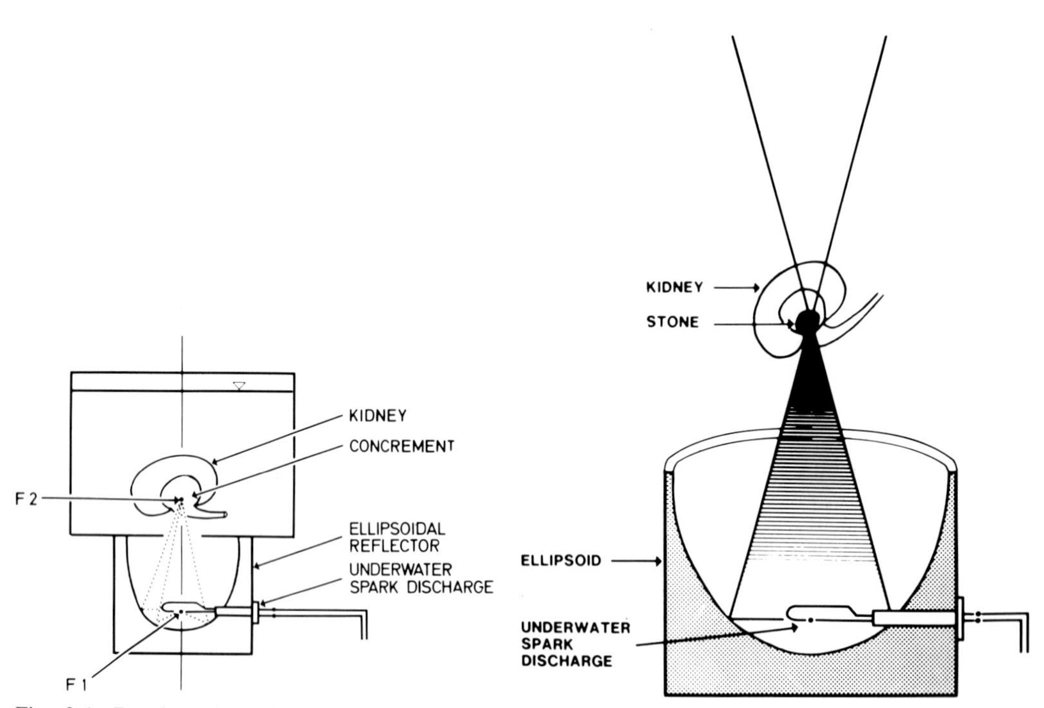


Fig. 2-1. Dornier schematic of shock wave propagation; shock wave is created at F1, reflected by the ellipsoid, and concentrated at F2 onto the calculus.

United States, experiments were undertaken at the University of Florida to further study shock waves. In addition, the use of microexplosive stone fracturing techniques (MEL), similar to those used by Watanabe,[7] have been investigated. This chapter outlines some of this work as well as discusses some of the physics and geometry pertinent to ESWL.

## SHOCK WAVE EXPERIMENTS

The goal of our initial investigation was to describe shock wave pressures as a function of lithotripter input generator voltage, electrode (shock plug) wear, and the distribution of the pressure waves around the discharging electrodes. Our pressure measurements were made under experimental conditions in the laboratory or in the ESWL suite. The water in each lithotripter bath or experimental situation was prepared as if it were to be used for a typical patient treatment. New electrodes were used in all experiments.

To measure the shock wave pressures, two commercially available piezoelectric crystal pressure transducers were used. These were made by Piezotronics of Buffalo, NY and possessed a sensing range up to 100,000 psi and a response time of approximately 1 μsec.

Correct calibrations were performed with the assembled equipment before each experiment. Pressure transducers were connected to two independent digitizers through a shielded cable to prevent electrical interference. The recorded signals from one digitizer were stored in digital form with a Techtronics computer system. SAS software was used to construct various plotter graphs showing maximum pressure distribution.

Two transducers were used to map the field of shock wave pressures around the focus of an extracorporeal lithotripter. One transducer was positioned near the open edge of the elliptical reflector, and served as a reference transducer. The second transducer was attached to the patient positioning chair (gantry), thereby allowing precise positioning of the transducer for each measurement with a nominal precision of ≤1 mm in all three directions. Location of the second transducer was calculated from the Dornier coordinates using a transformation computer program.

EHL measurements were performed in a Plexiglas container that could be filled with water, thus making it possible to photograph the shock wave electrical discharge. EHL shock probes made by the Richard Wolfe Company were used for the pressure measurements because single pulses could be initiated on their equipment. Shock plugs made by Northgate Research were used for the photography.

The pressures generated by EHL were recorded at distances of 0.25, 1.25, and 1.5 cm from the probe tip on low and high power settings. A high-speed 16 mm movie camera was used to visualize the underwater electrical discharge and the gas bubbles created by the spark plug. Each photographic frame represented 1/4,000 sec.

## RESULTS

The underwater electrical discharge of the ESWL probe was photographed. The electrical spark caused breakdown of the water in the area of the probe, and this created an initial shock wave and a gas bubble (Fig. 2-2). The photographs demonstrated that cracks appeared in the calculus before the bubble collapsed. The bubble was actually seen in two photographic frames, corresponding to a bubble lifetime of around 500 μsec.

The peak EHL pressure was approximately 2,000 psi measured 1.25 cm from the end of the probe, and this increased to approximately 8,500 psi at 0.25 cm. Increasing the power setting had little effect on the pressure measurements except at distances less than 1.2 cm. A typical pressure tracing recorded near the EHL probe showed an initial peak pressure followed by oscillations. These oscillations were probably caused by vibration artifacts originating in the transducer, because they appeared to correspond to the resonant frequency of the transducer in the positioning apparatus. However, we could

Fig. 2-2. **(A)** Electrical discharge of EHL plug (top) creates a plasma and dielectric breakdown of water nearby. (Courtesy of Mohammed Nassr, Ph.D., University of Florida, Gainesville, FL.) **(B,C)** Calculus becomes fragmented as bubble forms and enlarges around the EHL plug (at top).

not rule out the possibility that they were due to bubble oscillations. In each case the peak pressure lasted between 2 and 3 μsec, giving a pressure wave shape similar to that created by the extracorporeal lithotripter (Figs. 2-3 and 2-4).

In ESWL experiments, very long time intervals were recorded so as to observe the time course of the pressure around the focal point F2. Three pressure pulses were prominent on the tracings. After an initial triggering delay, a small pressure impulse was seen followed by two much larger pressure peaks, approximately 30 and 600 μsec later. The first two pulses were probably created by the electrical underwater discharge with a third pulse resulting from either the collapse of the gas bubble or secondary cavitation occurring from shock wave focusing. The initial small pressure impulse represented the unreflected portion of the initial shock wave created by the discharge. The pressure was small because only part of the initial shock wave energy was unreflected by the ellipse and its amplitude further decreased as it traveled from F1 to F2. The second shock wave represented the focused portion of the initial shock wave and accounted for most of its total energy. This peak had an average pressure in the experiments of approximately 10,500 psi; its pulse had apparent durations of approximately 2.5 μsec. In addition, the post-impulse resonance artifacts seen with EHL probe measurements were also seen with these measurements (Figs. 2-3 and 2-4).

The approximately 30 μsec delay between the unreflected and reflected portion of the initial shock wave was due to the different distances

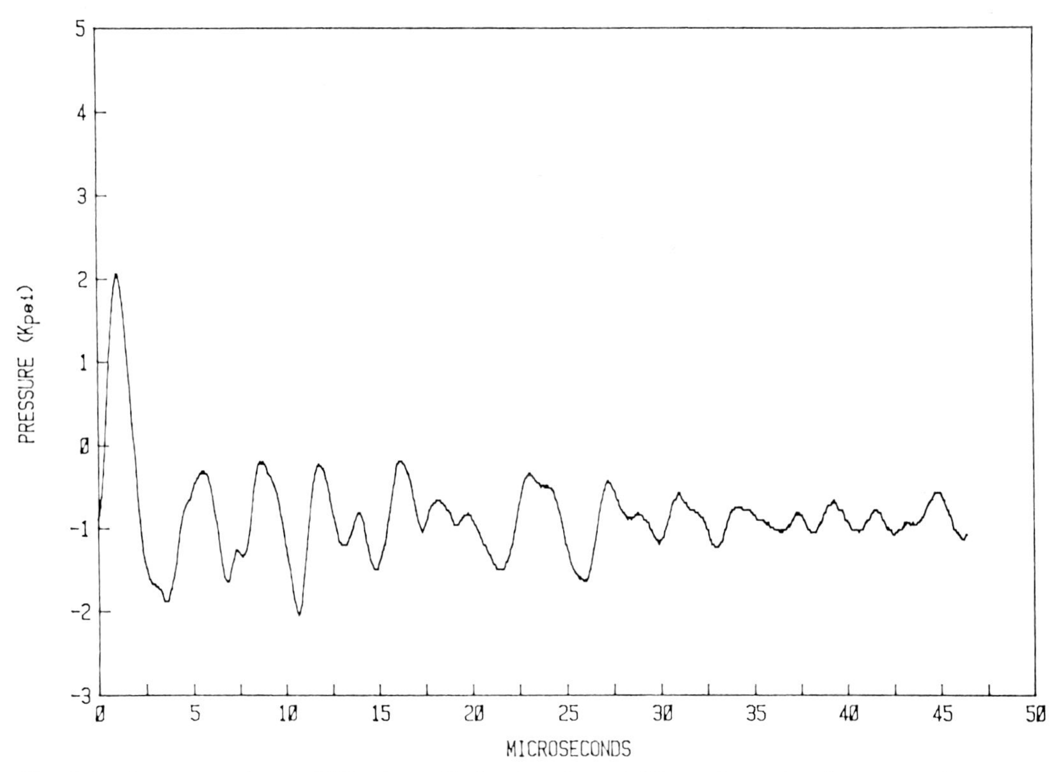


Fig. 2-3. Pressure tracing of EHL plug with the transducer 1.25 cm from plug tip; note peak pressure allowed by oscillations. (Reprinted by permission of the publisher from Weber W, Madler C, Keil B, Pollwein B, Lavbenthal H: Cardiovascular effects of ESWL. In Gravenstein JS, Peter K (eds): Extracorporeal Shock Wave Lithotripsy—Technical and Clinical Aspects. London: Butterworths (Publishers) Ltd., 1986.)

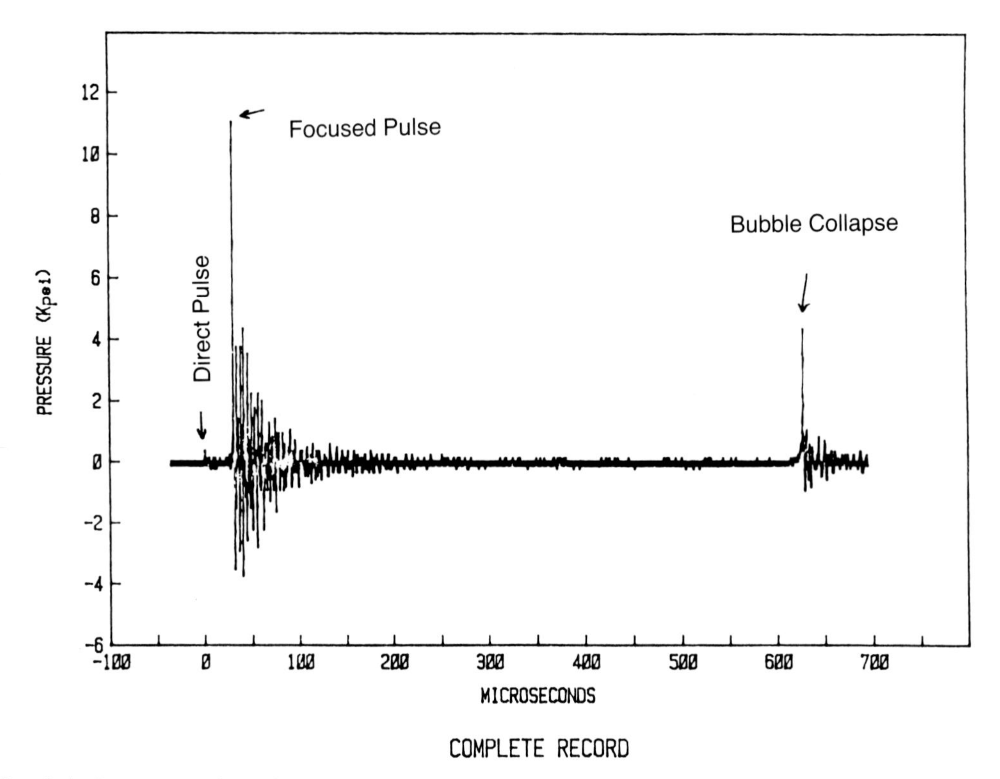


Fig. 2-4. Pressure tracing of ESWL with transducer located at geometric second elliptical focus; note peak pressure followed by oscillations. (Reprinted by permission of the publisher from Weber W, Madler C, Keil B, Pollwein B, Lavbenthal H: Cardiovascular effects of ESWL. In Gravenstein JS, Peter K (eds): Extracorporeal Shock Wave Lithotripsy—Technical and Clinical Aspects. London: Butterworths (Publishers) Ltd., 1986.)

traveled from F1 to F2. This corresponded to average shock velocity of approximately 1,700 m/sec over that distance.

To map the pressures in the area around F2, measurements were made following discharge of each shock wave from a new shock plug. Each pressure was normalized against a reference transducer. A rotational matrix was developed to convert all of the x, y, and z coordinates from the lithotripter tank into coordinates relative to the axis between the two foci. SAS isobar software was used to develop regression analyses in an attempt to explain the pressure maps and pseudo-three-dimensional plots created (Figs. 2-5 through 2-7).

Each irregular contour represented a relative pressure isobar in the horizontal plane perpendicular to the axis of the focus at F2. (This is a technique similar to that used to depict land elevation with contour maps.) The centermost contour represented the smallest area containing the highest relative pressure and each larger contour represented a radial increase of only a few millimeters in space. In this map, the relative pressure fell to less than 20 percent of the highest pressure in the center at a radial distance of 1.5 cm (Figs. 2-5 through 2-7).

## Discussion of Experimental Work

Although both EHL and ESWL employ underwater electrical discharges to create shock waves, to our knowledge, the simple physics have not been previously discussed in the urology literature. In experiments conducted at the University of Florida, it was discovered firsthand that a gas bubble, as well as a shock wave,

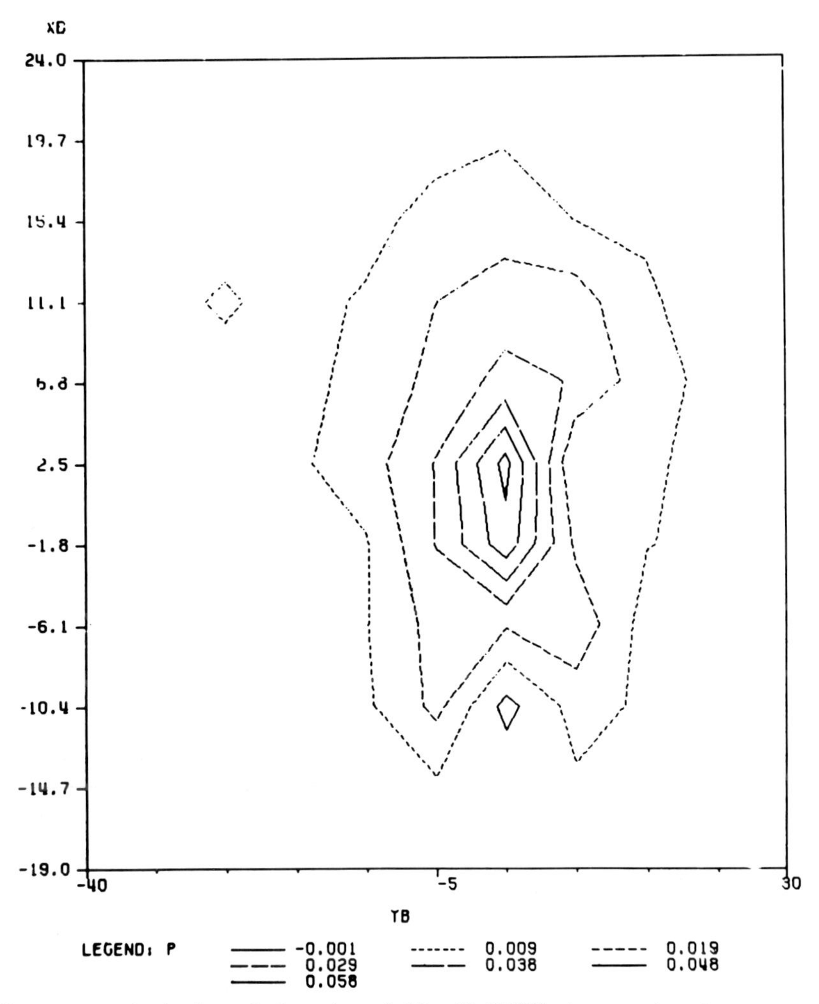


Fig. 2-5. Pressures on the horizontal plane through F2 with ESWL; innermost contours represent areas of highest pressure. (Reprinted by permission of the publisher from Weber W, Madler C, Keil B, Pollwein B, Lavbenthal H: Cardiovascular effects of ESWL. In Gravenstein JS, Peter K (eds): Extracorporeal Shock Wave Lithotripsy—Technical and Clinical Aspects. London: Butterworths (Publishers) Ltd., 1986.)

is created with each underwater electrical discharge during EHL. In the EHL photographs taken, there was evidence that the initial shock wave created by the electrical discharge cracked the calculus before the gas bubble collapsed. The gas bubble was seen in two more successive photographic frames after the first cracks appeared in the target calculus. This corresponded to a bubble life of around 500 μsec and was similar to the bubble lifetime discovered in the ESWL experiments. It is reasonable to suggest that the mechanisms of shock wave generation for EHL and ESWL are similar. In either case the initial shock wave would have more than enough time to repeatedly traverse a 1 cm calculus which could cause multiple stresses at different locations. These stresses in turn would have time to cause multiple cracks in the calculus before the bubble had time to collapse and create a second shock wave. Alternately, if a second

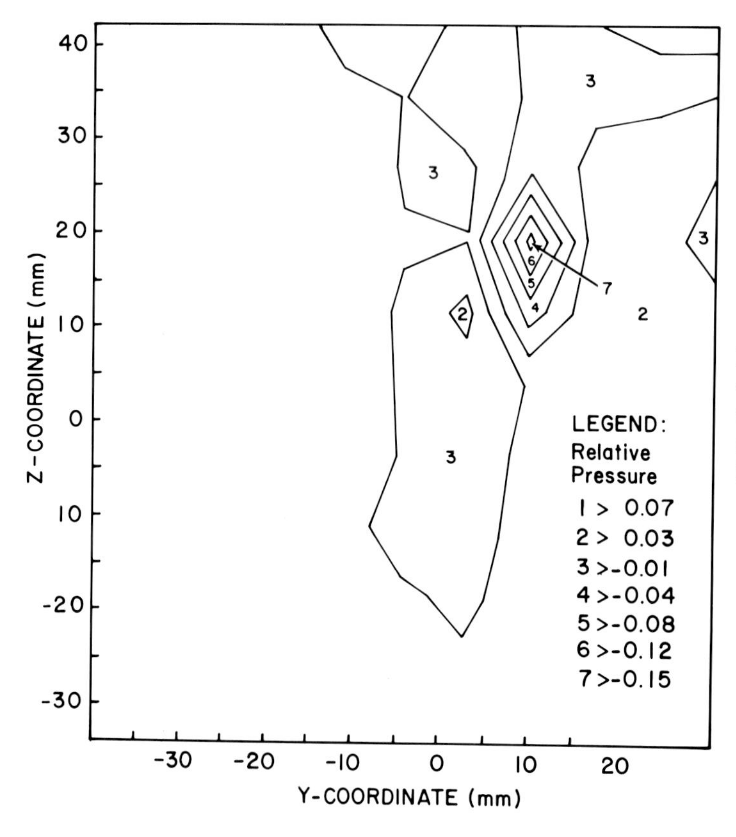


Fig. 2-6. Pressures along the foci axis depicted vertically show diamond shaped focal area. Highest relative pressures with ESWL are located in innermost contours.

shock wave were created by cavitation around F2, additional fracturing would also take place. However, it was believed that the secondary shock wave created by these phenomena (bubble collapse or cavitation) possessed much lower pressures when compared to the initial shock wave pressures and therefore, it was postulated that the initial shock wave was responsible for most of the destruction.

The shock wave shape was characterized by a rapid onset and a relatively gradual relaxation time in the microsecond range. Unfortunately, the pressure transducers used were probably too slow to record the actual peak pressure; that is, the shock wave had come and gone before the transducer had sufficient time to rise to its peak. Therefore, the actual pressures may have been even higher than those measured. Though the major difficulty in these investigations was centered around finding a durable and sensitive pressure transducer with a rapid enough response time, the experiments were designed so that a relative pressure could be used to compare findings with other pressure transducers. Therefore, the significance of most of the findings was not dependent upon the absolute, but rather on the relative pressure measurements. Accordingly, it was reassuring to find many similarities between this work and that reported by Chaussy.[3–6] Because of the similarities, the experimental work cited above will be used to add to the discussion of some of the geometry and shock wave physics pertinent to ESWL.

## ESWL Objectives

ESWL becomes possible only when at least three objectives can be achieved. First, a shock wave of sufficient strength must be created and transmitted to the targeted calculus safely and efficiently. This requires that the shock wave be created in a medium, such as water, which has an acoustical impedance very close to that

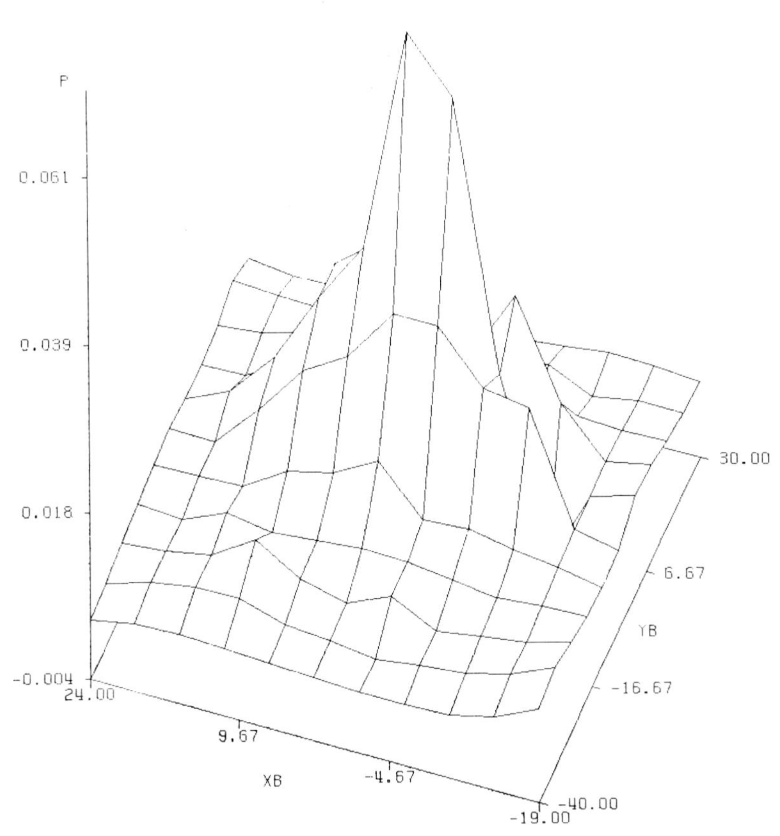


Fig. 2-7. Pressures on the horizontal plane at F2 with ESWL; highest relative pressure represented as a vertical tall peak along foci axis.

of human tissue. Second, the energy of the shock wave must be concentrated so that the appropriate tensile and compressive stresses can be created within the targeted calculus to cause fragmentation. Third, the patient must be reliably positioned so that periodic imaging of the calculus is possible. The Dornier lithotripter HM-3 reliably achieves these objectives. Since these objectives are not unique to the lithotripter, but rather to the technique itself, any future lithotripter technology must also successfully achieve these goals.

To meet the first objective, it is necessary to have a coupling between the body and the medium in which the shock waves are created. Either a water bath or a water-filled balloon may be used to achieve this goal. Because each shock wave created within water has a velocity very similar to tissue velocity, little energy is lost at the water body interface. Fortunately, water has many other properties necessary to generate and transmit shock waves reliably and efficiently.

Although the coupling effect of the water bath

or the balloon are important first steps, the focusing device used by each lithotripter determines its effectiveness and safety. Focusing requires precision and future lithotripters will have specific characteristics resulting from their own focusing device or shock wave reflector.

## Shock Wave Focusing

Preliminary geometrical considerations for the Dornier lithotripter are used as an example (Fig. 2-8). The Dornier lithotripter uses a truncated ellipsoid reflector with a shock plug positioned at one of the foci of the ellipse, F1. The general equation for the ellipse is

$$x^2/a^2 + y^2/b^2 = 1$$

If the center of the ellipse has coordinates (0,0), the length of the ellipse in the x axis is $a^2$ and the length of the ellipse in the y axis is $b^2$. The position of each focus is determined by the equation $a^2 - b^2 = c^2$; the foci are located at $(+c,0)$ and $(-c,0)$. In the Dornier lithotripter, $a = 14$ cm, $b = 7.98$ cm, and the ellipse is truncated 10 cm from $(+c,0)$; the foci are located at $(+c,0)$ for F1, and $(-c,0)$ for F2. These foci are separated by 23 cm.

Rotating the truncated ellipse around the x axis generates a three-dimensional approximation of the ellipsoid reflector used by Dornier. The distance from the truncated ellipse to the first focus is 12.65 cm. By using this radius as the radius of curvature of the unreflected shock wave as it emerges from the ellipse, the percentage of the initial shock wave that exits unreflected can be calculated. The area of the sphere with this radius of 12.65 cm is 2,010.9 $cm^2$. The area of the sector at the mouth of the ellipse is 210.63 $cm^2$. Therefore, assuming no losses due to transmission and reflector irregularities, approximately 10 percent of the initial shock wave energy created exits the ellipsoid unreflected. This unreflected shock wave may not contribute significantly to the fragmentation process because geometrical divergence diminishes the shock wave amplitude as it travels from F1 to F2. If little or no energy is lost in the reflector, then the remaining 90 percent of the initial shock wave created at F1 and focused by the ellipse emerges as a converging crescent-shaped wave front. The total energy is spread out over the entire crescent and, therefore, there is very little energy density at any one point where this shock wave arc enters the body. However, as it converges, all of the energy becomes concentrated within the 1 to 2 cm focal area at F2. This phenomenon of geometrical conver-

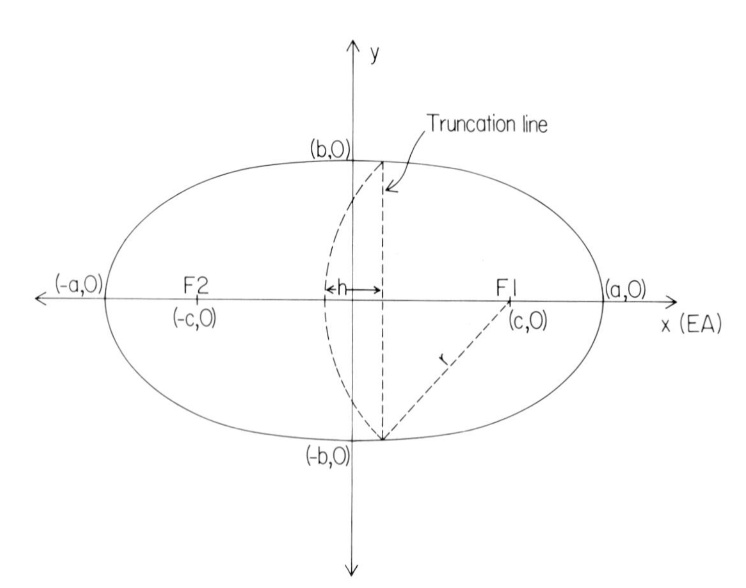


Fig. 2-8. Diagram of theoretical ellipse of ESWL; truncation line represents the opening of the brass ellipsoid reflector in the lithotripter. (Reprinted by permission of the publisher from Weber W, Madler C, Keil B, Pollwein B, Lavbenthal H: Cardiovascular effects of ESWL. In Gravenstein JS, Peter K (eds): Extracorporeal Shock Wave Lithotripsy—Technical and Clinical Aspects. London: Butterworths (Publishers) Ltd., 1986.)

gence spares the interposed tissues unnecessary injury and focuses the energy onto the targeted calculus.

Even though a theoretical point exists at F1, it is not possible to create a reflector that will give a sharp enough focusing at the second focus to create a point; therefore, a focal *area* results. Although it is believed that shock waves used for ESWL obey acoustical laws in this focal area, nonlinear considerations may be necessary to explain some of the phenomena that occur around F2 and are beyond the scope of this discussion.

Another theoretically interesting phenomenon occurs in the focal area. Immediately in front of the plane perpendicular to the axis of the focus at F2, the compressive shock wave converges in a crescent shaped arc. On the opposite side of this plane the shock wave diverges and therefore becomes tensile in its direction and its nature with respect to F2. Therefore, one could expect the area immediately behind F2, along the ''blast path'' of the shock wave, to have a more destructive effect on a calculus since these tensile forces are more effective than compressive forces in causing calculus fragmentation. In fact, preliminary experimental and clinical investigations performed at the University of Florida (Finlayson, personal communication) have demonstrated this to be true.

## Shock Wave Generation

Shock waves used for ESWL appear to be created with a mechanism similar to that used for electrohydraulic lithotripsy (EHL) and microexplosive lithotripsy (MEL).[2,3,7,8] For ESWL and EHL, an underwater spark creates the shock wave. The spark is generated by a sudden discharge of electrical energy through an appropriate circuit which contains a spark gap submerged in water. The spark gap initiation builds up a conductive plasma and a vapor between the two electrodes of the gap. The rapid expansion of this plasma creates an explosive impact and transmits very high pressures (shock waves) to the water. The plasma vaporizes and creates an expanding gas bubble. Similarly, an underwater explosion also creates a gas bubble with very high internal temperatures and pressures. Because water is isotropically compressible, the initial shock wave is transmitted radially. The shock wave configuration close to the electrode gap or the explosive depends on the shock plug or explosive configuration. However, at radial distances greater than the gap width or the dimensions of the explosive, the wave front becomes spherical and decays owing to geometrical divergence and frictional losses to the transporting medium. The decay may be more rapid than would be expected from geometrical considerations alone because of the entropic losses.

As the bubble expands, the initial shock wave is emitted, and the density in the bubble gradually diminishes. Because the surrounding water has inertia secondary to its mass, the water continues to move outward as afterflow, even after the gas bubble expansion ceases. Ultimately, a negative pressure differential emerges, causing the bubble expansion to cease and the bubble begins to contract at an increasing rate. The bubble finally collapses until there is an abrupt reversal of the process by compression of the gases, thereby creating a second shock wave. Theoretically the cycles may be repeated successively by further bubble oscillations. In our experiments the first oscillation created a shock wave with an amplitude of approximately 20 to 40 percent of the initial shock wave (Fig. 2-9). Successive oscillations were expected to become progressively weaker, but were not measured. With underwater explosions, the first bubble oscillation creates a shock wave lasting relatively longer than the initial shock wave and emits a pressure impulse that can be delivered.[8] It would appear that the time constant for these tertiary shock waves is not appropriate for stress induced fractures to be created within a calculus.

In addition to bubble oscillation and collapse related shock waves, cavitation related shocks could also cause stone fragmentation. If cavitation does occur near F2, then secondary shock waves are released near F2 after the focused

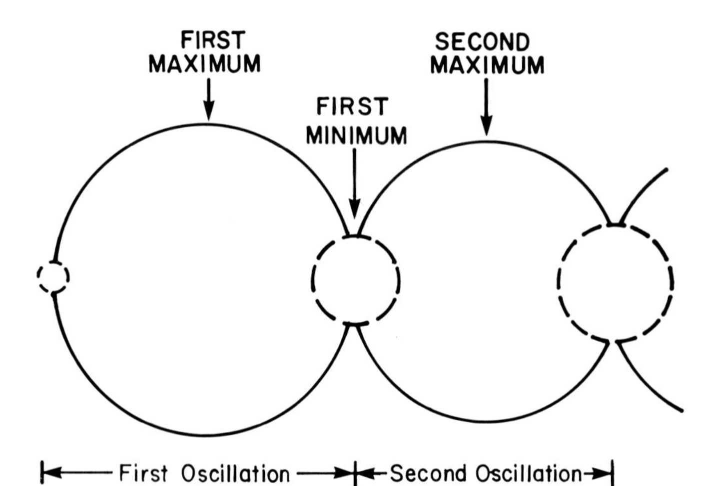


Fig. 2-9. After electrical or laser discharges, or an explosion underwater, a bubble is created with the initial shock wave. Subsequently, bubble collapse releases a second, less forceful, shock wave.

wave front has passed. In recent unpublished preliminary work, Finlayson and Nassr suggested that the third pressure wave measured in our initial experiments was due to cavitation and that the shock wave resulting from the bubble collapse actually occurred approximately 500 μsec later. Further work is needed to validate this new information and to clarify whether cavitation induced shock waves contribute to the calculus fragmentation process associated with ESWL.

## Shock Wave Transmission

Because shock waves, like sound, are compressive waves they can be transmitted through water, body tissue, and a calculus. They also obey acoustical laws linearly at low energy levels. Therefore, if one understands some of the basic physics surrounding sound, one can understand those of other compressive waves, including shock waves.

Qualitatively, it is important to remember that the velocity of all compressive waves, including sound, ultrasound, and shock waves, increases as the compressibility of the transmission medium decreases. In the case of ultrasound, the pressure and the density of the medium remains nearly constant throughout the process of wave transmission. Therefore, all parts of the wave are transmitted at the same velocity so that a sinusoidal wave remains sinusoidal indefinitely during its propagation. However, with shock waves, one can no longer assume that pressure and density remain constant within the transmission medium. Shock waves cause the transmission medium to become more dense, thus decreasing its compressibility and increasing the wave's velocity. It is this change in velocity at each point along the wave that changes the configuration of the half-sinusoidal wave to the sharp rise and gradual decay typical of a shock wave pressure tracing (Figs. 2-9 through 2-11).

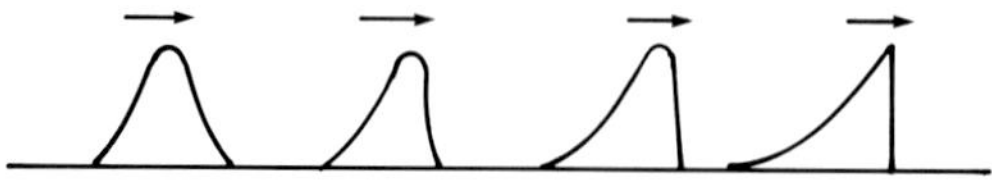

Fig. 2-10. As the lead edge increases its velocity, the half-sinusoidal wave transforms into the typical shock wave form.

To understand more completely how this change in velocity results with shock waves, consider the following example. Suppose a compressive wave which is an intense sinusoidal half-wave is created. At the beginning of the wave, the water is in the low amplitude range of pressure and therefore the velocity near this area is that of sound. Toward the middle of the wave, the pressure amplitude of each successive point becomes greater, which increases the density of the medium and also the velocity. With continued propagation of the wave, the pressure peak gradually increases its speed

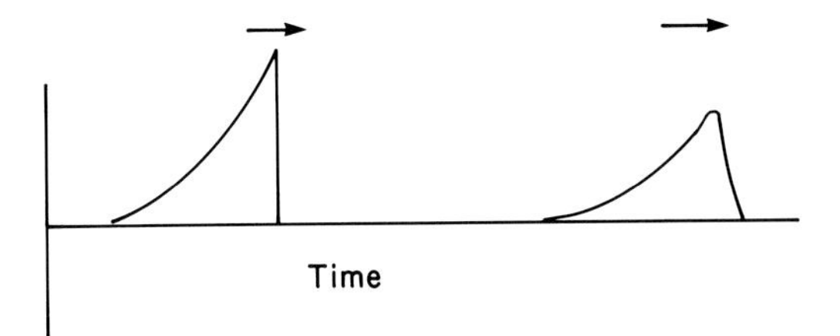


Fig. 2-11. As shock wave propagates, the lead edge slows and wave form degrades towards a half-sinusoidal wave.

enough to overtake the front end of the wave, thus altering its shape. Eventually its shape becomes transformed into a sudden and discontinuous increase in high-amplitude pressure followed by a gradual decay of pressure. This results in the typical profile of the shock wave shown in Figure 2-10. Once formed, this typical wave form does not propagate indefinitely through a medium without also undergoing gradual energy loss.

Initially the shock wave has a very sharp rise in pressure at its front, but as transmission continues, energy losses reduce this peak pressure until the shock wave degenerates into an ordinary sound wave (Fig. 2-11). In addition to this gradual energy loss, there are secondary tensile or rarefaction waves present that can overtake the main compressive wave and also reduce its peak pressure.

Unlike ultrasound, shock waves are composed of several different frequency components and have a spectrum; ultrasound has only one frequency. Usually, higher frequencies compose the sharp portion of the wave front and lower frequencies contribute to the shape of the remainder of the wave form.[8] It is the frequency distribution of a shock wave that determines its tissue penetration, its destructive forces on the calculus, and possibly the size of the particles that result during fragmentation.

## Shock Effects on Calculi

Shock waves undergo reflection and refraction when passing between two media at any interface.[8] Both reflections and refractions depend on differences between the acoustic impedances of the media at the interface. The more the acoustic impedances differ, the more the wave will reflect and refract at the interface. The large difference in acoustic impedance between a calculus and the surrounding water is thought to cause calculus fragmentation in ESWL. It is also believed that it is the small difference between the acoustic impedance of tub water and tissue that prevents tissue damage.

When the shock wave enters the calculus from the surrounding water, a reflected wave on the outer surface of the stone is created. The stone surface is compressed, and fragmentation may result.

Nevertheless, because the compressive strength of a urinary calculus is several times greater than its tensile strength, the compressive forces may not be large enough to break the calculus.

When a shock wave has traveled completely through the dense calculus, it meets the water interface on the back side of the calculus and is reflected again (Fig. 2-12). This reflection causes a rarefaction or tensile wave to occur. If the tensile strength of the calculus is overcome by this tensile wave alone, then a fracture occurs.

In addition to these forces acting on the calculus, even during propagation, each shock wave simultaneously creates additional stresses in two different directions inside the calculus. Compressional stresses are created in the direction of wave propagation and tensile stresses are created perpendicular to this direction. When shock waves are strong enough, these tangential stresses cause fragmentation to occur *within* the calculus along both the direction of propagation and the line of reflection (tension). Fortunately, because a calculus is heterogeneous and reflections occur at many different points, cracks are produced in many places with each shock wave. When enough stress is applied, a fracture will occur, regardless of the direction of the applied stress. Even as a fracture has begun to form, the rest of the shock wave will be reflected at this newly created interface so that a series of parallel fractures may actually be produced from

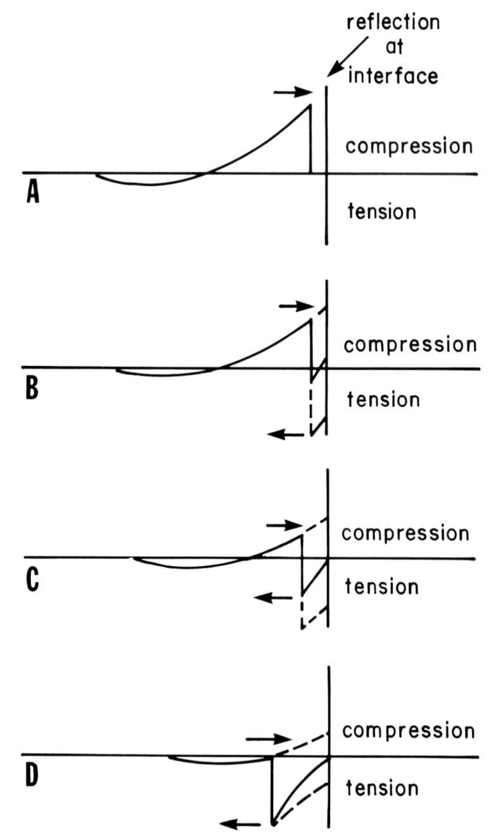


Fig. 2-12. **(A)** Initially, only compressive forces work on the stone reflective surfaces to cause fragmentation. **(B,C)** After reflection begins, the forces, a short distance from the reflecting surface, create a tensile force pulling in the direction opposite to the compressive forces in the latter part of the shock wave. Together these opposing forces cause stone fracturing. **(D)** Still later, the tensile (reflected) wave propagates through the stone away from the reflecting free surface.

a single shock wave. Things get much more complicated with each subsequent shock wave. Therefore, repeated shock waves usually result in multiple small fragments, as with ESWL.

Since human calculi behave as brittle materials under shock wave stress, in a fashion similar to other well-characterized brittle materials, it is believed that repetitive shock waves create small fragments because of their relatively short time constant.[8] Because a shock wave that possesses an amplitude sufficient enough to crack a calculus lasts only a microsecond or less, the resultant cracks do not have much time to spread. That is, the crack could not spread more than a millimeter or so before the shock wave pulse passes and the stress on the calculus is removed. Instead of long running cracks creating large fragments, a large number of separate fractures occur which gradually join, resulting in multiple small fragments being created.

## CONCLUSION

From the basic discussion above, it would seem that the key to ESWL lies in selecting the appropriate reflector to optimize the physical forces of the shock waves. Thus interactions between the calculus and shock wave occur, and small fragments result. Whether or not more efficient reflectors and coupling devices can be devised remains to be seen.

Although simple geometrical and physical considerations can be used to explain ESWL, it is understood that this discussion is an oversimplification of many complex physical laws.

ESWL is only possible when shock waves of sufficient magnitude can be created in a medium and transmitted via that medium through the body into the calculus. This requires a water bath or a water-filled balloon coupling device and an ellipsoid or similar reflector.

The shock wave is generated by an electrical discharge under water. This discharge creates a shock wave and a gas bubble. The initial shock wave emerges from the ellipsoid reflector, partially reflected and partially unreflected. Because of geometrical convergence, the reflected portion of the shock wave contains most of the energy and does most of the work on the targeted calculus. Although the gas bubble creates secondary shock waves through several oscillations, it is questionable whether these have sufficient energy to contribute to calculus fragmentation.

Because shock waves traveling from the water into the body do not encounter a large difference in impedance during their transmission, little energy is absorbed and most of the shock wave energy is delivered to the calculus. Because the calculus behaves as a brittle material under stress, multiple fractures occur within the calculus due to the combination of compressive and tensile forces. Fragmentation occurs in the focal area and may be more efficiently performed by positioning the targeted calculus behind F2 in the shock wave "blast path."

Recently preliminary work (International Biomedical, Seattle, Washington) has shown that energy from pulsed lasers may be concentrated with lenses to create shock waves of sufficient magnitude to fragment calculi in animals. Elsewhere, arrays of piezoelectric crystals (EDAP, Lyon, France) have been pulsed with large amounts of energy to also fragment calculi in animals and in some patients. Therefore, it is believed that future advances will be made in not only the shock wave generation efficiency, but also in coupling devices, calculus imaging, and changes in the shape of the focal area.

The current Dornier lithotripter is a marvelous and major advance in medical technology. Fortunately, it is based on simple physics and geometry, and further advances seem inevitable. Hopefully these advances will be as important to the clinical urologist as the advances in magnetic resonance technology and computed tomography screening have been to the radiologist.

## ACKNOWLEDGMENT

Supported in part by National Kidney Foundation, NIH Grant AM 20586 and Urotech Management Corporation.

## REFERENCES

1. Bulow H, Frohmuler HGW: Electrohydraulic lithotripsy with aspiration of the fragments under vision—304 consecutive cases. J Urol 126:454, 1981
2. Nassr M: A study of the effects of electrohydraulic lithotripsy. Masters research 1982–83. Thesis Dept. of Biomedical Engineering, University of Florida, Gainesville, FL
3. Chaussy C, Schmiedt E, Jocham D et al: First experience with extracorporeally induced destruction of kidney stones by shock waves. J Urol 127:417, 1982
4. Chaussy C, Schmiedt E, Jocham D et al: Extracorporeal shock wave lithotripsy. New aspects in the treatment of kidney stone disease. 1st Ed. Karger, Basel, 1982
5. Chaussy C, Schmiedt E: Shock wave treatment for stones in the upper urinary tract. Urol Clin N Amer 10:743, 1983
6. Chaussy C, Schuller J, Schmiedt E et al: Extracorporeal shock wave lithotripsy (ESWL) for treatment of urolithiasis. Urology 23:59, 1983
7. Watanabe H, Watanabe K, Shiino K, Oinuman S: Microexplosion cystolithotripsy. J Urol 129:23, 1983
8. Cole RH: Underwater Explosions. Ch. 13. Dover Publications, New York, 1965

# 3

# Biologic Effects of Shock Waves

The use of high-energy shock waves in the human body has raised many questions from physicians, physicists, and patients. Can a wave of this power pass through the body without affecting any human tissue or organ? If shock waves do affect the patient's body as well as his stone, are the effects detectable, and are the changes temporary or permanent? How can we use shock wave lithotripsy both effectively and safely, minimizing these changes?

Shock waves are high-energy pressure amplitudes generated in air or water by an abrupt release of energy in a small space. They propagate according to physical laws of acoustics and are transmitted through media with low attenuation. For example, when an atomic bomb explodes in air, a shock front representing a moving wall of highly compressed air is generated. Another familiar compression shock wave, a sonic boom, is created when an object, such as a supersonic aircraft, moves through a medium (air) faster than the speed of sound. Although unfocused, the audible wave (sonic boom) and mechanical wave (which breaks windows) are detectable evidence of this high energy form.

Inside the Dornier lithotripter, a rapid, high-voltage underwater spark discharge within an ellipsoid reflector generates a shock wave which can be focused and transmitted through water. This high-energy wave travels at supersonic speed through body tissue (a medium similar in acoustic impedance, i.e., density, to water) with slight attenuation. At its focus point, the impact of the wave against the stone liberates short-term, high-energy mechanical stresses. This stress overcomes the tensile strength of the calculus and causes disintegration. A summation of wave impacts should pulverize the calculus into sand. The voltage across the electrode determines the strength of each shock wave delivered and can be varied from 18,000 to 24,000 V.

Unlike high-frequency ultrasound, these low-frequency, positive-pressure compressive amplitudes have good tissue penetration without significant reflection. Only at the focal point (1.5 $cm^3$ in volume) are maximal compression and distractive energies developed. Pressures within the F2 focus vary between 800 and 1,200 bar and approach 14,000 psi.

Since histologic examination of normal human renal tissue after shock wave lithotripsy is not possible, various imaging techniques and indirect methods have been used to investigate and detect the effects of shock waves on human tissues.

The following discussion of the biologic effects of ESWL is divided into two sections.

*R.A.R.*

# Clinical and Experimental Effects Associated with ESWL

Robert C. Newman

## EFFECT ON SERUM PARAMETERS

The initial report of the United States extracorporeal shock wave lithotripsy (ESWL) experience, as compiled by Drach,[1] summarized information from the first 2,501 patients treated in the United States. This study reported no change in pretreatment and posttreatment creatinine, prothrombin time (PT), and partial thromboplastin time (PTT).

An upward trend was noted in the white blood cell count (WBC), lactic dehydrogenase (LDH), serum glutamic oxaloacetic transaminase (SGOT), creatinine phosphokinase (CPK), and bilirubin. A downward trend was seen in amylase, platelets, alkaline phosphatase, and hematocrit.

The slight decrease in hematocrit noted may be caused by hemolysis or by perirenal hemorrhage,[2,3] or more likely by hydration states. Although subclinical hematomas have been noted, very few patients (< 1 percent) have needed transfusion after ESWL at the University of Florida and at New York Hospital–Cornell Medical Center. To minimize the risk of hemorrhage, it is recommended that all anticoagulants, including aspirin, be discontinued at least 10 days before lithotripsy. Blood for clotting studies (PT, PTT, and platelets) should be drawn just before ESWL to confirm the absence of abnormalities.

## HEMATURIA

Gross hematuria is virtually universal after ESWL. The urine may be slightly pink or dark red. Unpublished canine studies done at the University of Florida suggest that the hematuria may be related at least in part to urothelial disruption in the pelvocaliceal system. The urine will usually clear within 24 to 48 hours, much as one might expect after a transurethral prostatic resection. If the urine is particularly dark, the administration of 20 to 40 mg of furosemide intravenously should diminish colic secondary to the passage of clots.

## INTRAVENOUS PYELOGRAPHY

An intravenous pyelogram (IVP) can be used to monitor post-ESWL function. It is usually done prior to ESWL to define the stone's position and the anatomy of the upper tract. During the early experience with ESWL in the United States, the Food and Drug Administration (FDA) required a pyelogram as a part of the post-treatment protocol.

At the University of Florida, urography was performed on a series of 100 patients 24 to 72 hours after lithotripsy.[4] Usually a kidney–ureter–bladder (KUB) film and 15-minute post-contrast injection film were obtained. Prone and erect follow-up films were taken up to 24 hours after injection in those patients demonstrating delayed excretion in the treated kidney and ureteral or pelvocaliceal dilation on the 15-minute film. Ureteral obstruction was arbitrarily classified into three grades: mild (contrast medium present distal to the stone fragments, preserved form of calyces); moderate (small amount of contrast medium seen to pass the stones; calyces beginning to lose their papillary appearance); and severe (markedly delayed excretion, no contrast medium distal to the stone fragments, dilation and clubbing of calyces). Twenty-one patients had complete stone disintegration followed by rapid fragment passage (Table 3-1). In three patients, no appreciable effect on the stones was noted following lithotripsy. Seventy-six had abdominal films demonstrating stone break-up without complete passage of the fragments within 24 to 72 hours.

Twenty-seven patients had ureteral obstruction ranging from mild to severe. When obstruction was present, the KUB usually demonstrated stone fragments in the ureter. The fragments often congregate, forming a steinstrasse or "street of stones" (Fig. 3-1). It is noteworthy that only 9 of 27 (9 percent overall) of these individuals required intervention. In these patients post-ESWL fragment manipulation was necessary because of intractable pain, nausea and vomiting, or fever associated with lack of fragment passage. Intervention consisted of retrograde ureteral catheter placement, stone basket manipulation, or percutaneous nephrostomy. Eighteen of the 27 patients with ureteral obstruction required no intervention and went on to pass their fragments spontaneously.

**Table 3-1. IVP Findings 24 to 72 hours after ESWL**

| Observation | Frequency ($n$ = 100) |
|---|---|
| KUB film | |
| Stones fragmented but not passed | 76 |
| No stones or fragments seen | 21 |
| No change in stone | 3 |
| Pyelogram | |
| No obstruction | 70 |
| Mild to severe ureteral obstruction | 27[a] |
| No or mild ureteral obstruction; delayed function or no dye excretion in treated kidney | 3 |

[a] Intervention was required in only 9 of 100 patients; all patients requiring manipulation demonstrated mild to severe ureteral obstruction.

More significantly, three patients with no or mild ureteral obstruction either did not excrete contrast on the delayed side or had severely delayed function suggesting possible parenchymal damage. In these three patients, magnetic resonance imaging (MRI) and quantitative radionuclide renography (QR) were performed. All three had renal enlargement presumably representing intrarenal edema, as imaged by MRI. One patient had evidence of subcapsular hemorrhage (Fig. 3-2); one had hemorrhage into a cyst; and the third had changes consistent with a perirenal fluid collection. All showed a pattern of total parenchymal obstruction on QR. In two of the three patients (the ones with subcapsular hemorrhage and hemorrhage into a cyst), repeat studies at 6 weeks and 6 months were normal. The patient demonstrating a perirenal fluid collection was lost to follow-up.

Overall, 70 percent of the patients had satisfactory stone fragmentation without complication. In this series, there was no definite evidence that stone size was related to obstruction. (Data obtained from a larger series studying 1,469 stone-containing renal units showed a definite correlation between stone size and the likelihood of obstruction ($P < .005$).[1] This indicates that patients with larger stones do have a higher probability of becoming obstructed.)

The KUB proved to be the most useful part of the examination in the 24 to 72 hour period after lithotripsy. This film assists the urologist in determining the adequacy of stone fragmentation and passage. A decision regarding the need

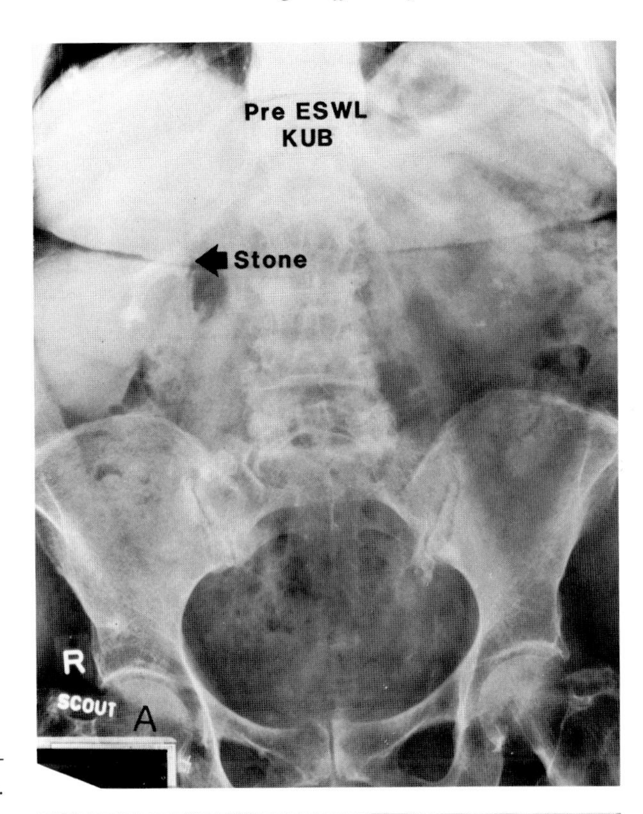


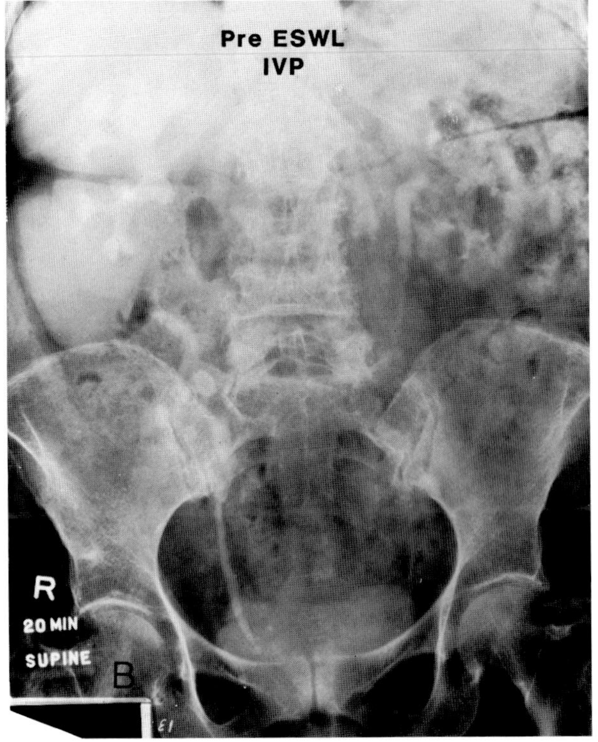


Fig. 3-1. **(A)** Pre-ESWL KUB film demonstrates stone in lower pole of right kidney. **(B)** Pre-ESWL pyelogram shows no obstruction. (*Figure continues.*)

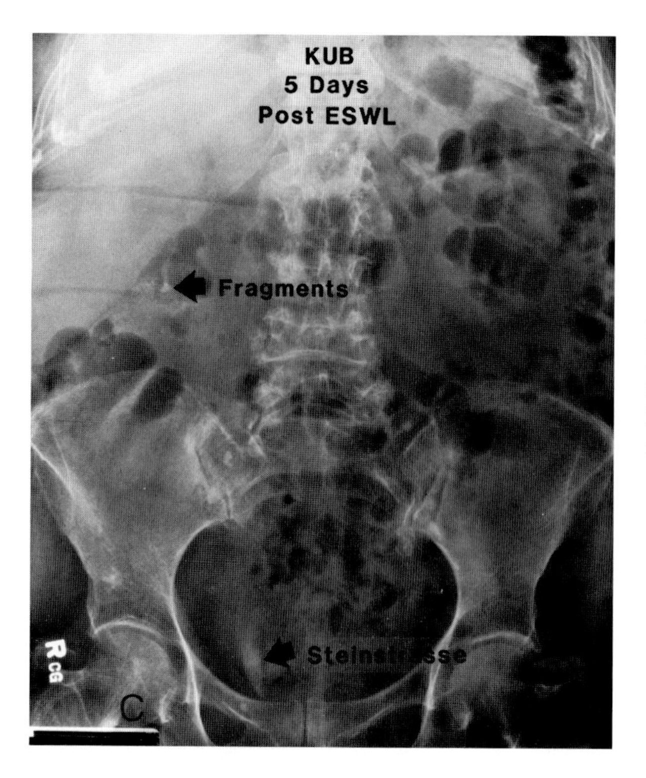


Fig. 3-1 (*continued*). **(C)** Post-ESWL KUB film (5 days) shows good fragmentation with steinstrasse in distal right ureter. Manipulation was not necessary as the fragments passed spontaneously.

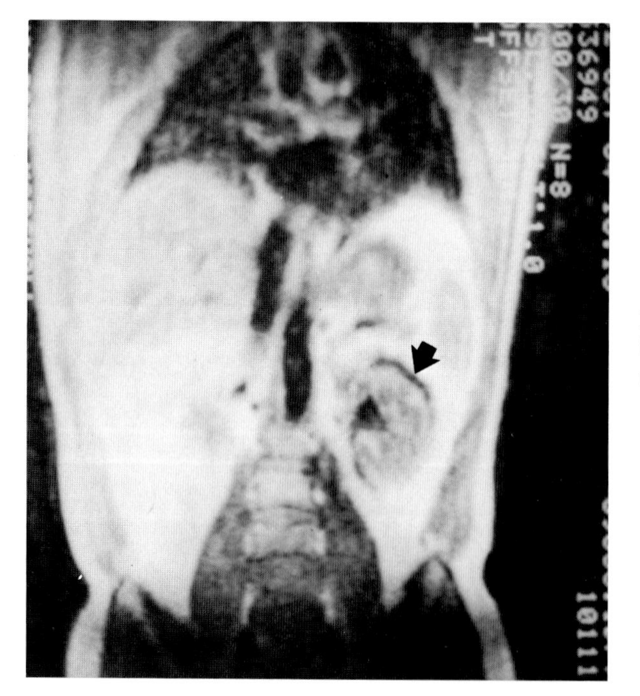

Fig. 3-2. MRI demonstrating a subcapsular hematoma in the upper pole of the left kidney.

for fragment manipulation (e.g., ureteral catheter placement or percutaneous nephrostomy) was generally made based on clinical criteria without regard to the pyelogram appearance. Based on this series and the associated clinical experience, it is felt that urography is generally unnecessary immediately after ESWL. If fragments remain on the post-ESWL KUB, the patient should be followed with serial KUB films as long as fragments remain. An IVP or similar study should be obtained 4 to 6 weeks after treatment to rule out the possibility of an unrecognized obstruction. Additional studies should be done in a timely fashion based on clinical indications.

## RADIONUCLIDE RENOGRAPHY

[$^{131}$I]iodohippurate renography has been used to study patients before and after ESWL. Chaussy's group followed 50 patients for a period of 2 years. They found a significant enhancement in function at 1.5 years; function had improved to 110 percent of the value obtained prior to lithotripsy.[5,6] At New York Hospital–Cornell Medical Center, Yankelevitz and Riehle (personal communication) studied 81 patients with [$^{131}$I]iodohippurate to determine effective renal plasma flow (ERPF), renal activity at 25 minutes, and the post-maximum renal clearance half time ($T_{1/2}$) 1 day before and 3 months after ESWL. No significant differences were found between the pre- and post-therapy value of these paremeters using the paired t-test ($P = 0.05$). They found no adverse renal effect of ESWL as measured by radiohippurate renography at 3 months; however, they suggest that such studies may be useful in monitoring selected patients (e.g., those patients with compromised renal function) before ESWL.

In the FDA study,[1] a group of 494 patients was reviewed. At 3 months, individual patient renogram clearances showed improvement when compared to studies done prior to treatment. Again, the paired t-test demonstrated statistical validity. For left-sided stones, $P < .0005$; whereas for right-sided stones, $P < .005$.

In one smaller series of 33 patients, a decrease of more than 5 percent in ERPF was seen in 10 patients (30 percent); six kidneys were treated twice. This change was statistically significant by the paired student t-test ($P = 0.025$) and the Wilcoxson signed rank test ($P = 0.003$). In this same group of patients, QR demonstrated partial (15 of 24 kidneys) or complete (9 of 41 kidneys) parenchymal obstruction.[7]

The improved function in some patients may be due to relief of the obstruction. It is worth noting that, in some patients, renal function worsened. Decreased function may be related to the number of shocks administered, the total power delivered, damage to the renal units per se, or the presence of obstruction due to passing fragments. Additional data are needed to clarify these points.

## RENAL SONOGRAPHY

Ultrasonic evaluation does not provide direct information about renal function, but does give the urologist useful information about the presence or absence of stone fragments, hydronephrosis or perirenal hematoma. At the University of Florida, J. V. Kaude and associates (personal communication, unpublished data) obtained abdominal radiographs and renal ultrasound examinations on 286 patients (321 treated kidneys) within 24 hours after ESWL.

Of 321 kidneys, 286 (89 percent) had evidence of stone fragments at 24 hours after ESWL. One patient had stone fragments seen on KUB but not on sonography, whereas another patient had stones evident on sonography but not seen radiographically. Seventeen percent had dilated collecting systems. Six percent showed findings consistent with intrarenal or subcapsular hematoma or perirenal fluid collections; these findings were not clinically significant. For comparison, Chaussy and Schmeidt noted sonographic findings consistent with subcapsular hemorrhage in approximately 0.6 percent of their patients.[8]

Information about stone disruption and position can generally be obtained via conventional abdominal radiography. For the occasional patient with lucent stones, ultrasound provides a noninvasive means of following the patient. In general, dilation of the collecting system does not necessarily mean that fragment manipulation or a secondary procedure will be required.[4] Even if dilation (hydronephrosis) is present soon after ESWL, the patient can be followed easily on an outpatient basis with periodic KUB films and sonographic examinations. If the patient's clinical course does not correlate with sonographic findings, more definitive radiographic examinations (e.g., IVP or MRI) will usually resolve unanswered questions.

## MAGNETIC RESONANCE IMAGING

Kaude and associates reported on a group of 33 patients undergoing MRI within 24 to 48 hours after ESWL.[7] In this group 38 kidneys were shocked; 6 kidneys received two treatments each. From 800 to 2,400 shocks were delivered per treatment (mean 1,808 shocks). Of the 38 treated kidneys, 14 (37 percent) were normal by MRI following ESWL. The following changes were noted in the remaining 24 kidneys: subcapsular hemorrhage in 9 of 38 (24 percent), hemorrhage into a renal cyst in 2 kidneys (5 percent), loss of corticomedullary demarcation in 9 (24 percent), perirenal fluid collection in 12 (32 percent), and unexplained changes in 3 (11 percent). In the last category of three patients, one individual had high-signal-intensity $T_1$ and $T_2$ weighted images consistent with, but not diagnostic of, fat. The other two patients had smaller and more numerous high-signal-intensity objects scattered through the renal parenchyma 6 weeks after ESWL. These high-signal-intensity changes were not evident on the immediate post-ESWL scans and their etiology and significance were unclear.

The other changes seen may have been related to obstruction secondary to fragment passage, the number of shocks administered, or parenchymal damage. Follow-up MRI examinations were not done in all of these cases. However, it appears that most of these image changes are reversible. Future studies should help to define the clinical significance of these types of findings.

MRI appears to be more sensitive than some other techniques (e.g., sonography) in detecting subtle changes. Therefore, computed tomography (CT) and MRI may be helpful in defining abnormalities in cases where findings on more commonly used and less expensive studies (e.g., KUB, IVP, QR, and sonography) do not correlate with the patient's clinical course and additional data are needed for decision-making in patient management.

## ANATOMIC CHANGES IN HUMAN KIDNEYS AFTER ESWL

For obvious reasons, the opportunity to examine pathologically human renal units that have undergone ESWL does not occur frequently; however, at the University of Virginia, two patients died at 1 and 12 days after lithotripsy,[3] one of metastatic gallbladder cancer and the other of a myocardial infarction. The three treated kidneys examined at autopsy showed perirenal bleeding (100 to 170 ml). No other gross or microscopic abnormalities were seen.

## ANIMAL PATHOLOGIC STUDIES AFTER ESWL

Early experimental work in Munich was done with a dog model.[9] Seventeen dog kidneys were implanted with human calculi. Five hundred shocks were delivered to the stones, and the animals were sacrificed 14 days after lithotripsy. None of these kidneys demonstrated pathologic changes.

In preliminary canine studies done by Brendel, shock waves in three different "therapeutic" doses were delivered.[10,11] Renal hemorrhage ranging in size from small but macroscopic to greater than 0.5 cm were noted. Petechiae were seen on the renal capsule. Pre-

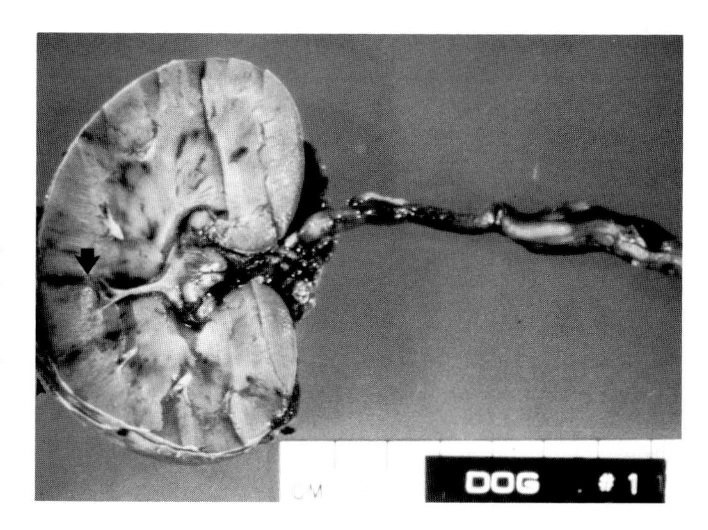


Fig. 3-3. This canine kidney was shocked 4,500 times. Note the perirenal bleeding (arrowhead) and radially oriented corticomedullary hemorrhage (arrow). Similar abnormalities were seen in kidneys given 1,776, 6,000, and 8,000 shocks. (Courtesy of Ray Hackett, M.D., Department of Pathology, University of Florida, Gainesville, FL.)

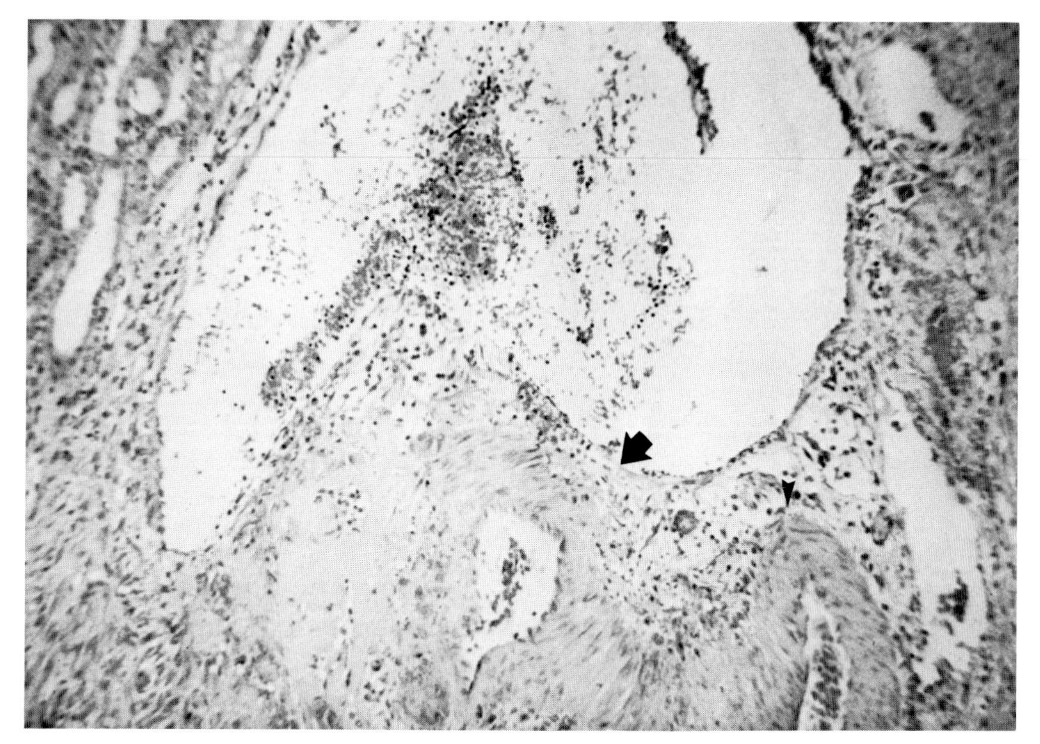

Fig. 3-4. Microscopic section taken through an arcuate area of a canine kidney that received 4,500 shocks. Note the venous endothelial disruption (arrow) and hemorrhage; the arcuate artery (arrowhead) remains intact. (Courtesy of Ray Hackett, M.D., Department of Pathology, University of Florida, Gainesville, FL.)

liminary data suggest that the number but not the size of bleeding sites depended on the number of shock waves administered.

In unpublished work at the University of Florida, normal in situ dog kidneys were shocked from 1,600 to 1,800 times using 18 to 24 kV. The animals were killed and autopsied at 48 hours or 30 days after shock wave delivery. High frequency jet ventilation was used during anesthesia. The predominant findings were macroscopic perirenal hematoma and radially oriented corticomedullary hemorrhage in the area shocked (Fig. 3-3). Microscopic damage to renal venules and tubular dilation was observed 24 to 48 hours after lithotripsy (Fig. 3-4).

Renal units examined 30 days after shock wave delivery showed evidence of fibrosis in the shocked area (Fig. 3-5). The kidneys in this group received either 1,600 or 8,000 shocks. It is worthy of note that the kidney receiving 8,000 shocks was viable and functioning; the scar tissue was confined to the area that received the shocks and, though this is not a quantitative assessment, the scarring at 8,000 shocks appeared to be far less than five times the scarring resulting from 1,600 shocks.

Correlating the results of various studies is difficult because all the studies involved a relatively small number of kidneys. Chaussy's study used fewer shocks than are usually administered to human kidneys. Brendel administered "therapeutic" numbers of shocks, whereas several animals in the Florida study received supraconventional numbers of shocks. Though the studies differ in some aspects, it has been demonstrated that as few as 1,600 shocks can cause both reversible and permanent changes in canine kidneys. It appears likely that the extent of damage

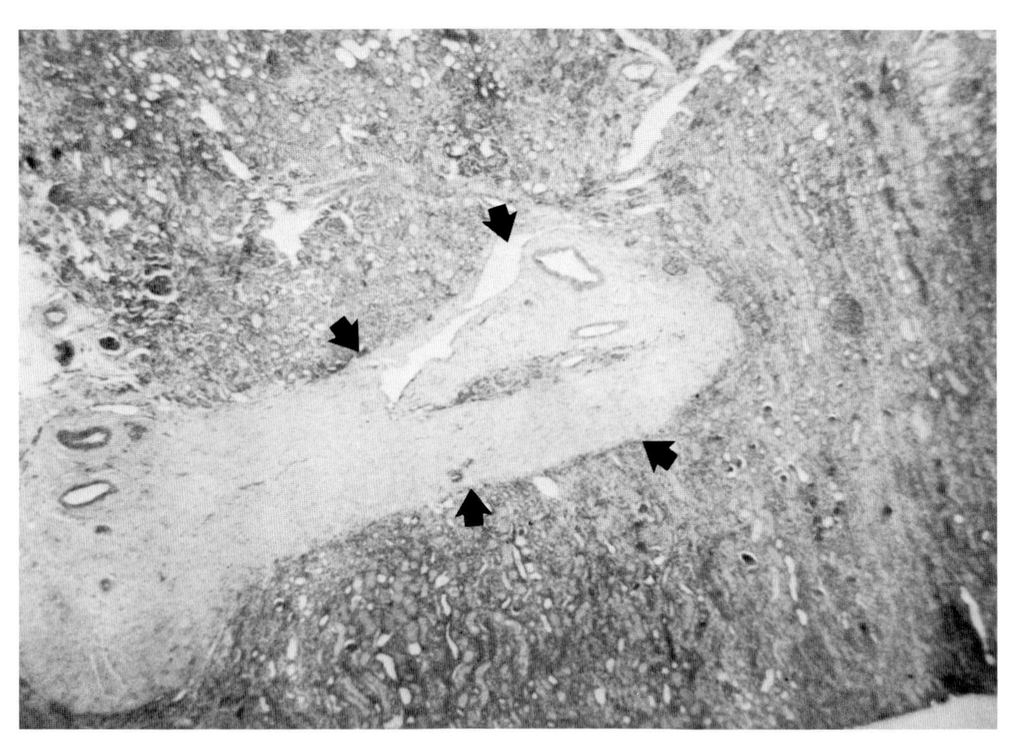

Fig. 3-5. Low-power view through shocked area of a canine kidney that received 8,000 shocks; the animal was killed 30 days after ESWL. The arrows outline an area of fibrosis (scar tissue) which is evidence of permanent change. (Courtesy of Ray Hackett, M.D., Department of Pathology, University of Florida, Gainesville, FL.)

is related to the number of shocks given and the power used; however, additional studies will be necessary to clarify this area.

## SUMMARY

It is extremely difficult to study the effect of high-energy shock waves on tissue, and indirect imaging techniques do not really substitute for pathologic and electron microscopic examination.

In most circumstances, quantitative radionuclide renography did not demonstrate significant differences on pre- and post-ESWL studies. QR, perhaps in conjunction with retrograde pyelography, is very useful in the patient with an iodine allergy. If renal function is compromised at the outset, use of this modality before and after ESWL will assist the urologist in monitoring the patient's progress or lack thereof.

Renal sonography is helpful in patients with lucent stones not visible without the use of contrast. Ultrasound should also demonstrate the presence or absence of hydronephrosis. If dilation is present, the patient should be monitored closely until fragments are passed and resolution occurs.

MRI is particularly sensitive in detecting intraparenchymal changes (Figs. 3-6, 3-7), whereas CT may demonstrate perirenal changes more distinctly (Fig. 3-8). Neither MRI nor CT is necessary on a routine basis; they should be used when patient's clinical course does not correlate with findings on more common (and less expensive) radiographic studies, and more information is needed to make clinical decisions.

Few data are available on the pathologic effects of shock waves in humans. Studies in dogs have demonstrated that ESWL can cause both reversible and permanent histopathologic changes in the kidney. It appears that damage is more likely when higher numbers of of shocks are used. Many questions remain unanswered.

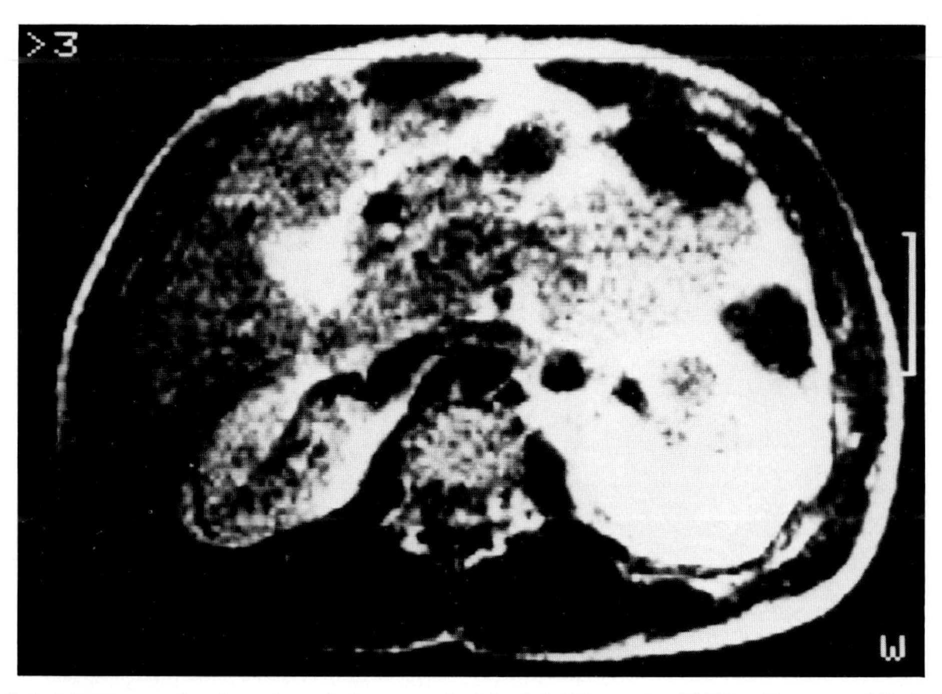


Fig. 3-6. MRI image showing edema (enlargement) of the left kidney after ESWL. (Courtesy of D. Jocham and J. Lissore, Ludwig-Maximillian University, Munich, F.R.G.)

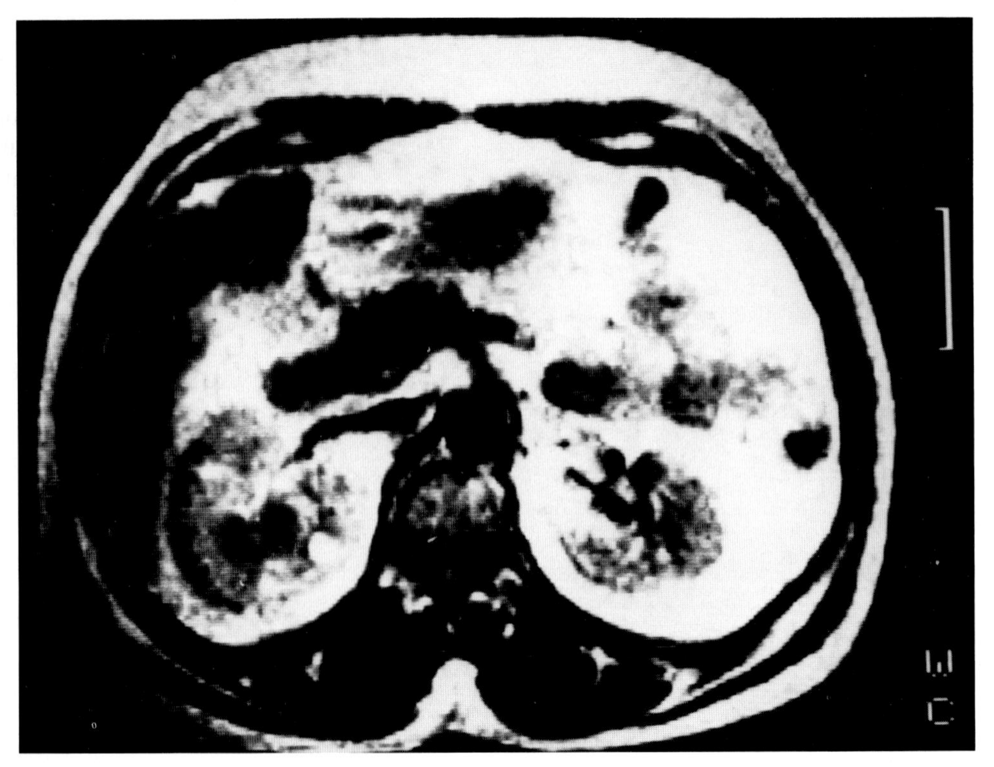

Fig. 3-7. MRI image showing subcapsular bleeding in the right kidney after lithotripsy. (Courtesy of D. Jocham and J. Lissore, Ludwig-Maximillian University, Munich, F.R.G.)

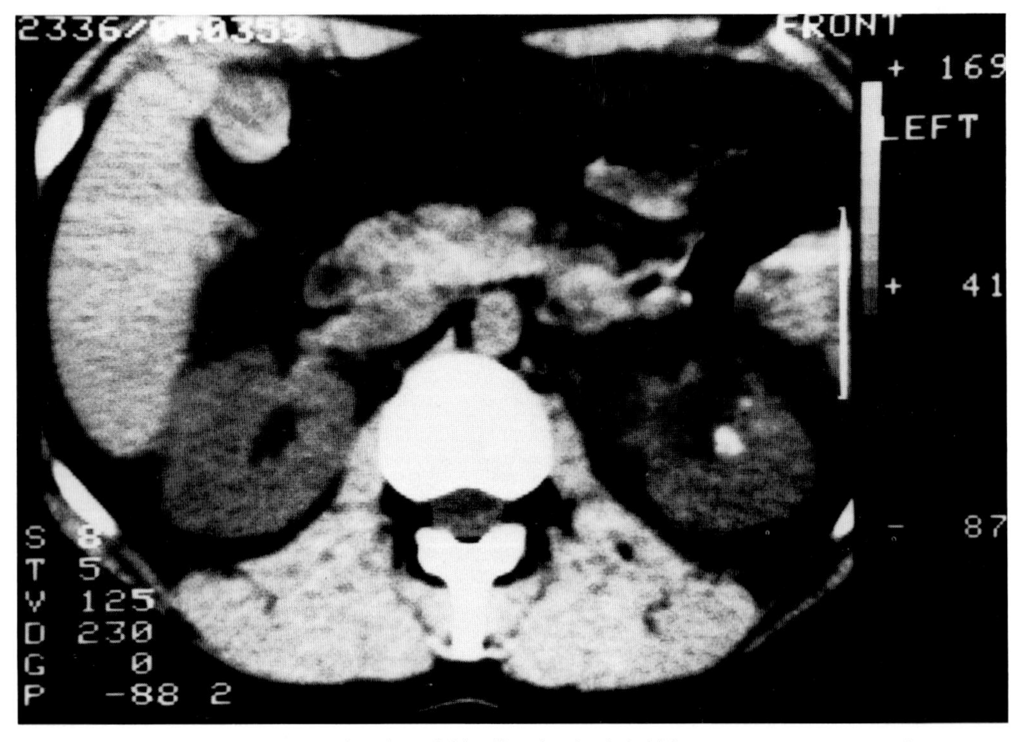


Fig. 3-8. Typical CT appearance of perirenal bleeding in the left kidney; operative intervention was not necessary in this case. (Courtesy of D. Jocham and J. Lissore, Ludwig-Maximillian University, Munich, F.R.G.)

If damage does occur, what are the long-term effects? Is it better to give 4,000 shocks in one session or to give 2,000 shocks followed by another 2,000 shocks 2 to 5 days later? Logic suggests that renal damage incurred after 2,000 shocks would not be totally repaired after 2 to 5 days. In short, the window of safety for shock wave lithotripsy is not presently known. Caution is in order until more data have been gathered.

## ACKNOWLEDGMENTS

This study was supported in part by NIH grant AM20586 and by a grant from Uro-Tech Management Corporation.

## REFERENCES

1. Drach GW et al: Report of the United States cooperative study of extracorporeal shock wave lithotripsy. J Urol 135:1127, 1986
2. Jocham D, Beer M, Rath M, Schmeidt E: ESWL-related changes in kidney imaging procedures. Syllabus from the Third World Congress in Endourology, New York, September 1985, E 8
3. Gillenwater J, Sturgill BC, Jenkins AD: Anatomic changes in kidneys recently treated with shock waves. Syllabus from the Third World Congress in Endourology, New York, September 1985, E 20
4. Grantham JR, Millner MR, Kaude J et al: Renal stone disease treated with extracorporeal shock wave lithotripsy: Short term observation in 100 patients. Radiology 158:203, 1986
5. Chaussy C, Schmeidt E, Jocham D, et al: Extracorporeal shock wave lithotripsy (ESWL) for treatment of urolithiasis. Urology (special issue) 23:59, 1984
6. Jocham D, Brandl H, Chaussy C, Schmeidt E: Treatment of nephrolithiasis. p. 35. In Gravenstein JS, Peter K (eds): Extracorporeal Shock-Wave Lithotripsy for Renal Stone Disease—Technical and Clinical Aspects. Butterworths, Stoneham, MA, 1986
7. Kaude JV, Williams CM, Millner MR, et al: Renal morphology and function immediately after extracorporeal shock-wave lithotripsy. AJR 145:305, 1985
8. Chaussy C, Schmeidt E: Extracorporeal shock-wave lithotripsy for kidney stones: An alternative to surgery? Urol Radiol 6:80, 1984
9. Chaussy C (Ed). Extracorporeal Shock-Wave Lithotripsy. New Aspects in the Treatment of Kidney Stone Disease. Karger, New York, 1982
10. Brendel W. Effect of shock-waves on canine kidneys, p. 141. In Gravenstein JS, Peter K (eds): Extracorporeal Shock-Wave Lithotripsy—Technical and Clinical Aspects. Butterworths, Stoneham, MA, 1986

# Clinical Studies Documenting Renal Change after ESWL

Dieter Jocham

The discharge of stone particles after extracorporeal shock wave lithotripsy (ESWL) is a dynamic process that requires radiologic monitoring. These imaging techniques may also be used to study the kidneys after shock wave treatment. Routine monitoring after ESWL has revealed a few isolated clinically significant changes caused by subcapsular and perirenal hematoma.[1–4,6] Concurrent reports by many ESWL users show overwhelmingly that even in the case of intrarenal hemorrhage, surgical intervention is rarely needed. For example, in a series of 2,800 patients treated in Munich, there was only one case of hemorrhage, from a nephrostomy tube tract in a patient who underwent percutaneous lithotripsy followed 5 days later by ESWL. A purse-string ligature of the bleeding source in the fistula was performed.

Sonography and single kidney–ureter–bladder (KUB) x-ray films serve as routine diagnostic procedures for ESWL patients. Follow-ups ($n$ = 100 patients) of renal function in ESWL-treated patients for more than 4 years at our clinic by means of [$^{131}$I]iodohippurate scintigraphy have confirmed that no impairment of renal function occurs after ESWL. Therefore, this examination need not be performed routinely. Only in individual cases at routine follow-ups may alterations in the kidney imaging be observed. Obviously, a specific explanation for these renal alterations is needed. Animal experiments were conducted at the Institute for Surgical Research at the Ludwig-Maximilians University, Munich, under the guidance of Professor Brendel. The histopathologic findings of the tissue in correlation with the results from various monitoring methods provided precise descriptions of the renal alterations.[1]

Because there is a possible correlation between the appearance of pathologic changes (e.g., edema, hematoma) and the applied shock wave energy, a prospective study of 20 patients having similar partial or total staghorn stone constellations was conducted at the Munich ESWL Center. The groups were treated with ESWL (16 patients), percutaneous lithotripsy (PL) (2 patients), or open operation (2 patients). The goal of these studies was twofold: first, to determine the relationship between ESWL or alternative therapies and subcapsular, parenchymal, and perirenal alterations; and second, to establish the value of monitoring techniques for the assessing these changes.

The goal of this study was to estimate the effects of different treatment modalities (ESWL, percutaneous nephrolithotomy, surgery) on human kidneys, and also to compare various monitoring techniques for evaluating renal changes. The sensitivity of several imaging techniques in detecting renal changes after ESWL was also

**Table 3-2. Protocol of Studied Monitoring Methods as Applied to Modern Stone Therapies[a]**

| Exam Performed[b] | Before ESWL 1–5 hours | 24 hr after ESWL | 7 days to 3 months after ESWL | Equipment Used/Clinical Parameters |
|---|---|---|---|---|
| MRI | + | + | (+) | Siemens Magnetom 0.35 T SE TR 400 (35/90) 2000 (35/90) (TE) |
| CT | (+) | + | (+) | Siemens Somatom II scanner |
| Sonography | + | + | + | Toshiba Sono-Layer SAL 35-A |
| Renography | + | + | (+) | Ohio Nuclear Sigman 410 |
| IVP | + | + | − | Siemens Urograph |
| Clinical examination | + | + | + | Pain, confinement to bed |

[a] Various diagnostic procedures are outlined as well as the apparatus used; in addition, the patient's subjective impairment owing to various therapeutic procedures is evaluated.
[b] Examinations were performed after ESWL (16), open surgery (2), and percutaneous nephrolithotomy (2).

tested. The patients (well hydrated) were evaluated by a clinical examination, serologic evaluation, magnetic resonance imaging (MRI), renal sonography, and a nuclear renogram ([$^{131}$I]iodohippurate scintigraphy) (Table 3-2).

Twenty-four hours after the treatment, all examinations were repeated and supplemented with computed tomography (CT). In cases where pathologic alterations were noticed, studies were repeated periodically up to 3 months. Renal enlargement (see Table 3-3) corresponding to edema as seen from animal experiments can be seen after shock wave therapy. The MRI scan in Figure 3-9 shows an example of transient renal edema after ESWL. This type of edema can also often be seen sonographically as only a diffuse enlargement of the kidney. Such temporary parenchymal changes are typical after ESWL and were not found after either percutaneous nephrolithotripsy or open surgery. The severity of swelling after ESWL increases as the number of shock waves administered increases.

After ESWL, subcapsular bleeding (Fig. 3-10) was noted uniformly only after a large number of shocks. In these cases (Table 3-3) any diminished function of the treated kidney was found to be temporary and completely reversible. Figure 3-11 shows the typical CT appearance of perirenal bleeding after ESWL. Even in this case, operative intervention was not required. Table 3-3 summarizes the inci-

**Table 3-3. Post-ESWL and Post-Procedural Renal Alteration as Observed by Various Monitoring Techniques and Clinical Parameters[a]**

| Lesion | ESWL, 1,500 Shocks, Percent ($n$ = 11) | ESWL, 2,000 Shocks, Percent ($n$ = 5) | Open Surgery or Percutaneous Nephrolithotomy ($n$ = 4) |
|---|---|---|---|
| Intrarenal | | | |
| Edema (mild) | 54 | 60 | 0 |
| Edema (severe) | 27 | 40 | 0 |
| Hematoma < 1 cm | 37 | 40 | 50 |
| Hematoma > 1 cm | 0 | 20 | 25 |
| Perirenal | | | |
| Hematoma < 1 cm | 45 | 80 | 0 |
| Hematoma > 1 cm | 10 | 20 | 100 |
| Clinical | | | |
| Confined 24 hr to bed | 9 | 0 | 100 |
| Decrease of renal function (20%)[b] | 18 | 40 | 0 |

[a] Monitoring techniques used: QR, renal sonography, MRI, CT. For ESWL, the number of shock waves administered correlates directly with the number and severity of resulting renal changes. The subjective impairment of patients with ESWL is remarkably less than after percutaneous nephrolithotomy or open surgery.
[b] [$^{131}$I]iodohippurate scan before and after treatment.

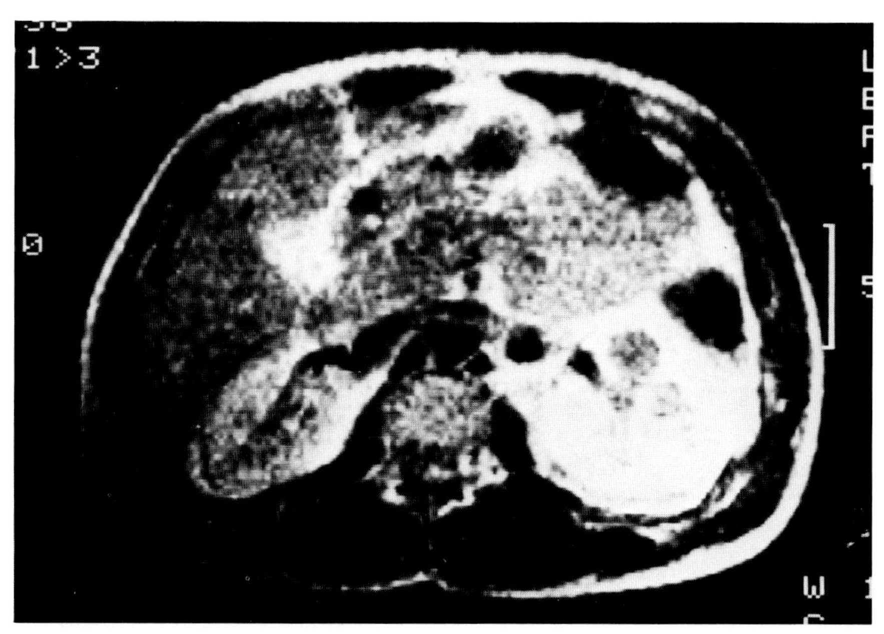


Fig. 3-9. MRI showing edema of the left kidney after ESWL. The kidney is remarkably enlarged and shows a nonuniform parenchymal structure. (Courtesy of the Department of Radiology (Professor Dr. J. Lissner, Director) of the Ludwig-Maximilians University, Munich, F.R.G.)

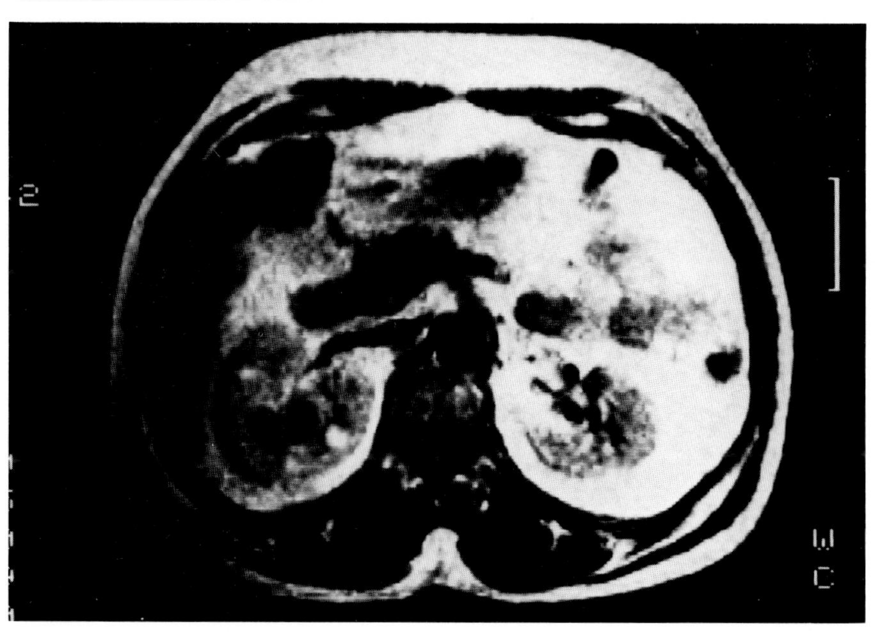

Fig. 3-10. MRI showing subcapsular bleeding of the right kidney. (Courtesy of the Department of Radiology (Professor Dr. J. Lissner, Director) of the Ludwig-Maximilians University, Munich, F.R.G.)

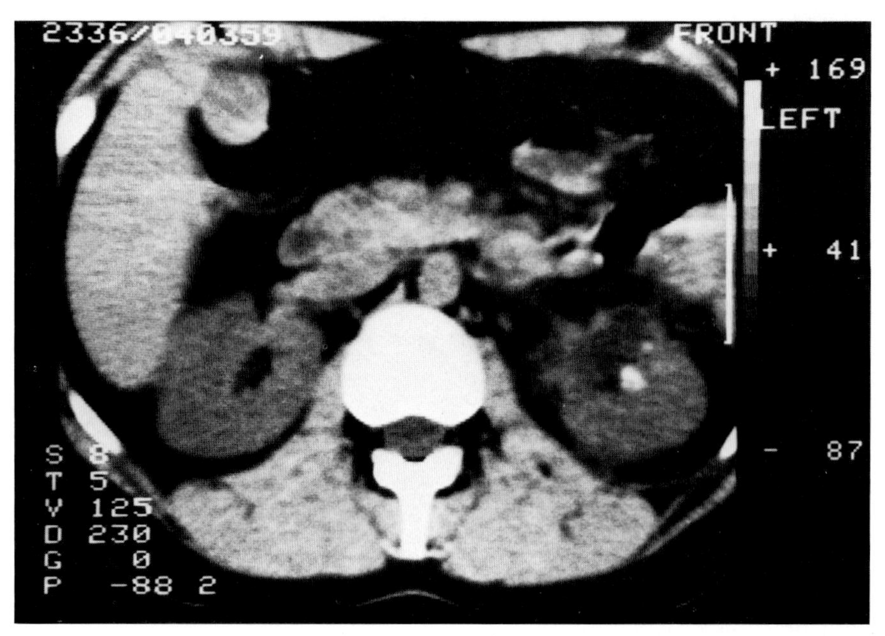


Fig. 3-11. CT image showing perirenal bleeding of the left kidney. The margin of the hematoma is clearly recognizable. According to our experience, even such a finding as this does not require surgical intervention. (Courtesy of the Department of Radiology (Professor Dr. J. Lissner, Director) of the Ludwig-Maximilians University, Munich, F.R.G.)

dence of iatrogenic changes after ESWL, percutaneous lithotripsy, and open surgery.

Temporary parenchymal edema is common after ESWL. The degree of edema correlates with the number of shocks administered (Table 3-3). Animal studies[1] have shown that edema and subcapsular bleeding can be induced even by few shocks (<500 individual shocks). In humans, parenchymal edema was found in 81 percent of patients treated with 1,500 shocks and in all patients after 2,000 shocks (Table 3-3).

The incidence of subcapsular bleeding, as detected clinically or radiographically, was greatest in patients treated percutaneously or by open surgery. However, subcapsular bleeding was also noted after lithotripsy using high shock wave energies (Table 3-3). Perirenal bleeding was found primarily after open operation and nephrolithotomy (Table 3-3). In 40 percent of cases, scintigraphic analysis of kidney function showed reduced function in the treated kidney after ESWL using higher shock wave energies; occasionally, the reduction was quite pronounced. Those patients treated with other techniques showed no postoperative decrease

**Table 3-4. Sensitivity of MRI, CT, and Sonography in Detecting Changes after Therapy for Renal Calculi[a]**

| | MRI (%) | CT (%) | Sonography (%) |
|---|---|---|---|
| Intrarenal alternations ($n$ = 14) | 100 | 71 | 78 |
| Perirenal alternations ($n$ = 15) | 73 | 100 | 40 |

[a] Therapies used: ESWL ($n$ = 16), percutaneous nephrolithotomy ($n$ = 2), and open surgery ($n$ = 2). Intrarenal changes (edema or hematoma) are reliably detected by MRI; perirenal alterations are better detected by CT. Sonography is inferior to MRI or CT in detecting discrete perirenal changes. However, sonography suffices as a routine diagnostic procedure because it normally detects clinically relevant renal changes.

in kidney function. In all cases, the decreased function was completely reversed as measured by renography by the time the changes detected by kidney imaging were no longer visible.

In addition to establishing the identity of the radiographic changes discernible after therapy, these studies also provided an indication of the sensitivity of the different imaging techniques (Table 3-4). MRI was superior to CT and sonography in detecting intraparenchymal changes. Perirenal changes, however, were best seen with CT.

## COMMENTARY

In spite of the fact that the size of our series was limited by cost, we noted that ESWL did induce subcapsular changes (edema) which were not observed after open surgery or percutaneous nephrolithotomy. These typical ESWL changes were transitory, dosage-dependent, and significantly increased after 1,500 shocks. In our study, edema usually resolved by 1 week after ESWL; whereas evidence of hematoma was absent after 6 weeks. The changes occasionally coincided with a diminution of renal function as determined by [$^{131}$I]iodohippurate scan.

Based on this study, we feel ESWL patients can be followed in a clinically effective, cost-efficient manner by obtaining periodic sequential abdominal flat plates and by renal sonography. Only when there is a discrepancy between these radiographic findings and the patient's clinical course is CT or MRI called for. To reduce ESWL-related risks, the Munich group recommends using less than 2,000 individual shocks at 18 kV for treatment of stones in a single kidney.

## ACKNOWLEDGMENT

Support for this study was provided by M. Beer, M.D., E.A. Moser, M.D., M. Rath, M.D., and Dornier Medical Systems, Inc.

## REFERENCES

1. Brendel et al: Personal notes on results from animal experiments. Institute for Surgical Research, Ludwig-Maximilians University, Munich, Klinikum Grosshädern, FRG 1985
2. Jocham D, Schmiedt E, Walther V et al: In Chaussy C (ed): Extracorporeal Shock Wave Lithotripsy—New Aspects in the Treatment of Kidney Stone Disease. Karger Vlg., Basel, 1982
3. Jocham D, Brandl H, Chaussy C, Schmiedt E: Treatment of nephrolithiasis. In Gravenstein JS, Peter K (eds): Extracorporeal Shock-Wave Lithotripsy for Renal Stone Disease—Technical and Clinical Aspects. Butterworths, London, 1986
4. Kaude JV, Williams CM, Millner MR, Finlayson, B: Magnetic resonance imaging of the kidney after ESWL. In Gravenstein JS, Peter K (eds): Extracorporeal Shock-Wave Lithotripsy for Renal Stone Disease—Technical and Clinical Aspects. Butterworths, London, 1986
5. Meeting report of the Annual Conference of the German ESWL Users in Mainz, West Germany, October 1, 1985

4

# Design Considerations for the ESWL Unit

Thomas W. Schoborg

The concept of treating symptomatic stone-forming patients by referral to a stone center has evolved over the last two decades.[1,2] Therefore, in discussing the necessary design and equipment of an ESWL unit, the concept of complete patient care, whatever the individual modality, must be considered. The overall successful management of renal and ureteral calculi is not a matter of extracorporeal lithotripsy alone, but often involves multiple endoscopic and percutaneous modalities. Facilities for accurate preoperative radiologic assessment, auxiliary percutaneous or cystoscopic procedures, ureteroscopic manipulations, chemolysis, ultrasonography, and renal scans must be available.[3] Subsequent outpatient visits to determine any underlying metabolic derangement should be anticipated,[4] and proper office space must be designed for administrative support personnel.

All activities of the stone center can be coordinated through one central office, which should encompass a Stone Registry—a computerized record constituting a database consisting of patients' sociologic and insurance information, stone type, treatment modalities, successes of treatment, and medical evaluation. If feasible and consistent with the goals of the institution, a research facility with clinical and biochemical divisions would complete the stone center concept.

## STONE CENTER CONCEPT: CONSIDERATIONS IN DESIGN

The ideal concept in design would consist of the ESWL unit as an integral part of the comprehensive stone center. Within the stone center, all requirements and necessary facilities for the successful treatment of urinary (and possibly biliary) calculi would be available and readily accessible within the same physical area. At a minimum, it is necessary to have the cystoendoscopy suite next to the ESWL suite. Ancillary facilities such as x-ray and endourology suites, reception, and patient recovery rooms should be nearby, but not necessarily adjacent to this core facility if existing installations are efficient and adequate.

### Design: The ESWL Suite

Since only one type of lithotripter is currently available, designs for ESWL suites throughout the world have been fairly standardized. Inpatient and outpatient units may have slightly different requirements in terms of recovery room beds, ambulatory patient facilities (lockers, dressing rooms, etc.), ambulance entrances, and anesthesia induction rooms. Of course, inpatient hospital beds are best situated as close as possible to the lithotripter. The lithotripter suite must accommodate the patient, anesthesiologist, urologist, and technical support staff. Spatial and economic limitations of each facility will dictate individual design considerations. The teaching responsibilities of a training center must be anticipated, and a variable number of operators with their own patients must be served. It seems appropriate to design the ESWL suite in such a way that the operator can perform several activities simultaneously as efficiently as possible. This would mean that the x-ray view box, telephone, dictaphone, observation window (if present), and control module should be centralized within reach of the urologist. At the same time, proper visualization of the patient, anesthesiologist, monitors, image intensifiers, and personnel would be facilitated. An attempt to achieve this optimum design has been made at the Atlanta Stone Center of Georgia Baptist Medical Center (Figs. 4-1 and 4-2). If a visitor's observation room is present, blinds should be installed for situations requiring privacy as requested by the operator or patient.

Since endourology and shock wave lithotripsy are often both used in the complex stone patient, facilities for both procedures should be close together. The design of the ESWL suite must, therefore, include plans for one or preferably two endoscopic (endourology) suites. One would be used mainly for cystoscopic, ureteroscopic, and ureteral stone manipulation procedures, and the other for percutaneous nephroscopic instrumentation. The two would be equipped slightly differently, as described later.

In addition, a nearby surgical suite, although not mandatory, would be helpful when operative intervention is necessary. A separate room equipped for overhead x-rays for examining pretreatment and posttreatment radiographs as well as intravenous pyelograms would also be a useful addition.

Finally, to complement the treatment facility of the stone center, patient reception, prepara-

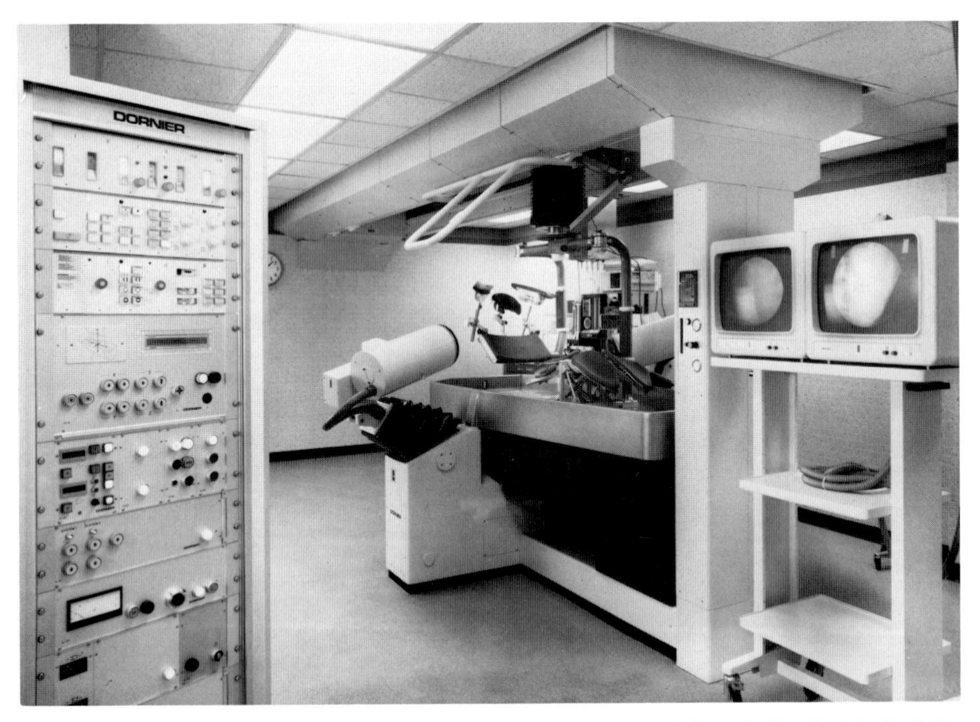


Fig. 4-1. View afforded operator and observer (Atlanta Stone Center of Georgia Baptist Medical Center, Atlanta, GA).

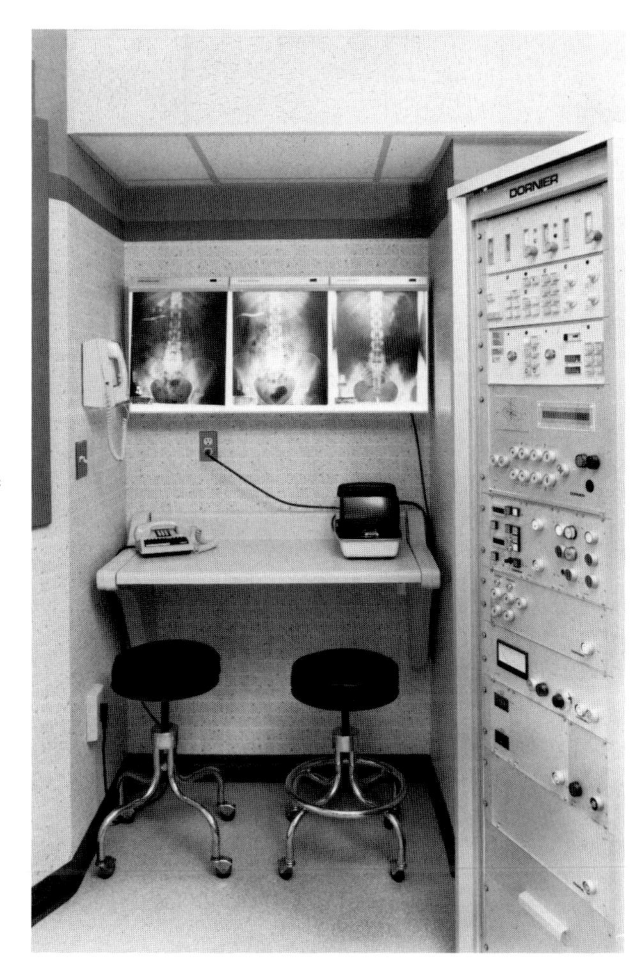


Fig. 4-2. Operator station (Atlanta Stone Center of Georgia Baptist Medical Center).

tion and recovery areas, including a floor for admitted patients, could be considered. Medical management in the form of chemolysis could be conducted on this urologic wing of the center, as well as inpatient metabolic evaluation of the recurrent stone-forming patient. Outpatient metabolic evaluation programs can be conducted through the office of the urologist or through a separate laboratory facility.

Coordinating all these activities would be a central office where medical and nonmedical personnel could organize and record procedures and clinical research. This central office would, of necessity, include a stone registry containing separate records of patients with a diagnosis of urolithiasis as well as the computer capability for recording and producing information regarding details of individual procedures and clinical research projects. The central office would also have reception and discharge capabilities in the event the number of outpatient procedures increased.

A separate research facility investigating biochemical and physiochemical aspects of urolithiasis would coordinate its activities with the stone center, depending on the academic and pharmaceutical affiliations afforded the center.

In summary, various levels of development of the ESWL suite are possible, considering the limitations imposed by physical and budgetary constraints. Starting with the Dornier-recommended ESWL suite and progressing to the more complex and comprehensive stone center, the following list gives the options in design of the lithotripsy suite or stone center (Figs. 4-3 through 4-5).

1. ESWL suite as proposed by Dornier Medical Systems, Inc. (Fig. 4-3).
2. ESWL suite and adjoining endourology suite as separate in-hospital facility with spatially separate surgical suites, x-ray facilities, recovery room, and urology floor (Fig. 4-4).
3. ESWL suite adjoining endourology and x-ray suite; other facilities spatially separate as available (Fig. 4-4).
4. Comprehensive stone center: ESWL suite, two adjoining endourology suites (ureteroscopy and percutaneous nephroscopy), adjoining x-ray facility, patient reception, preparation, and recovery rooms, and adjacent urology floor with central office for research, records, and data control; outpatient services available (Fig. 4-5). Furthermore, potential additional space allocation would be available for newer kidney stone treatment modalities as well as newer technology for the treatment of gallstones. All facilities would be in close functional proximity to an operating room.

Although specific architectural features of the ESWL suite are discussed later in this chapter, Alder[5] appropriately lists the basic features:

1. Insulation and sound absorptive materials in walls, floor, and ceiling

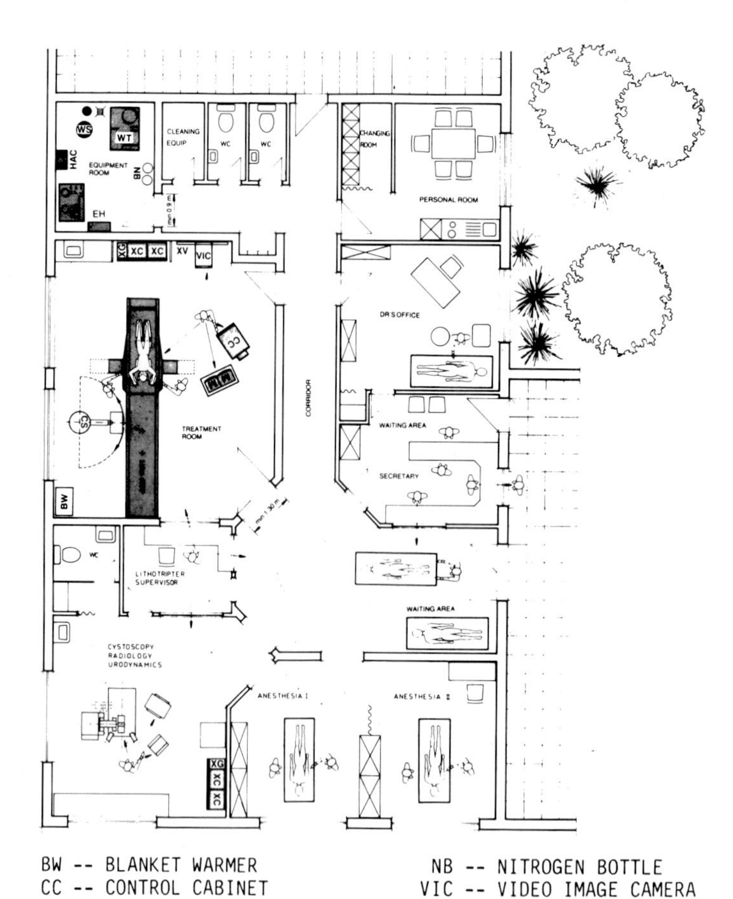


BW -- BLANKET WARMER
CC -- CONTROL CABINET
CS -- CEILING-MOUNTED SUPPLY UNIT
EH -- ELECTRIC BOX/HYDRAULIC
HA -- HYDRAULIC AGGREGATE
HAC -- HYDRAULIC ACCUMULATOR
M -- MONITOR
NB -- NITROGEN BOTTLE
VIC -- VIDEO IMAGE CAMERA
WS -- WATER SOFTENER
WT -- WATER TREATMENT
XC -- X-RAY CABINET
XG -- X-RAY GENERATOR
XV -- X-RAY FILM VIEWER

Fig. 4-3. ESWL suite design recommended by Dornier. (Courtesy Dornier Medical Systems, Inc., Marietta, GA.)

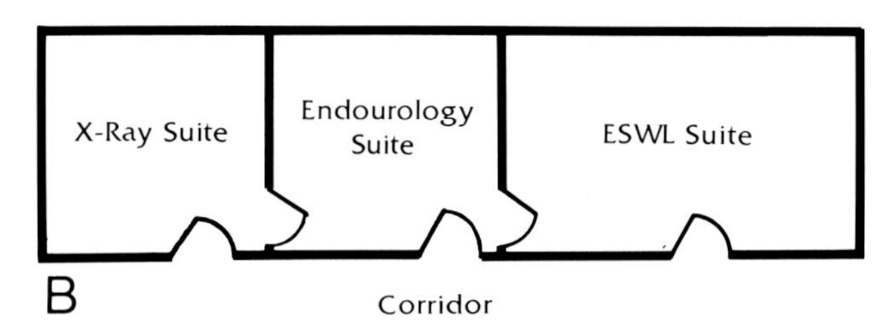


Fig. 4-4. Hypothetical floor plans (excluding restrooms, offices, and storage). **(A)** ESWL and endourology suites as separate facilities (endourology suite also serves as patient preparation suite). **(B)** ESWL suite with adjoining endourology and x-ray suites as separate facility (endourology suite serves as patient preparation area).

2. Wall coverings of vinyl or fabric to absorb sound
3. Floor stress/load bearing proportionate to equipment weight (specifically to water-filled bath)
4. Solid wall construction on bearing plate wall
5. Ceiling height at least 10 ft, 4 in. in the treatment room above the bath
6. Several levels of available lighting with rheostat or separate controls
7. Possible removal of several inches of concrete to allow fitting of base plate supplied by manufacturer
8. Equipment room should be adjacent and insulated to reduce noise from the hydraulic system and water degassifier
9. Absolute minimum space for treatment area, 400 ft$^2$ (ideal: 450 to 600 ft$^2$)
10. Absolute minimum size of equipment room, 150 ft$^2$

## THE FUTURE: SECOND GENERATION LITHOTRIPTERS

Although not directly related to present design of ESWL suites, it must be remembered that alternate lithotripters are currently being developed and evaluated. The specifics describing

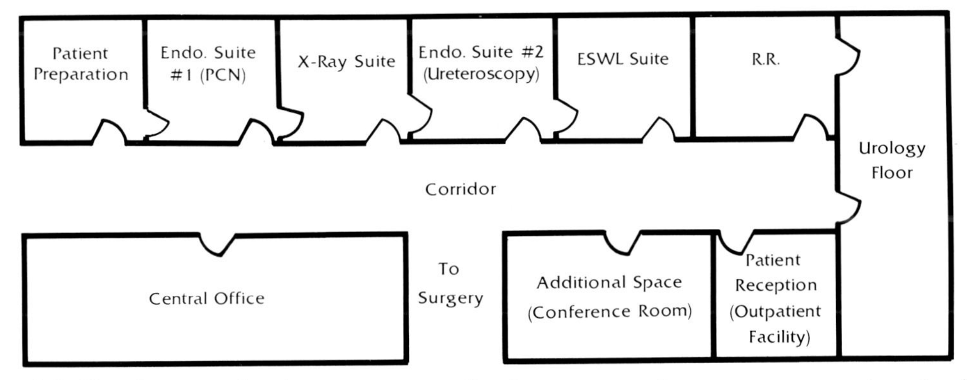


Fig. 4-5. Plan for comprehensive stone center; all suites necessary for stone management are in close continuity.

alternate technologies are available in a recent report of the American Urological Association ad hoc committee to study the safety and clinical efficiency of current technology and noninvasive lithotripsy.[6] In the future, different methods of imaging, patient positioning, shock wave generation, and anesthesia will have to be accommodated in the suite design. If it becomes possible to incorporate a tubless lithotripter into a multipurpose fluoroscopy table, then the ESWL room will have to be equipped for other procedures as well. In addition, new localization techniques may necessitate room temperature control and ventilation for computer components.

In addition to alternative second-generation lithotripters, Dornier Medical Systems, Inc. is developing a mobile lithotripter unit (personal communication, Thomas Brooks, Dornier Medical Systems, Inc., Marietta, GA, November 8, 1985). Installed in a truck, the lithotripter unit with accompanying x-ray fluoroscopy units, water degassifying and softener facilities, and self-contained power generator will couple with the receiving facility (hospital) which would supply the connections for auxiliary power, water, and drainage. Obviously, the design for such an ESWL suite must be modified to assure ground floor location, proximity of access, and relative isolation from daily hospital or outpatient traffic. The constant factor in any present or proposed unit is a well-designed and equipped endourology suite close to the lithotripter to allow preoperative preparation of the stone patient for lithotripsy and immediate postoperative management of any associated problems.

## SPECIFIC ARCHITECTURAL REQUIREMENTS AND RECOMMENDATIONS

*J. David Mullins*

Once patient load and flow have been determined for a given institution, they must be translated into a physical reality. The addition of a lithotripsy unit to a hospital may be in the form of renovating existing space, a newly constructed addition, or a freestanding facility.

### Functional Location and Support

The siting and placement of a lithotripsy unit is first determined by its functional needs in the context of the hospital and its operations. A location near existing fluoroscopic and cystoscopic rooms is desirable, and can save expense and expedite patient flow.

The suite must have an area adequate for the lithotripter treatment room, direct support rooms, and ancillary rooms specific to the hospital's needs. The basic rooms consist of the treatment room, the equipment room, a patient preparation/recovery room, and storage. Dornier recommends a room size of 7 × 7 m for the treatment room; however, a size of 6 × 6.5 m is considered more optimal. An adjacent patient preparation room can be 3 × 4 m. If it is decided to combine patient preparation with a patient recovery room, a minimum size would be 5 × 5 m. The equipment room must contain enough area for all support equipment. Its minimum width must be 2 m to adequately contain and service all equipment. The actual room configuration depends upon the specific floor plan.

### Ancillary Rooms

A cystoscopy room adjacent to the patient preparation room has proven to be helpful in performing percutaneous procedures in combination with lithotripsy. If provided, a standard-sized cystoscopy room (5 × 5 m) should be equipped with C-arm x-ray and an overhead x-ray with an endourologic table. Shielding can be minimized if the cystoscopy room is located on a common wall with the lithotripter treatment room.

An observation room, positioned behind the lithotripter control panel, provides a view of the control panel monitors, patient and tank, and anesthesiologist. Though this room should be no smaller than 2 $m^2$, the actual size depends on whether it is used as a physician consultation area or as a teaching facility, or both.

The following rooms should be considered for inclusion, if space and budget allow:

1. Physician's office—dictation, charting, and consulting space is often needed for both staff and consulting physicians
2. A dedicated x-ray room and darkroom which allows improved patient care and workflow within the suite
3. Storage room for supplies and spare parts

## Design Parameters for the Lithotripter Treatment Room

As in all health care occupancies, a lithotripter suite design must meet all building and safety codes. In the United States, the Joint Commission on Accreditation of Hospitals recognizes the Life Safety Code of the National Fire Protection Association (NFPA 101 and NFPA 220). All design aspects of the lithotripter suite in regard to egress from the suite, fire safety, and construction must conform to NFPA and local building code standards.

The most demanding physical design parameters are those for the lithotripter treatment room. Physical considerations include the following:

1. Weight of the lithotripter versus available structure
2. Area of room and ceiling height
3. Materials appropriate for the conditions
4. Sound transmission
5. Radiation
6. Lighting and electrical requirements
7. Mechanical requirements

While specific requirements depend upon the actual installation, basic parameters of these conditions must be rigidly observed.

## Structural Requirements

The Dornier lithotripter tub contains 850 L of water when filled. When this is combined with the weight of the machine and the patient, the live load adds up to 70 lb/ft$^2$. Patient care areas in most hospitals often exceed this figure, having a load capacity of 100 lb/ft$^2$ or greater; however, standard building codes require a live load capacity of only 40 lb/ft$^2$ in patient ward areas. This load-bearing capacity often dictates that the lithotripter unit be located on the ground floor on a slab-on-grade, which solves structural requirements as well as reducing sound transmission and improving accessibility to the suite for installation of large pieces of the lithotripter equipment. A slab-on-grade prevents noise transmission to any space below and with proper isolation, can prevent sound transmission via vibration to adjacent spaces.

## Area and Volume Requirements

Dornier recommends a treatment room size of 7 × 7 m; a more optimal size is 6 m wide × 6.5 m long. Rooms slightly less than 5 m wide have proved satisfactory; however, this generally results in a treatment room with a less-than-optimum anesthesiologist's work space, cramped personnel circulation, and small work/charting areas.

The treatment room should have adequate space for the tub, control panel, monitors, electrical panels, anesthesiologist's cart and column, nurses' charting area, doctors' view box and counter, holding area for stretchers, blanket warmers, and supplies, and adequate circulation space for personnel and patient. All requirements translate to a length of 7 m and a width of 6 m.

The height available for the treatment room is equally important. The main ceiling minimum height requirement is 2.55 m, and the required distance from the finished floor to ceiling above the lithotripter lift guideway is 3 m. An additional 20 cm is required for clearance between the ceiling and any cable trays running above the guideway ceiling. This, plus space between the cable trays and the deck above, could result in a clear height between floor and deck of 3.4 m. Because of the extensive plumbing and electrical requirements in the ceiling plenum area, this clear height should be considered a minimum. Site selection in a renovated space should carefully observe available height clearances.

## Materials

Material selection should conform to all code requirements. For example, if the treatment room shares a common wall with a corridor, the wall materials must conform to the fire rating requirements of the corridor wall. Materials should be selected for function, durability, and patient comfort.

The floor must be durable, easy to clean, impervious to water, and nonslip. Various non-skid vinyl floorings are available (such as Mipolam by Dynamit Nobel) which are easily cleaned to maintain infection control standards.

Walls can be of any material normally found in a hospital. Glazed ceramic tile block can be used because it is easily cleaned and reduces airborne sound transmission; however, this material increases sound reverberation within the room and has a "cold" feel. A staggered metal stud wall with fiberglass sound batts and gypsum wallboard significantly reduces sound transmission and when covered with a class A vinyl wallcovering is moisture resistant and can be soothing to an anxious patient.

## Miscellaneous Finishes

Ceilings should be lay-in acoustical tile with acoustical sound batts on top of the ceiling. Doors may be standard hospital doors, but the recommended width of all doors into the treatment room is 4 ft 0 in. The main access door should have a minimum clearance width of 4 ft 3 in. to allow installation of the oversize equipment. Should the door require a fire rating, the problem may be solved by a 4 ft 0 in. leaf and a 1 ft 0 in. leaf with an astragal. All doors and frames into the treatment room must be lead-lined.

## Sound Transmission

Sound transmission through air and solids is an important design consideration in a lithotripter treatment room. Airborne readings at the tub can exceed 95 decibels. Decibel readings on the slab from 3 to 5 ft from the tub decrease from 73 dB to 64 dB, respectively (approximately the sound of a slamming door). As discussed in previous sections, transmission of sound to the exterior of the room can be attenuated by fiberglass sound batts in the ceiling and walls, attention to special wall construction, and isolation of the floor slab. Additional sound transmission prevention methods can include thresholds under entry doors and isolation hangers on duct work and light fixtures. A combination of heavy wall materials (i.e., concrete block or ceramic tile block) and sound dampening wall materials (i.e., sound attenuation blankets in a gypsum wallboard and metal stud wall) can reduce sound levels by 25 dB to approximately 51 dB outside the treatment room (a level at which further sound reduction methods would be considered unnecessary).

## Radiation

Since the Dornier lithotripter employs the use of two standard x-ray tubes, the treatment room requires standard radiologic shielding. While a radiation physicist must determine each facility's specific shielding requirements, shielding is generally required from the finish floor to 7 ft above the finish floor. Lead shielding thickness is generally 1.6 mm (1/16 in.) to 0.8 mm (1/32 in.). All doors and frames into the room require lead shielding. If an observation window faces onto the room, the glass must be lead glass and the frame lead lined. Finally, a movable lead glass shield is recommended for the treatment room personnel.

## Electrical and Lighting Requirements

Specific electrical requirements for a lithotripter suite depend on the size, services, and design of the given facility. Basic electrical service required for the lithotripter proper is a 480 V, three-phase, 50 amp service. Other normal electrical loads, as well as 220 V service for x-ray equipment, are usually well within the capacities of most hospital facilities. Nearby location of electrical panel boards is highly desirable.

The large number of cables required for and by the lithotripter and access to these cables demands the use of cable trays in the ceiling plenum. The amount of available vertical space for cable trays should be determined as early as possible and coordinated with other disciplines.

Lighting required for the treatment room should be versatile. High illumination (>70 foot-candles) is required for patient preparation and location, monitoring, charting, and general illumination; low levels of lighting (<40 foot-candles) are preferable for film and monitor viewing. This duality can be resolved by providing either fluorescent cove lighting or standard fluorescent troffer fixtures along both sides of the room for general illumination, and high intensity spot lights for charting, patient monitors, controls, and anesthesiologist's work areas. Variable light levels can be achieved with fluorescent lights with variable dimmer switches, or considerably less expensive multiple switching of the fixtures. Placement of microspot lights on variable dimmer switches is desirable for charting areas, anesthesia machine locations, and patient monitoring. Thus, should the physician require even complete darkness to view the monitors, the anesthesiologist can still view and monitor the patient and the nurse can continue charting.

### Mechanical Requirements

Heating and air conditioning required for the suite are well within norms found throughout a typical hospital patient care area. No special equipment should be needed within a lithotripter suite, as long as the existing equipment can easily maintain a temperature of 24° to 28°C with <85 percent humidity. Heat gain from lithotripter electrical panels and support equipment can be handled by most in-place mechanical systems. A minimum of six air changes per hour with a maximum of ten air changes per hour is required. Care should be taken with placement of supply and return air vents and grilles. Incorrect placement can result in inadequate cooling of equipment areas and exposing a wet patient to direct air currents.

Plumbing requirements include nearby adequate drain and water supply systems. The lithotripter tub requires placement of a floor drain (70 mm diameter minimum) and a minimum hospital water supply pipe of 28 mm. Hot water (~ 50°C) is required to be supplied to the equipment room.

The hydraulic patient chair requires the use of seamless stainless steel piping and tubes. A specialty supplier will be needed for this piping and related welds.

## EQUIPMENT

*Thomas W. Schoborg*

In discussing the equipping of an ESWL unit, the concept of a stone center is relevant. A stone center provides for a close partnership between endoscopic urology (percutaneous and ureteroscopic) and extracorporeal lithotripsy, ensuring the successful, efficient, and safe treatment of urinary calculi.[7] For this reason, the equipping of an ESWL suite will be discussed in terms of the individual components or suites of the stone center (Table 4-1). It must be emphasized that many types of equipment are already available, and, as with any rapidly changing technology, a rapidly changing set of accessories will appear in the future.

### Endourology Suite

As discussed above, the endourologic management of urinary calculi, as an adjunct to extracorporeal lithotripsy, would be best accommodated in two suites: one for percutaneous endoscopic procedures and one for cystoscopy and ureteroscopy.

Referring to recent publications on equipping endourologic suites,[7,8] many types of endoscopic instrumentation and accessories are avail-

**Table 4-1. The ESWL Unit: Equipment**

I. Existing stone centers
 A. Endourology suite(s)
  1. Table
   a. Portable
   b. Fixed
  2. X-ray units
   a. Overhead
    i. Portable
    ii. Fixed
   b. Fluoroscopy
  3. Endoscopic equipment
   a. Ureteroscopic
   b. Percutaneous
   c. Cystoscopic
  4. Accessory catheters
 B. ESWL suite
 C. Separate x-ray facility
 D. Office of stone center
  1. Personnel
  2. Computer for database—stone registry
  3. Reception
 E. Research
  1. Clinical—stone registry
  2. Biochemical
 F. Ancillary services
  1. Outpatient services
  2. Recovery room
  3. Physicians' and nurses' rooms
II. Future stone centers

able as well as various fluoroscopic instruments and endoscopic tables.

Regarding the percutaneous procedures, my preferred instrument for rigid nephroscopic removal of calculi is the 24 F sheathed nephroscope with ultrasonic or electrohydraulic energy sources to fracture calculi (Wolf, ACMI). The 15 F flexible nephroscope (Olympus, Pentax, ACMI) is also frequently used. Both are now primarily used in the well established method of percutaneous debulking of large staghorn calculi prior to ESWL therapy.[9] Accessory graspers such as the four-wire 4 F grasper (ACMI) are frequently needed.

At the end of the procedure, a newly designed post-percutaneous neprostomy, pre-ESWL 18 to 24 F nephroureterostomy catheter (Cook Urological) is placed after the percutaneous procedure. (Figs. 4-6 through 4-8). This design serves a three-fold purpose: first, to maintain position of the nephrostomy tube after the percutaneous procedure by means of the 5 F ureteral stent; second, to allow possible irrigation of the renal pelvis following percutaneous neprostomy (PCN) via the inner 5 F ureteral stent portion of the catheter; and third, to allow easier passage of stone debris after ESWL therapy once the inner 5 F removal ureteral stent portion of the catheter is removed. Of course, the standard (Cook) 18 to 22 F nephroureterostomy stent is also acceptable.

In equipping the cystoureterscopic suite, both standard and offset (integrated) ureteroscopes are required. Generally, for stone manipulation a 9,5 F ureteroscope with electrohydraulic lithotripsy or the 11.5 F ureteroscope with ultrasonic probe lithotripsy is used. Newer flexible ureteroscopes are now available with (Reichert) or without (Olympus, ACMI) operating channels. The Reichert nonarticulated flexible ureteroscopes are the FUS-7 (7 F) with a 1.5 F operating channel, the FUS-9 (9 F) with a 3.5 F operating channel, and the FUS-10 (10 F) with a 5 F operating channel. Newer ureteroscopic accessories (3 F) are currently available to extract stones and catheters from the ureter (Fig. 4-9). Ureteral balloon dilators and double pigtail ureteral catheters with attached 4–0 sutures for easy outpatient removal should be on hand as well.

In attempting to treat impacted ureteral calculi with ESWL, it is important to either dislodge the stone back into the kidney[10,11] or by-pass the stone with a ureteral catheter.[12] However, it is impossible to dislodge or by-pass some severely impacted ureteral stones with conventional ureteral catheters. To deal with this problem, we designed new 4 F, 5 F, and 6 F ureteral catheters for use with ESWL (Fig. 4-10). They have a 20-degree angle tip at the end of the catheter that allows the operator to "torque" the catheter around the stone, seeking a passage for the 0.035 inch floppy-tip guidewire; also, a beveled or tapered tip eases passage of the catheter around the stone once the guidewire has been passed. Since the introduction of these catheters, more stones than previously have in fact been bypassed and subsequently disintegrated with ESWL (Fig. 4-11). Specifically, in a recent series of 35 patients with ureteral stones, 94 percent of stones were bypassed,

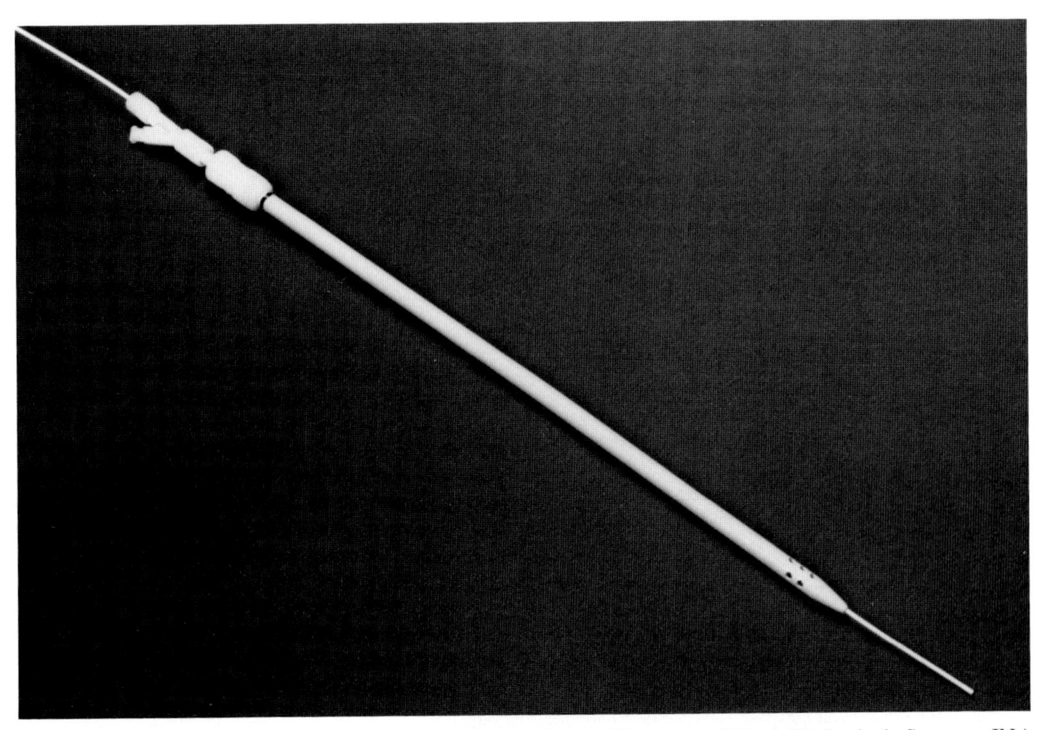

Fig. 4-6. Post-PCN, pre-ESWL nephroureterostomy catheter. (Courtesy of Cook Urological, Spencer, IN.)

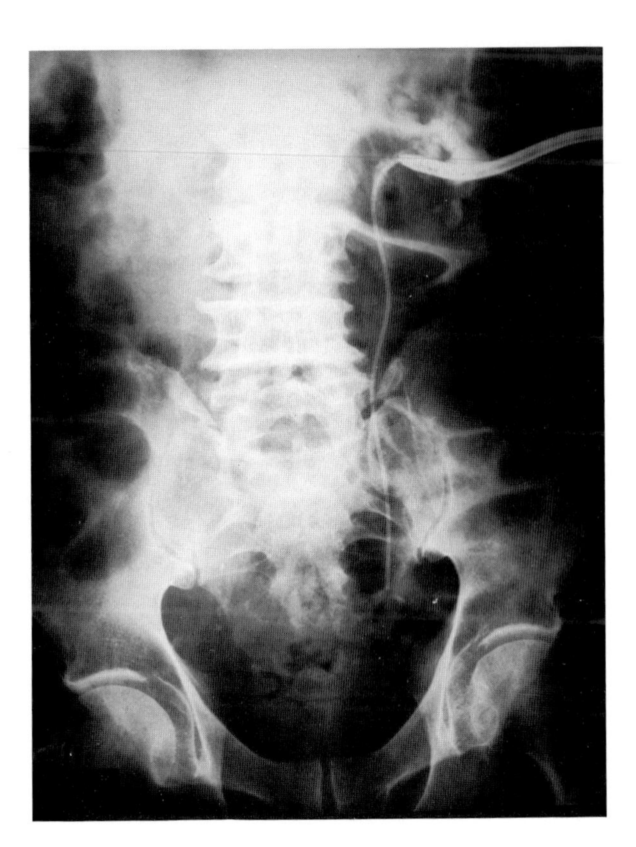

Fig. 4-7. Radiograph of nephroureterostomy catheter immediately after percutaneous nephrostolithotomy.

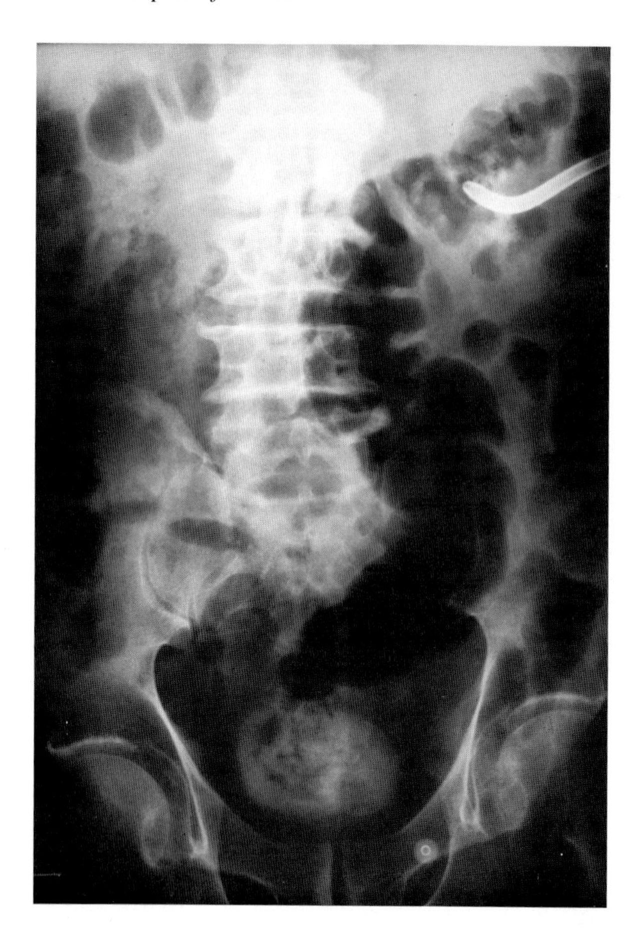

Fig. 4-8. Radiograph of nephroureterostomy catheter following ESWL.

and 91 percent of these stones were treated successfully. Smaller (4 F) catheters with 0.028 inch guidewires are also available for particularly difficult cases. Alternatively, other ureteral catheters such as the 6 F open-end whistle-tip catheter or a UPJ occlusion balloon catheter may be preferred to manipulate ureteral calculi.

## Endoscopic Tables

The portable AMSCO Endo-Uro 1100 table (Fig. 4-12) functions both as an endoscopic table with overhead film and C-arm fluoroscopic capabilities and as a portable "stretcher" that can be used to transfer the patient from the endourology to the lithotripsy suite. Alternatively, an AMSCO operating room table with C-arm extenders will suffice for percutaneous procedures or, if necessary, for ureteroscopic manipulation. If a fixed pedestal unit is preferred, the Hydrajust table will accommodate both cystoscopy and ureteroscopy. It has fixed overhead fluoroscopy and permanent film capability; a high-quality kidney–ureter–bladder (KUB) film can be obtained because a Bucky cassette unit with a moving grid and more limited focal range can be used.

Fluoroscopy in the endourology suite is best accomplished with any standard C-arm or U-arm equipment. Some centers prefer the additional capability of permanent films (8 × 10 in.). Ceiling-mounted fluoroscopy can be used, but it requires a special steel suspension system and is not easy to store against the wall. It is always preferable to have a separate x-ray facility with x-ray tubes fixed to the table for superior-resolution KUB films.

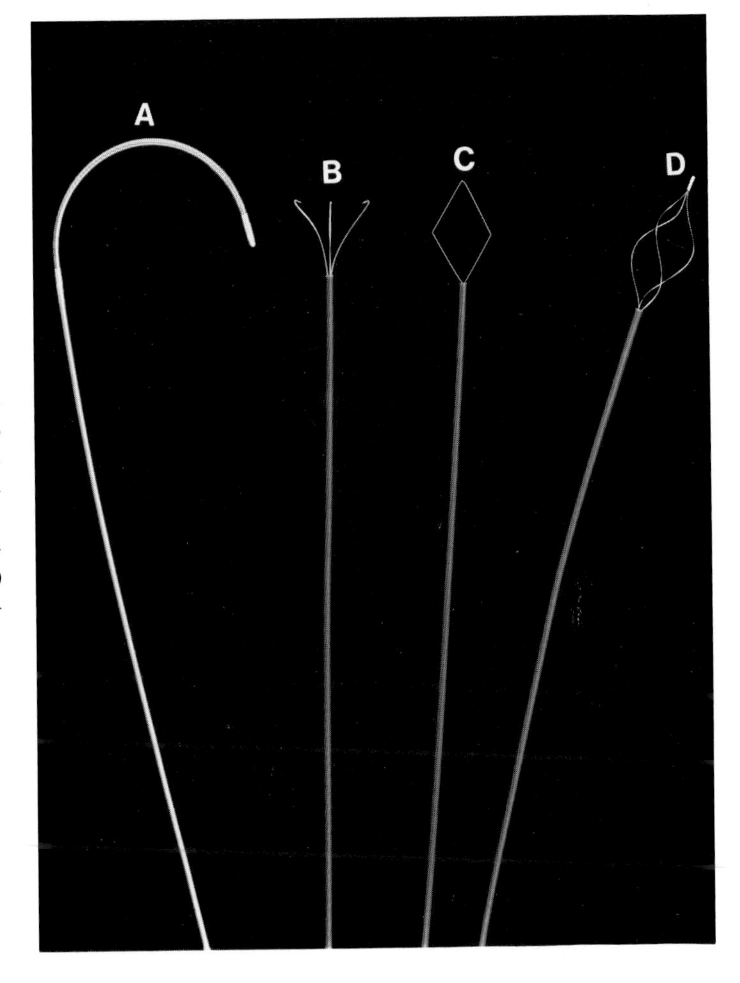


Fig. 4-9. **(A)** 0.038 inch deflecting wire guide used with ureteroscope to facilitate passage or to "tease" calculi in difficult locations. Also used to deflect nonarticulated flexible ureteroscopes. **(B)** 3 F three-prong grasping forceps. **(C)** 3 F loop retriever. **(D)** 3 F three-wire helical stone extractor.

## ESWL Suite

Radiation exposure of personnel is slight even in the immediate vicinity of the lithotripter. However, protective lead aprons should probably be worn by the anesthesiologist and all those working close to the tub. A movable protective glass screen should also be available for observers.

Some centers, including the author's, have been very successful in replacing the post-ESWL KUB film with "manually" performed underwater spot kidney films. For this technique, the x-ray technician holds an 8 × 10 in. gridded cassette film (Kodak) just beneath the appropriate image intensifier in such a way that the grid lines line up with the x-ray beam. Adjustments in the control module are made, appropriate KV and MAS readings are calibrated (according to the patient's girth), and the "snapshot" button pushed. In the author's first 600 cases, this technique was performed in approximately one-third of the patients with results comparable to standard KUB films (Fig. 4-13). This technique makes it possible to minimize the number of shocks administered to the patient and also decreases the number of post-procedure films needed. At present, in conjunction with several companies, our group is attempting to design attachable cassette holders that may be

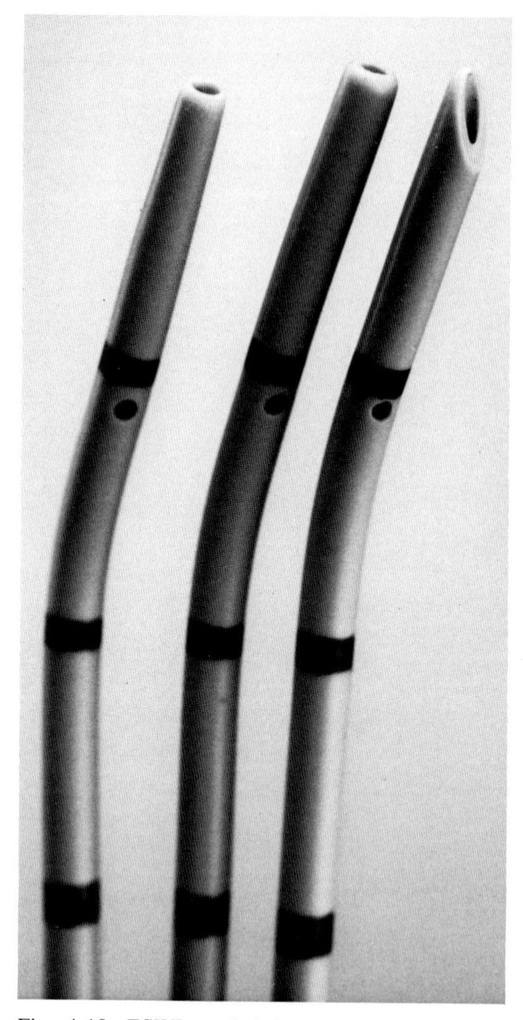

Fig. 4-10. ESWL angled-tip ureteral catheters, tapered or beveled, through which guidewires are passed.

placed on the image intensifier (Fig. 4-14). Virginia-Mason in Seattle has reportedly designed such a cassette holder which is commercially available (personal communication, November, 1985, R.P. Gibbons, Mason Clinic, Seattle, Washington).

The noise background created by the spark gap discharge is sometimes bothersome, but not dangerous to patients or personnel. Audiograms performed on personnel associated with lithotripsy at the author's facility have failed to reveal any significant changes. Nonetheless, the patient is always offered ear protection during treatment, and some technicians wear earmuffs when treating at a high voltage.

Anesthesia requires the usual equipment for induction and maintenance of general and epidural anesthesia, as well as the necessary equipment for defibrillation and cardiac resuscitation.

Other equipment helpful in the ESWL suite includes mobile stools (as opposed to fixed chairs, since space is usually at a premium). Blanket warmers are a necessity, as well as an unlimited supply of towels. Flotation devices for the arms are now routinely used in many centers.

## FUTURE STONE CENTERS

Future stone centers possibly will be equipped for gallstone as well as kidney stone lithotripsy. Experimentation in the extracorporeal fragmentation of gallstones in animals and recently in humans is underway.[13] Like the therapy of kidney stones, gallstone management will require multiple modalities, such as chemolysis,[14] endoscopy, extracorporeal lithotripsy, percutaneous techniques, and metabolic evaluation with subsequent medical therapy and prevention.

Stone centers in the future should offer advanced endoscopic techniques, endoscopes with smaller outer diameters, and more functional capabilities for the retrograde or percutaneous extraction of stones in either the kidney or the biliary tract. These centers should be directed and operated primarily by surgeons (urologic and gastrointestinal) as only they can appropriately select cases and manage all possible complications (including surgical intervention). Subspecialization in both fields is likely to continue.

## ACKNOWLEDGMENT

Special appreciation is expressed for the contribution of Mr. Terry Bennett, Chief Radiologic Technician, Georgia Baptist Medical Center, in the development of the technique of spot kidney films in the lithotripter.

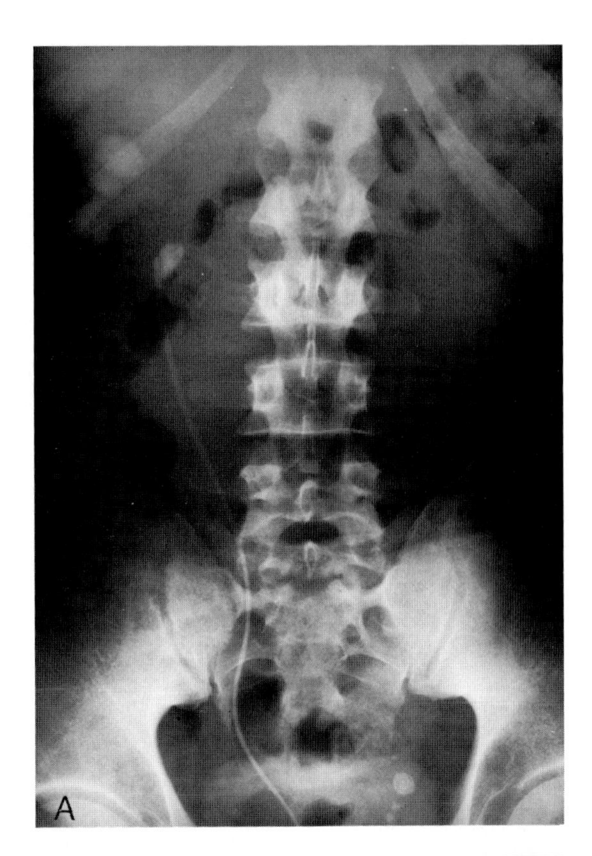


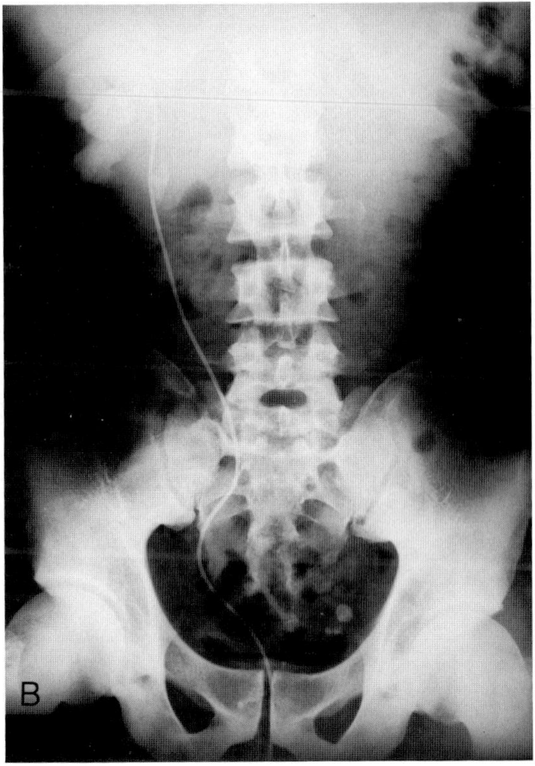


Fig. 4-11. **(A)** Radiograph after ESWL (failed), following unsuccessful passage of conventional ureteral catheter. **(B)** Radiograph of same patient following second ESWL after successful passage of new ESWL angled-tip ureteral catheter. (Courtesy of Cook Urological, Spencer, IN.)

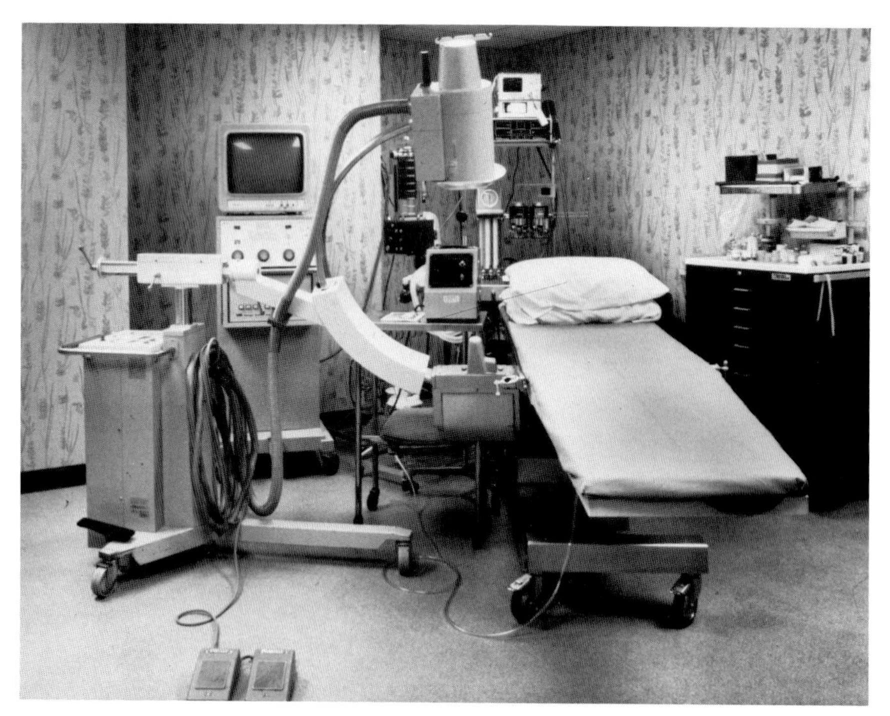

Fig. 4-12. AMSCO Endo-Uro 1100 table in endourology suite (Atlanta Stone Center of Georgia Baptist Medical Center, Atlanta, GA).

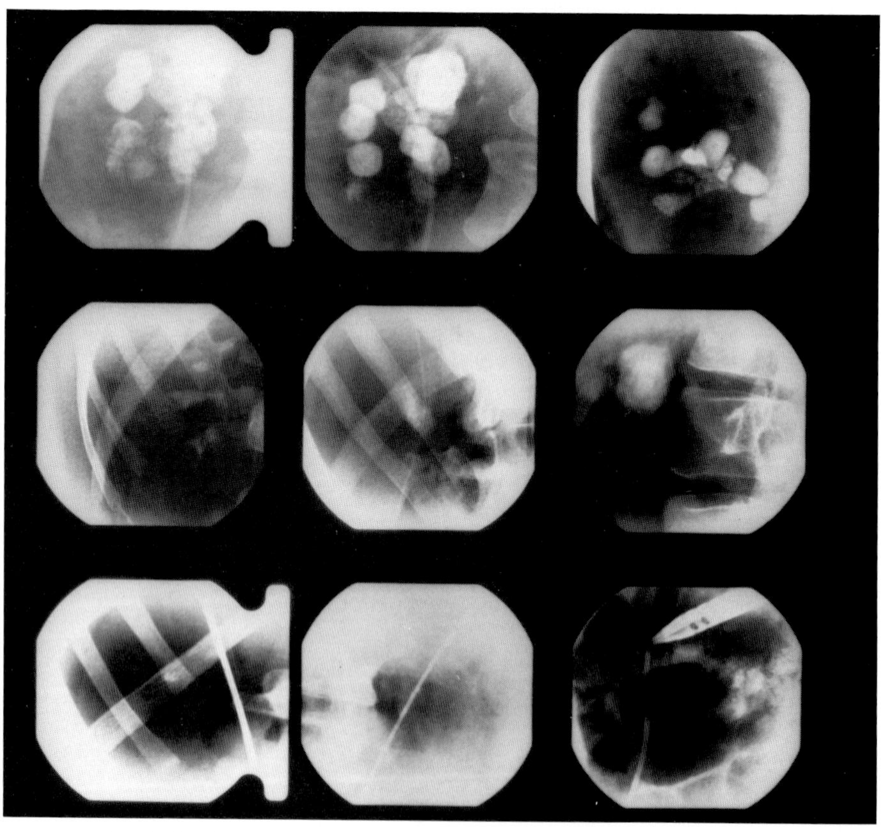

Fig. 4-13. A series of 8 × 10 inch underwater spot kidney radiographs.

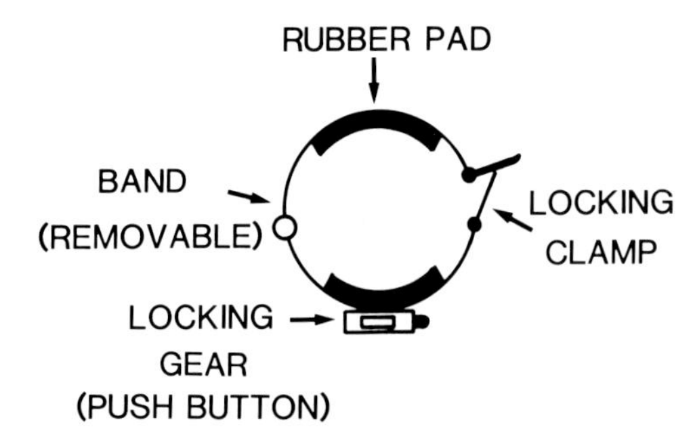


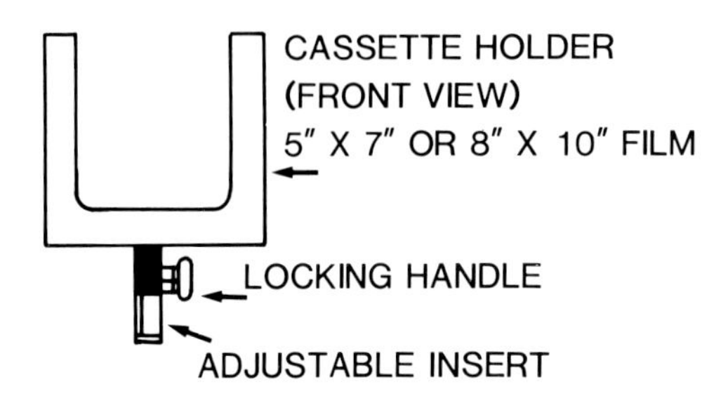


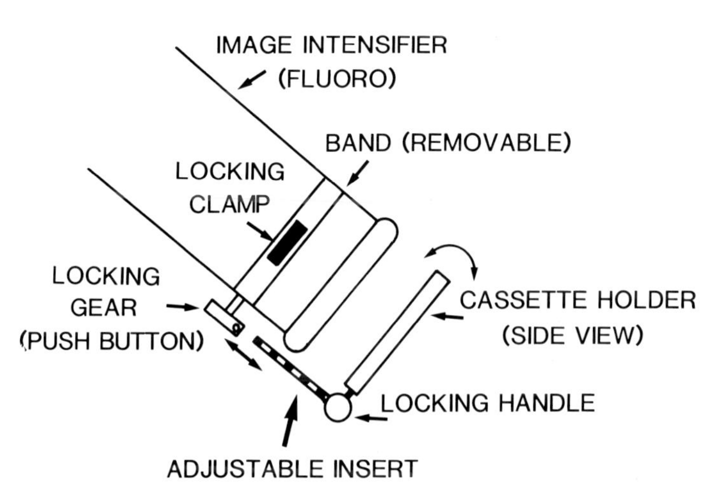


Fig. 4-14. Schematic representation of removable cassette holder designed for underwater spot kidney radiographs.

## REFERENCES

1. Chaussy C, Schmiedt E: Shock wave treatment for stones in the upper urinary tract. Urol Clin North Am 10:750, 1983
2. Chaussy C, Schmiedt E, Jocham D, et al: Four years experience with extracorporeal shock wave lithotripsy (ESWL) in Munich (abstract). J Urol 131:246A, 1984
3. Clayman RV, Castaneda-Zunega W: Appendix 1. Chemolysis. p. 357. In: Techniques in Endourology: A Guide to Percutaneous Removal of Renal and Ureteral Calculi. Year Book, Chicago, IL, 1984
4. Pak CYC, Button F, Peterson R et al: Ambulatory evaluation of nephrolithiasis: Classification, clinical presentation and diagnostic criteria. Am J Med 69:19, 1980
5. Alder HC: Lithotripters: Noninvasive devices for the treatment of kidney stones. A.H.A. Hospital Technology Series (I55MO735–4681), 4:40, 1985

6. Study of safety and clinical efficiency of current technology of percutaneous lithotripsy and non-invasive lithotripsy, AUA Ad Hoc Committee, Dr. DL McCullough, Chairman, Baltimore, May 16, 1985
7. Clayman RV, Castaneda-Zunega W: Appendix 4: instrument list. p. 369. In: Techniques in Endourology: A Guide to the Percutaneous Removal of Renal and Ureteral Calculi. Year Book, Chicago, IL, 1984
8. Stone LO: Instruments and supples for endourology suite. p. 21. In Schoborg, TW (ed): Operating Room Manual of Endourologic Procedures. Frix Publications, Atlanta, GA, 1985
9. Lingeman JE, Kahnoski RJ, Coury TA: The removal of staghorn calculi using percutaneous nephrostolithotomy and extracorporeal shock wave lithotripsy (abstract). J Urol 133:321A, 1985
10. Dretler SP: Combination treatment: ureteroscopy and extracorporeal shock wave lithotripter (ESWL) (abstract). J Urol 133:311A, 1985
11. Wilbert DM, Muller SC, Thuroff JW et al: Our experience with extracorporeal shock wave lithotripsy and special aspects in more than 1,000 cases (abstract). J Urol 133:310A, 1985
12. Lithotripsy Users Conference with Christian Chaussy at annual meeting of AUA, Atlanta, Georgia, May 14, 1985
13. Brendel W, Enders G: Shock waves for gallstones: Animal studies. Lancet 1:1054, 1983
14. Schmeller T, Chaussy C: Combination treatment of renal calculi with shock wave lithotripsy and chemolysis (abstract). J Urol 131:291A, 1984

# 5

# Patient Selection and Education

Alan D. Jenkins

Patients with upper urinary tract calculi have been actively seeking extracorporeal shock wave lithotripsy (ESWL) since its introduction into the United States in 1984. Although most patients with renal or ureteral calculi can be successfully treated with ESWL, this may not be the optimal form of therapy for some patients. The treatment of each patient's surgically active stone disease must be individualized; attention must be given to the specific characteristics of the stone, urinary tract anatomy, and associated illnesses or physical handicaps.

Patients have become more educated health care consumers, and will investigate the purchase of health care as diligently as the purchase of a new automobile. They will not base their decision solely on the advice of their urologist, but will consider information obtained from family, friends, relatives, business associates, and the press. Some patients inform their urologists when and how they wish to be treated. Urologists at ESWL centers are being asked to treat patients with classically asymptomatic stones who would not be considered for traditional surgical removal. ESWL can pulverize the vast majority of upper urinary tract calculi, but should all patients with radiopaque material in their kidneys be treated? If patients are symptomatic, *when* should they be treated?

These questions will not be answered in this chapter, but the factors pertinent to the selection of patients for ESWL and the timing of ESWL will be outlined. A few experimental and controversial areas will also be mentioned, and the aspects of patient education unique to ESWL will be discussed.

## PATIENT SELECTION

### Surgically Active Urolithiasis

The distinction between metabolically active urolithiasis and surgically active urolithiasis is of utmost importance when treating patients with renal stone disease.[1] Metabolic activity is related to the actual crystallization of stone-forming salts, whereas surgical activity is related to the signs and symptoms produced by stones that have already formed in the urinary tract. Surgically active urolithiasis has been defined as renal colic, upper tract obstruction, or a urinary tract infection associated with the stone material. Examples of surgically active urolithiasis are a 2 cm renal pelvic calculus that intermittently obstructs the ureteropelvic junction or a 1 cm midureteral stone that is responsible for severe, but asymptomatic, hydronephrosis. Until recently, patients with such clinical presentations would have required an open surgical procedure for stone removal. The introduction of percutaneous nephrostolithotomy (PNL) and ESWL has permitted the less invasive removal of these stones. If ESWL treatment can successfully treat such surgically active stones, then few physicians can argue that it is not the treatment of choice.

The commonly accepted indications for open surgical stone removal can also be applied to the selection of patients for ESWL treatment. If a patient has a stone that might otherwise need percutaneous or surgical removal then this patient is a potential candidate for ESWL.

## Initial Clinical Criteria

During the initial German and American trials with the Dornier lithotripter, patients were carefully selected for treatment.[2] The first 50 patients treated at each investigational center in the United States had to have solitary pelvic or caliceal stones less than 2 cm in diameter (Food and Drug Administration category A) (Table 5-1). The urine was required to be sterile, the ureter distal to the stone patent, and the stone radiopaque. No significant aortic or renal artery calcification could be present. Finally, the patients had to be in class 1 or class 2 of the physical status classification system adopted by the American Society of Anesthesiologists.[3] These patients could have had only mild or moderate systemic disorders, such as slightly limiting organic heart disease, mild diabetes, or essential hypertension.

**Table 5-1. FDA Patient Classification**

Criteria for category A patients
1. Solitary, densely opaque pyelocaliceal stone
2. Stone 2.0 cm or less in axial length
3. Sterile urine
4. Absence of obstruction distal to stone
5. Normal body habitus
6. Creatinine less than 3 mg/dl
7. Absence of significant aortic or renal artery calcification

Criteria for category B patients
1. Multiple pyelocaliceal stones
2. Stones larger than 2.0 cm in axial length
3. Upper ureteral stones
4. Radiolucent stones (localization facilitated by exogenous contrast material)
5. Infection stones (partial or full staghorn, positive urine cultures)

When it became apparent that healthy patients with solitary small stones could be safely and effectively treated with ESWL, the indications were expanded (FDA category B). Patients with stones larger than 2 cm in diameter, multiple stones, or upper ureteral stones were considered suitable candidates for ESWL (Fig. 5-1). Radiolucent uric acid stones could be localized using radiocontrast agents injected intravenously or, preferably, through a ureteral catheter. Patients with infection stones could be safely treated if antibiotics were administered for 24 to 48 hours before the procedure. Patients with complete staghorn calculi, however, were excluded from the first phase of the American clinical trials. Before FDA approval in December 1984, patients at high risk from the anesthesia were also excluded from ESWL treatment.

## Anesthesia Management

The risks associated with ESWL are primarily related to the anesthesia. The anesthesia risks are related to the physical condition of the patient and the method of anesthesia. The American Society of Anesthesiologists has adopted a classification of physical status used to assess the relative anesthetic risk in a particular patient (Table 5-2). Although this practical system is not scientifically precise, greater safeguards should be employed for patients in a poorer class.

Anesthetic techniques that have been used for ESWL include local with intravenous sedation, regional (subarachnoid block or continuous epidural), general endotracheal anesthesia, and high frequency jet ventilation (HFJV). Overall, the specific anesthetic technique is not as impor-

**Table 5-2. American Society of Anesthesiologists Physical Status Classification**

| | |
|---|---|
| Class 1. | Patient generally healthy without serious systemic problems |
| Class 2. | Patient has mild or moderate systemic problems |
| Class 3. | Patient has serious systemic disturbances |
| Class 4. | Patient suffering from life-threatening but not necessarily terminal conditions |
| Class 5. | Moribund patient who has little chance of survival |

From Dornier Lithotripter Patient Report Form. Rhode Island Health Services Research, Inc. (SEARCH), Providence, RI, 1984, p. 9.

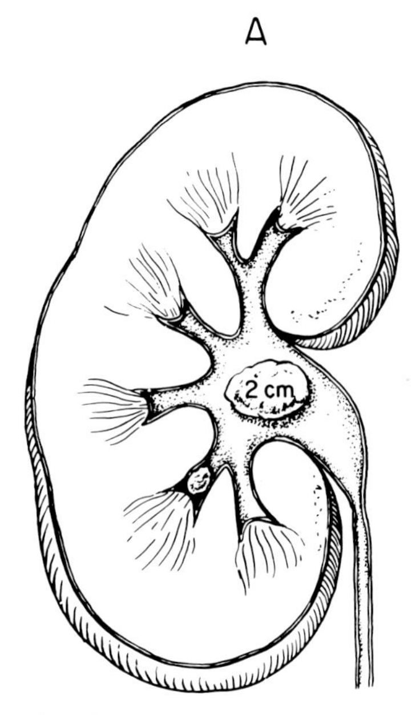


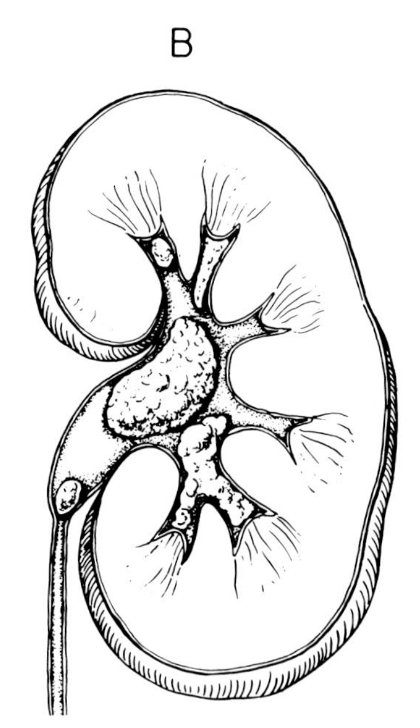


Fig. 5-1. FDA categories A and B. Note that category A includes only solitary stones, either caliceal or renal pelvic.

tant as the general health of the patient. Both of these considerations are secondary to the care with which the anesthetic is administered.

## Radiation Exposure

The use of x-rays is integral to the performance of ESWL using the Dornier lithotripter. Fluoroscopy is used for initial localization and aiming, and for monitoring the progress of pulverization during treatment. Pregnant patients are not suitable candidates for ESWL. Furthermore, the effects of shock waves on placental integrity and fetal well-being are not known. Temporizing measures should be used in pregnant patients with symptomatic urinary calculi. ESWL can safely be used after delivery.

## Body Habitus

The distance between the top of the ellipsoidal reflector and the second focal point is fixed in the Dornier lithotripter. If this distance is less than the distance between the patient's stone and the skin of the posterior flank (thickness of the patient), then the stone cannot be accurately placed at the second focal point. Accurate localization can usually be achieved in patients weighing less than 135 kg and having less than 30 percent body fat. Although small children have been treated using ESWL, the patient support was designed for patients between 4 ft 0 in. and 6 ft 6 in. in height.

In very large patients, stones in the renal pelvis or calices can be localized more readily than stones in the ureter. If an obese patient presents with a stone in the ureteropelvic junction, an attempt should be made to manipulate the stone back into the kidney with a ureteral catheter before proceeding with ESWL.

## Disorders of Blood Coagulation

One of the known adverse side effects of ESWL is intrarenal or perinephric bleeding which has been associated with heavy aspirin use in some patients,[5] but may also occur in healthy patients taking no medications.[6]

ESWL can safely be performed in patients

with congenital disorders of blood coagulation, including hemophilia, if they are treated as if they were undergoing a major surgical procedure.[7] A detailed history should reveal any abnormal bleeding tendencies meriting a more extensive hematologic evaluation. Specific factors should be replaced before and during treatment, and the patient should be watched closely during the initial 24 hour recovery period.

Patients who have been anticoagulated with warfarin can be treated if the anticoagulant has been discontinued and the clotting studies have returned to normal levels. Anticoagulation can be reinstituted soon after ESWL treatment, if the physician is confident that additional procedures, such as repeat ESWL or ureteroscopy, will not be needed to aid particle passage.

## FDA Approval for Clinical Use

The Appendix gives the text of the label included with each Dornier Model HM-3 lithotripter.[8] This particular lithotripter has been approved for the treatment of *upper* tract stones—renal caliceal stones, renal pelvic stones, and upper ureteral stones. Contraindications include pregnancy, presence of a cardiac pacemaker, urinary tract obstruction distal to the stone, and renal artery calcification on the treated side. Caution must be exercised in patients with a high risk of heart failure, with pneumonia, with very low diaphragms, or with coagulation disorders. Hydronephrosis, presence of ureteral stents or nephrostomies, prior renal nephrostolithotomy or surgery, previous ureteral surgery or reimplantation, horseshoe kidneys, ileal conduits, and solitary kidneys are not contraindications.

## Developments in Selection Criteria

Experimental studies have been undertaken in three categories since FDA approval in 1984: (1) distal ureteral stones, (2) patients with cardiac pacemakers, and (3) children.

### DISTAL URETERAL STONES

Distal ureteral stones are probably best treated either by pushing them proximally and then disintegrating them by ESWL, or else by extracting them ureteroscopically. If the stones are impacted, however, they can safely be treated in situ with ESWL. Good pulverization can be achieved if the stone is distal to the caudal edge of the sacroiliac (SI) joint. A successful outcome is less likely if the shock waves must pass through the SI joint to reach the stone. Stones overlying the SI joint are effectively pulverized if the patient is placed in the prone position. No adverse skeletal effects have been observed.[9]

### CARDIAC PACEMAKERS

A few patients with cardiac pacemakers have been safely and successfully treated with ESWL at the University of Virginia. All of the pacemaker generators have been located in the upper chest. The EKG electrodes have been placed such that the spark discharge is triggered by the patient's R wave, not the pacemaker spike. No significant arrhythmias have been seen.

### CHILDREN

When the Dornier lithotripter was approved by the FDA in December, 1984, there had been no experience with ESWL in children less than 7 years old. This was primarily due to the minimum height (4 ft 0 in.) requirement for placement on the gantry. Since then, individual ESWL centers have temporarily modified the chair with hammock-like devices, and children as young as 1 year old have been treated at the University of Virginia.

In a preliminary review of five centers in the United States employing ESWL for pediatric patients, 20 patients with an age range of 12 months through 16 years were reported to have undergone ESWL.[10] Most required general anesthesia. The mean number of shocks was less than that required for an adult population. All stones were fragmented and most patients were stone free at the time of last follow-up. Additional procedures such as cystoscopy with ureteral catheterization and percutaneous nephrostomy were required in some patients. As with the adult population, serious perioperative complications were rare.

With children, care must be taken to shield the lungs with an acoustic barrier, such as a sheet of Styrofoam, to prevent the potential complication of a pulmonary contusion.

## Selection Considerations

### STONE COMPOSITION

Most upper urinary tract calculi are composed of calcium oxalate or a mixture of calcium oxalate and calcium phosphate. Most of these stones are easily pulverized with ESWL, but the evidence is that very old calculi or calcium oxalate monohydrate stones are more difficult to fragment.

Many struvite or magnesium ammonium phosphate stones are finely pulverized by shock waves, although some infection stones are very resistant to ESWL. This resistance may be related to the age of the stone or the proportion of carbonate-apatite or matrix in the stone.

Radiolucent uric acid stones can be treated with ESWL, but localization requires the use of a radiocontrast agent injected intravenously or through a retrograde ureteral catheter. Shock waves will fragment uric acid stones, but it is more difficult to monitor the progress of pulverization. Although the increased surface-to-volume ratio permits rapid dissolution of many residual particles with medical therapy (alkalinization, fluids, and allopurinol), the kidney must remain decompressed for effective chemolysis. A nephrostomy tube or external ureteral stent can be used for irrigation until dye studies (antegrade pyelogram) reveal resorption of most uric acid particles.

Cystine calculi are the most resistant to ESWL. Assuming a strict medical regimen, cystine renal pelvic stones less than 2 cm in diameter can successfully be treated with ESWL, but higher energy and more shock waves will be needed than for a comparable calcium oxalate stone. Large cystine stone loads should be treated with primary percutaneous nephrostolithotomy and subsequent percutaneous chemolysis or secondary ESWL.[11]

Two patients with large (>1 cm) cystine ureteral stones have been treated by ESWL at the University of Virginia. No fragmentation was visible radiographically, so the stones were surgically removed. One stone was found to be completely intact, whereas the other was cracked in several places. The pieces of the cracked stone were held together by interlocking, like the pieces of a wooden puzzle.

### LARGE CALCULI

The care of patients with large staghorn calculi is one of the most challenging tasks for urologists. Although cystine and uric acid stones may have a branched configuration, the vast majority of staghorn stones are composed of struvite (magnesium ammonium phosphate). ESWL easily pulverizes most struvite stones, but the large quantity of sand does not readily pass through the ureter. Several combinations of ESWL, percutaneous lithotripsy, and chemolysis have been proposed for the treatment of large staghorn calculi. Some institutions still believe that large staghorn calculi are best removed surgically. A more detailed discussion of the role of ESWL in the treatment of staghorn calculi will be presented in Chapter 12.

### OUTPATIENT ESWL

During the investigational trials, almost all patients were admitted to a hospital the day before the procedure and were kept as inpatients for 24 to 48 hours after treatment. Since FDA approval, many patients have been admitted on the day of treatment, and certain patients have been permitted to return home after recovering from the anesthetic. Acute post-ESWL medical care would most likely be required for renal colic or sepsis. Approximately one-third of patients will require parenteral narcotics for pain relief. Although it is not possible to predict with complete accuracy who will have severe renal colic, the size and location of the stone are important factors. A patient with a 6 mm lower-pole caliceal stone is less likely to have renal colic than a patient with a 2 cm pelvic stone. Temporary double-J ureteral catheters may prevent ureteral obstruction by fragments and the occurrence of severe renal colic.

Patients with infected calculi and who may become septic after ESWL should be kept under close observation for 24 to 48 hours after ESWL. Patients with complicating systemic medical conditions, such as cardiovascular disease or diabetes mellitus, are best treated as inpatients.

### ASYMPTOMATIC STONES

Patients with asymptomatic small caliceal stones or small ureteral stones are usually not candidates for open surgical removal. Percutaneous manipulative techniques and ESWL, however, permit the safe and effective treatment of these small stones. There is anecdotal evidence that many isolated caliceal stones are not asymptomatic. These patients may have minor flank pain or constitutional symptoms that resolve after percutaneous extraction or shock wave pulverization of a small caliceal stone.

A patient's occupation or geographic location are important considerations. An aircraft pilot may be grounded if a small caliceal stone is present on an abdominal radiograph. Patients who spend considerable time in isolated geographic areas may not have ready access to medical care. ESWL treatment of stones with a small but finite chance of causing ureteral obstruction and renal colic can be of great benefit to these patients.

Ureteral stones less than 4 to 6 mm in diameter are likely to pass spontaneously,[12] but patients must frequently use oral narcotics to control the pain. Small stones in the upper two-thirds of the ureter can be easily treated with ESWL. Successful shock wave pulverization is less likely to occur if the stone remains in the ureter for more than 6 weeks or passes into the distal ureter. Prompt ESWL treatment of small upper ureteral stones can spare patients the misery of stone passage.

## PRE- AND POST-ESWL EDUCATION

Patient education may be the most important aspect of ESWL. Because this new technology is accompanied by more unfamiliarity than traditional open surgical procedures used for stone removal, post-treatment convalescence will be much smoother if patients understand the nature of ESWL and its associated side effects.

### Methods

At the University of Virginia, an educational program consisting of printed material, a videotape, and discussions with the medical and paramedical staff has been implemented. Most ESWL centers have booklets that briefly describe ESWL. When patients arrive at the hospital, they and their families are invited to watch a videotape that outlines the principles of ESWL, dramatizes an actual treatment procedure, and explains the post-treatment recovery period. Most of the routine questions are answered by a paramedical staff member, whereas the urologist responsible for the patient's care answers specific questions. New patients and their families have the opportunity to talk with formerly treated patients.

An anesthesiologist discusses the anesthetic choices with each patient. Although regional anesthesia is favored at the University of Virginia, general anesthesia is used if medically indicated or requested by a patient.

**COMPLICATIONS ASSOCIATED WITH EXTRACORPOREAL SHOCK WAVE LITHOTRIPSY (ESWL)**

The purpose of ESWL is to pulverize kidney stones to the extent that they may be passed spontaneously. ESWL does not actually remove the kidney stones. The sand-like particles created by the treatment must pass out of the kidney, through the ureter (the tube leading from the kidney to the bladder), and into the bladder.

1. 30% of patients will have pain as the particles pass through the ureter. This may be quite intense, but sufficient pain-relieving medication will be given. You must tell the nurses when you are having pain.

2. 10% to 20% of patients will require additional ESWL treatment sessions (at reduced cost) to adequately pulverize their kidney stones. Large stones are more likely to require multiple treatment sessions, but even small stones may require more than one treatment. If an additional ESWL treatment session is required, it will usually be done one or two days after the initial treatment. In some patients it may be prudent to wait one or two weeks before an additional treatment is given.

3. The adequacy of stone pulverization will be determined from x-rays taken the day after treatment. The numerous sand particles can obscure a larger fragment. A larger fragment may become apparent on radiographs taken several weeks later — after the smaller particles have passed. Another ESWL treatment may be needed to complete the pulverization of any larger fragments.

4. 10% of patients will require supplemental procedures. These include placement of a percutaneous nephrostomy (a tube inserted into the kidney through the back), cystoscopy with ureteral catheterization or basket extraction, and ureteroscopy. A single patient may need several supplemental procedures.

5. Three months after ESWL, 90% of patients will be stone-free, but 10% will still have sand-like particles in their kidney. Some of these particles may continue to pass, but others may remain in the kidney. The long-term effects of these retained particles are not known.

6. Bleeding around the kidney has occurred in some patients, but no lasting damage to an internal organ or its function has been observed.

7. ESWL will not prevent the formation and growth of new kidney stones. Metabolic evaluation and medical therapy will be required to achieve this goal.

- - - - - - - - - - - - - - - - - - - - - - - - - - - - - - -

PATIENT'S STATEMENT

I have read the preceding information and have received a copy of this sheet.

____________________ ______________________________

Date | Patient's Signature

Fig. 5-2. Complication sheet that each patient treated at the University of Virginia is asked to read and sign.

## DISCHARGE INSTRUCTIONS FOR LITHOTRIPTER PATIENTS

The purpose of your treatment with extracorporeal shock wave lithotripsy (ESWL) was to break your stones into pieces small enough to easily pass through your urinary system. Mild pain may occur as these fragments are passing. This should be relieved by the prescription (usually for Percocet) which you will be given.

Some patients are also given a prescription for an antibiotic, for the treatment of a urinary tract infection. If so, you should take all of the medicine and have a urine culture done by your family physician or urologist one week after you finish taking the antibiotic.

Unless you are told otherwise by your physician, you should be able to resume all of your normal activities. A liberal intake of fluids and mild exercise, such as walking, may aid the passage of the fragments.

Mild back pain may be relieved by soaking in a warm tub of water once or twice a day. Two Tylenol or aspirin tablets may also help relieve these aches.

We would like you to strain your urine for 1 to 2 weeks after leaving the hospital. Any fragments should be placed in the small plastic bag and mailed back to us in the envelope that we provide. No postage is needed. Some patients will not pass stone fragments immediately, but may begin to do so 4 to 6 weeks after discharge. If this should occur, we would like you to collect any fragments that you conveniently can and send them to us.

If you develop a high fever (greater than 101° F or 38.5° C) or severe pain that is not relieved by the oral pain medication, you will need to see a physician. If you decide to see a doctor in your hometown, please show him the sheet "GUIDELINES FOR UROLOGISTS FOLLOWING LITHOTRIPTER PATIENTS." If your physician has questions, or if you would like to talk with someone at the University of Virginia, please call 804-924-9548 (8 AM to 5 PM, Monday through Friday). At night or on weekends or holidays, you should call the page operator at the University of Virginia Hospital, 804-924-0211, and ask for the urology resident on call. You will need to leave your phone number. The urology resident will call you back.

---

For Office Use Only

Return appointment: Date ______ Time ______ Urologist ______________________

24-hour urine: Yes ______ No ______

Pain medication: Yes ______ No ______

Antibiotics: Yes ______ No ______

Reviewed with patient by: #1 ______________________________ Date ______

#2 ______________________________ Date ______

Follow-up: ____________ Hometown physician ____________ University of Virginia

Fig. 5-3. Discharge instructions used at the University of Virginia.

## GIVE THIS SHEET TO YOUR UROLOGIST

### GUIDELINES FOR UROLOGISTS FOLLOWING LITHOTRIPTER PATIENTS

1. Many patients require several weeks to pass all of their fragments. Ten percent of patients will have residual particles three months after ESWL, but even these may pass.
2. Please obtain an excretory urogram (or a plain abdominal radiograph and renal ultrasound) three to four weeks after lithotripsy. Your patient may not be stone-free at that time, but the presence or absence of high-grade, asymptomatic obstruction will be determined.
3. If your patient has a high-grade obstruction, we recommend the placement of a percutaneous nephrostomy. Retrograde passage of a ureteral catheter through the sand in the ureter is usually not possible.
4. If a plain abdominal radiograph shows a column of stone material and sand in the distal ureter, the first step should be a ureteral meatotomy (and perhaps gentle agitation of the leading particles with a pigtail ureteral catheter). This should initiate the passage of particles over the next several days. An attempt to remove all of the particles with a basket is usually not productive and may injure the ureter.
5. The passage of a column of sand may be impeded by a large leading fragment. Ureteroscopic extraction of such a fragment may permit the smaller particles to pass. Attempts to manually extract all of the smaller particles should be avoided.
6. Three months post-ESWL is a good time to assess the presence or absence of retained sand in the kidneys with a plain abdominal radiograph. Small particles that can easily pass should be followed while larger fragments may merit another ESWL treatment (at reduced cost).
7. We would greatly appreciate it if you would let us know the results of any follow-up studies and the development of any post-treatment problems.
8. If you have any questions, please call us:
   Monday–Friday, 8 AM–5 PM: 804-924-9548
   Nights, Weekends, Holidays: 804-924-0211 (Page the urology resident on call.)

Fig. 5-4. Standard guidelines for referring physicians.

## Complications

During their pretreatment evaluation, patients are asked to read and sign a form listing the complications associated with ESWL (Fig. 5-2). This form outlines the nature of ESWL treatment, the possible occurrence of renal colic, and the possible need for additional ESWL treatment sessions or supplemental procedures. To reinforce the importance of the information, patients are asked to sign the form. A standard University of Virginia Medical Center operative consent is also obtained. Permission is requested for ESWL treatment and possible cystoscopy, ureteral catheterization, or percutaneous nephrostomy.

## Post-ESWL Follow-Up

Three-fourths of our patients are discharged from the hospital the day after ESWL treatment. Each patient is given a discharge sheet of instructions (Fig. 5-3) and an instruction sheet for their referring urologist (Fig. 5-4). All patients are asked to strain their urine for 1 to 2 weeks and return the stone particles for analysis in addressed, postage-paid containers given them at the time of discharge. Our patients are also asked to obtain an excretory urogram or a plain abdominal radiograph and renal ultrasound, usually 3 to 4 weeks after treatment. The purpose of these studies is to exclude the presence of asymptomatic ureteral obstruction. The treating urologist writes the referring physician and outlines the exact procedures performed and the need for any special follow-up studies. Each referring physician is sent a copy of the stone analysis.

# CONCLUSION

Even more than percutaneous lithotripsy, ESWL has revolutionized the treatment of urolithiasis. Shock waves can pulverize virtually any upper urinary tract stone, but proper patient selection must be given the highest priority if success is to be maximized and morbidity minimized. Patient education is intimately associated with patient selection. Patients must understand the nature as well as the limitations of ESWL. ESWL is certainly not magic—stones are pulverized, not vaporized. The particles must still be passed, and the patient must be cooperative and patient.

Once the decision is made to proceed with ESWL, the patient's understanding of the side effects and potential complications should be reinforced. Difficult particle passage will occur in some people, and patients should be reassured that this side effect can be safely managed. With appropriate pretreatment education, the emotional impact of post-treatment problems will be minimized, and patients will remain satisfied with their decision to undergo ESWL.

# REFERENCES


1. Smith LH: Urolithiasis. In Earley LE, Gottschalk CW (eds): Strauss & Welt's Diseases of the Kidney, 3rd Ed. p. 893. Little Brown, Boston, 1979
2. Chaussy C, Schmiedt F: Shock wave treatment for stones in the upper urinary tract. Urol Clin N Am 10:743, 1983
3. Dripps RD, Eckenhoff JE, Vandam LD: Introduction to Anesthesia. W B Saunders, Philadelphia, 1982
4. Dornier Lithotripter Patient Report Form. Rhode Island Health Services Research, Inc. (SEARCH), Providence, RI, 1984, p. 9
5. Finlayson B, Thomas WC Jr: Extracorporeal shock wave lithotripsy. Ann Int Med 101:387, 1984
6. Kande JV, Williams CM, Millner MR et al: Renal morphology and function immediately after extracorporeal shock wave lithotripsy. AJR 145:305, 1985
7. Krieger JN, Hilgartner MW, Redo SF: Surgery in patients with congenital disorders of blood coagulation. Ann Surg 185:290, 1977
8. Operating Manual, Dornier Kidney Lithotripter. Dornier Medical GmbH, Marietta, Georgia, 1985
9. Jenkins A, Gillenwater J: ESWL treatment of distal ureteral stones, Third World Congress on Endourology, New York, 1985

10. Sigman M, Laudone V, Jenkins A et al: Initial experience with extracorporeal shock wave lithotripsy in pediatric patients. J Urol (in press)
11. Schmellen NT, Kersting H, Schullen J et al: Combination of chemolysis and shock wave lithotripsy in the treatment of cystine renal calculi. J Urol 131:434, 1984
12. Sandegard E: Prognosis of stone in the ureter. Acta Clin Scand Supple., 219:1, 1956

## APPENDIX

The following is the text of the "package insert" provided with the Dornier Kidney Lithotripter, Model HM-3.

### Clinical Information for Physicians

1. DESCRIPTION. This device is the Dornier Lithotripter, Model HM-3. It is manufactured by Dornier Medical GmbH, 810-B Franklin Court, Marietta, GA 30067.

A detailed description of the device is contained in the Operating Manual.

2. INDICATION AND USAGE. The Dornier Lithotripter, Model HM-3, is indicated for use in the disintegration of upper urinary system stones—i.e., renal calyx stones, renal pelvic stones, and upper ureteral stones.

3. INFORMATION FOR USE. Caution: Federal law restricts this device to sale, distribution, and use by or on the order of a physician. Detailed information for use is contained in the Operating Manual.

4. CONTRAINDICATIONS. The Dornier Lithotripter, Model HM-3, is contraindicated for treatment of gallstones, lower ureteral stones and bladder stones. The Dornier Lithotripter, Model HM-3, is also contraindicated when:

1. use of both general and peridural anesthesia is contraindicated
2. exposure to radiation is contraindicated—e.g., with pregnant women
3. the anatomy of the patient does not permit focusing of the device into the posterior flank in the area of the kidney stone, for example, in patients with severe curvature of the spine or excess body fat
4. the patient has a heart pacemaker
5. there is a distal urinary obstruction
6. there is renal artery calcification on the treated side.

Extreme caution must be used in treatment of patients with high risk of heart failure, pneumonia, very low diaphragms, or abnormalities of coagulation. Caution must be used in treatment of patients using drugs, including aspirin, which may affect coagulation. Caution must also be used with patients with neurogenic bladders.

There has been no experience with the Dornier Lithotripter, Model HM-3, with children under 7 years of age.

5. WARNINGS. The Dornier Lithotripter, Model HM-3, may be used only by medical specialists fully trained in its use in accordance with standards established by the manufacturer.

Specific warnings concerning use of the Dornier Lithotripter, Model HM-3, are contained in the Operating Manual.

6. PRECAUTIONS. Precautions concerning the use of the Dornier Lithotripter, Model HM-3, are contained in the Operating Manual.

7. ADVERSE EFFECTS. There have been several reported cases of subcapsular hematoma, including two associated with heavy aspirin use. There has been one case of pancreatitis which may have resulted from the treatment. No lasting damage to an internal organ or its function has been observed.

The most common side effects are blood in the urine for a few days and colic associated with the spontaneous passage of stone particles. This colic can be treated with conventional analgesic therapy. In about 15 percent of cases, obstruction by stone fragments may occur and will require adjunctive urological treatment.

# 6

# Principles of Treatment

Robert C. Newman
Robert A. Riehle, Jr.

Surgeons have often said that the outcome for the patient is usually determined during the operation itself. Proper technique and planning during the procedure often pave the way for a smooth postoperative course and an ultimately satisfied patient. The same may be said about shock wave lithotripsy. In order to successfully treat a patient with a stone using the extracorporeal shock wave lithotripter, the operator must master patient positioning, biplanar fluoroscopy, and operation of the console. After the gantry itself is maneuvered so that the stone is brought into focus using underwater fluoroscopy, the surgeon must supervise the delivery of shock waves to disrupt the concretion(s). The treatment strategy used depends on the stone size, location, and opacity.

The comments in this chapter relate to the HM-3 lithotripter; in some cases, the instructions indicated will have to be slightly modified for newer models.

## PATIENT POSITIONING

To perform extracorporeal shock wave lithotripsy (ESWL), the patient must be placed on a device called the gantry (Fig. 6-1). The gantry has adjustable head, arm, back, and leg supports which are adjusted according to the patient's height and girth, the location of the stone in the urinary tract, and the side on which the calculus lies.

The vertical support bars are rotated to the right or left so that they are located on the same side as the stone to be treated (Fig. 6-2). The locking levers at the top of each bar must be depressed to allow swiveling. If the patient is to have bilateral sequential treatments, it is usually necessary to remove the patient from the tub after the first lithotripsy, place him on a litter, rotate the vertical supports to the opposite side, and reposition the patient on the gantry. When the patient is of small stature and the stones are medially placed, personnel may reposition for treatment on the opposite side by adjusting the leg supports in the reverse direction and shifting the immersed patient from one side of the gantry to the other. This is best done by two technicians, one responsible for holding the patient underwater to prevent buoyancy while the other directs the placement of the support system.

Each patient's height may be accommodated by moving the leg support in a caudad or cephalad direction. A scale is etched in the side of the leg support (Fig. 6-3), and can be moved by pulling the lock pin out. The "0" setting on the scale is appropriate for an individual 5 ft, 6 in. tall. Changes may be made in increments

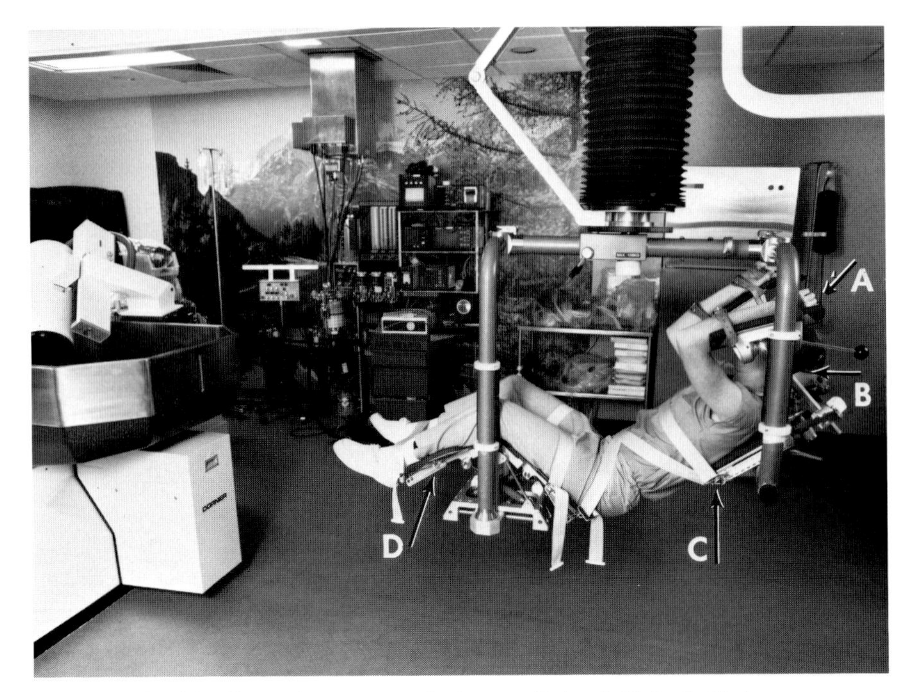


Fig. 6-1. Gantry (note supports): A, arm; B, head; C, back; D, leg.

of 2 in. By placing web straps under the patient for support, patients as short as 4 ft (1.2 m) have been treated; presumably, even shorter patients could be accommodated using this method. It is possible to treat some patients over 6 ft, 6 in. (1.98 m) with limits being defined largely by the individual's body build and weight (a 297.6 lb (135 kg) maximum has been recommended by Dornier).[1]

The calf and thigh pads are parts of the leg support. Upward movement of the pads results as the brass sleeves are twisted counterclockwise. Clockwise twisting lowers the thigh or calf, while a swivel mechanism allows side-to-side movement. The side of the body being treated should be canted 15 to 20 degrees so that the patient is actually rotated toward the side designated for treatment. This can be accomplished by adjusting the thigh pads. For example, for a right-sided stone, the right leg supports (calf and thigh) should be brought close together, while the angle between the left thigh and calf is increased. Placement of a pad or roll under the contralateral shoulder helps to facilitate torso rotation toward the side to be treated.

The backrest should be positioned so that the lower edge of the pad is between the axilla and nipple level. The headrest may be moved from side to side, moved up and down, and lengthened or shortened. It can be positioned either before or after the patient is anesthetized.

When the Dornier armrests are used (Fig. 6-1), care must be taken to place the arms in a position to avoid nerve compression or stretch. Alternatively, the arms may be allowed to float on the water or can be supported using Swimmies (Kransco, San Francisco, CA).

If ureteral catheter placement or stone manipulation is to be performed prior to ESWL, or if regional anesthesia is chosen, the patient will be placed on the gantry after anesthetic induction. Otherwise, the patient may assist to position himself/herself on the gantry. Once on the gantry, webbed straps are placed across the chest, abdomen, and thighs to secure the patient and help to counteract the effects of buoyancy in water. As a safety measure, straps are also

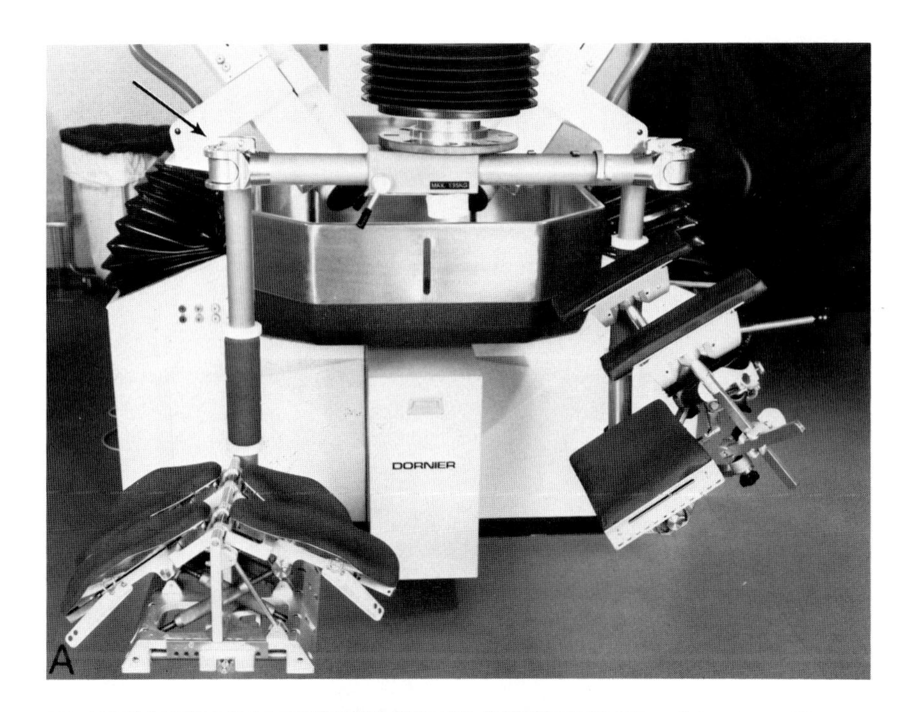


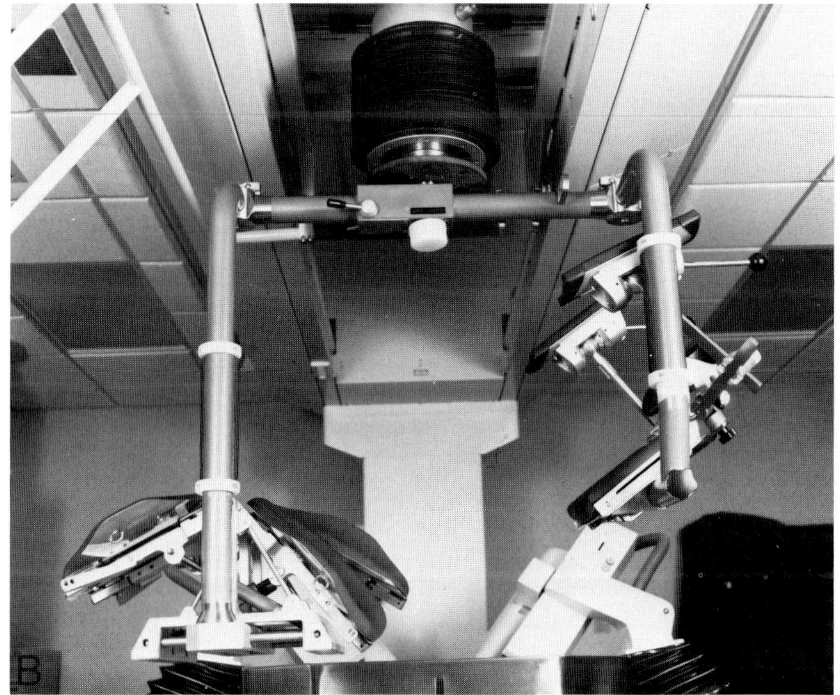


Fig. 6-2. **(A)** Gantry prepared for right-sided treatment. **(B)** Gantry prepared for left-sided treatment.

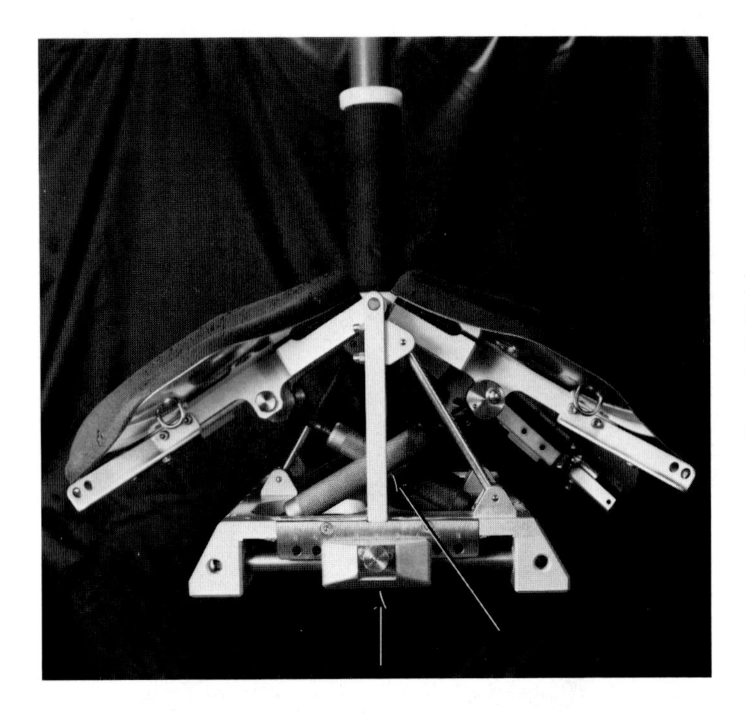

Fig. 6-3. Leg support: lock pin (short arrow) allows for height adjustment; brass cylinders (long arrow) under leg support may be adjusted according to body habitus and the side being treated.

placed beneath the patient. This keeps the patient's slippery body from jackknifing and falling to floor between back and leg supports while being removed from the tub.

If a child is being treated, a Styrofoam pad should be placed under the thorax with the lower edge just above the level of the stone. This should diminish the potential for lung damage induced by the shock waves.

## Gantry Positioning

### MOVEMENT INTO AND OUT OF THE TUB

The patient is usually placed in the gantry when the support device is perpendicular to the tub to allow more room for personnel to work. After the patient is safely and correctly positioned, the gantry itself is positioned so that it lies parallel to the tub (Fig. 6-4). The swivel mechanism makes this movement relatively effortless once the gantry lock is depressed.

The gantry is then raised hydraulically using controls on the console (control panel) (Fig. 6-5). The overhead lever is depressed while the gantry is pushed over the tub; it is then locked into position. Before lowering the patient into the tub, the ''centering'' button (four arrows on button) should be depressed (Fig. 6-6, Panel E, Module 19). This will position the gantry so that it will not hit the tub sides as it is lowered into the water. Once the ''centering'' button is depressed, the gantry continues to move until it is in proper position. The patient is then slowly lowered into the tub hydraulically. These steps are reversed to remove the patient from the tub.

### EMERGENCY SYSTEM

If necessary, the patient can be removed from the tub quickly by using the emergency buttons on the overhead lever. A patient can be emersed and placed on a litter in 30 to 45 sec using this system. Thus, resuscitative measures can be instituted in a timely fashion.

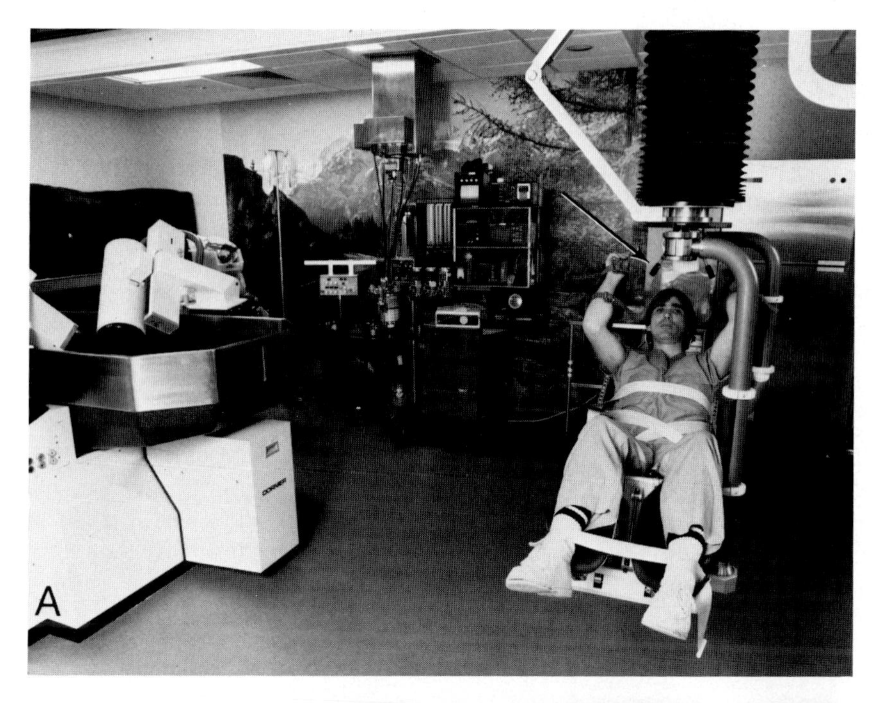


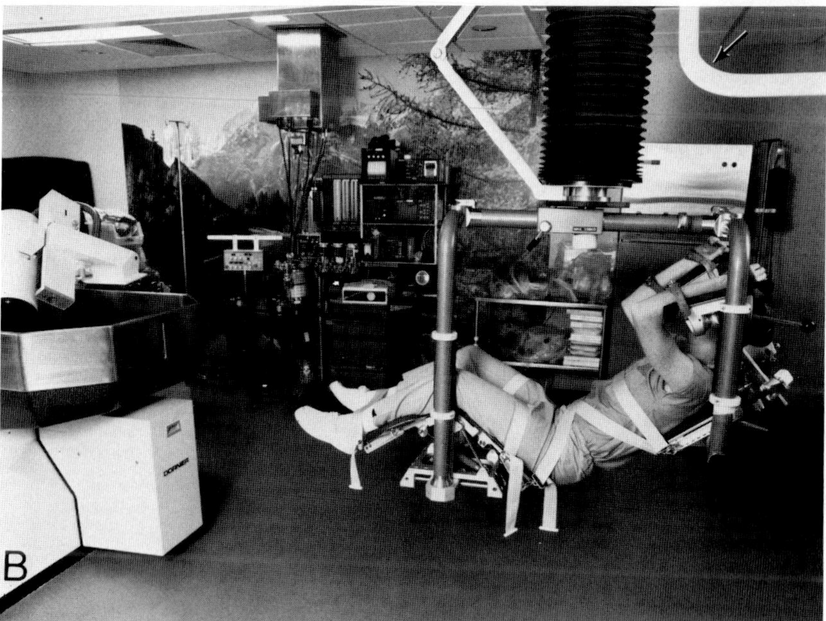


Fig. 6-4. **(A)** Gantry perpendicular to tub for patient positioning and anesthesia induction. Arrow points to gantry locking lever; pushing the lever up allows the gantry to swivel. **(B)** Gantry parallel to tub, ready to move into position in tub.

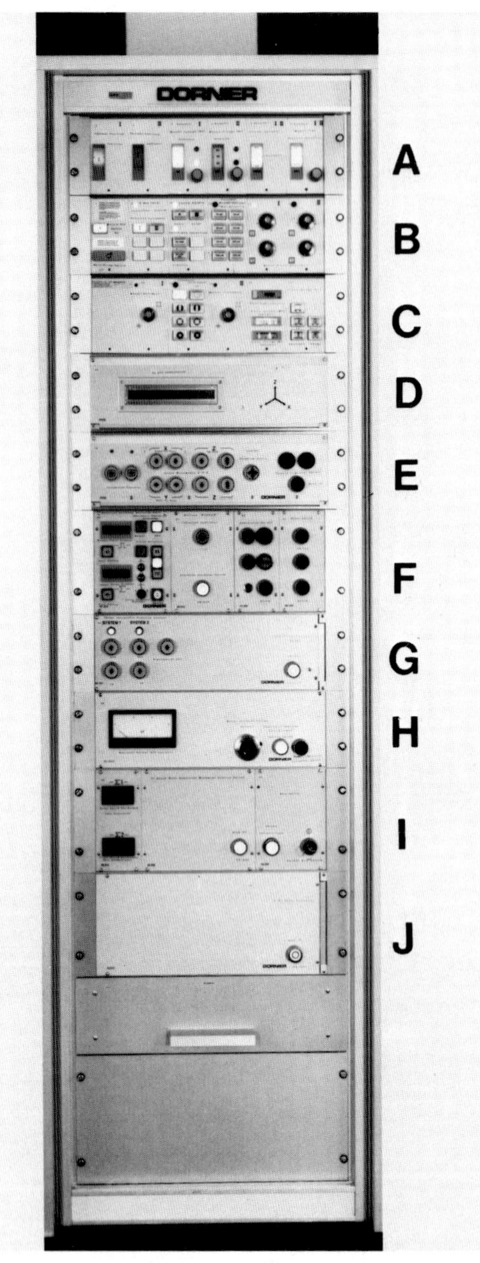


Fig. 6-5. Console. Note control panels A through I; one to six modules are contained in each panel.

## MOVEMENT IN THE TUB

Once in the tub, three-dimensional movement can be controlled via the console by the operator. The hydraulic system is activated when the green "pump" light is lit. If the system pressure is low, an alarm bell sounds and the green light goes out. Depressing the red alarm button stops the bell ringing; however, the red light remains on until the proper pressure level has been automatically reached.

Gantry movement is based on a XYZ system. The digital display indicates the distance in millimeters from the "home" or "center" position on each axis. When an "average" patient has been lowered into a position where the stone can be visualized—not focused—the indicator reads approximately X = 0, Y = 0, Z = 550 to 750 mm. The Z buttons move the patient up and down. Buttons with double arrows allow more rapid movement and are generally used to move the patient into or out of the tub. Buttons with one arrow allow finer, more precise movement. The X buttons move the gantry and thus the stone cephalad and caudad, and the Y buttons move it left and right.

It is important to realize that the XYZ axis orientation is different on the videoscreens than one would expect. The X and Z axes are oriented diagonally on the monitors, while Y and "integrated function" (see below) movement is along the horizontal plane on the screen (Fig. 6-7). This occurs because of the necessity of having a system which allows one to view actual three-dimensional movement in two planes on the monitor. Movement of the stone from the top to bottom on the screen or vice versa cannot be accomplished by depressing a single button. However, vertical movement, on the monitor, may be obtained by simultaneously depressing the X and Z buttons. The X and Z down buttons are pushed together so the concretion will move toward the bottom of the screen; pushing the X and Z up buttons have the opposite effect.

Diagonal movement of the patient (not on the monitor) is possible when using the integrated function buttons (Fig. 6-6, panel E, module 19). When using this function, movement along all three axes occurs simultaneously. As mentioned above, the operator will see movement from left to right on the monitor when the integrated function button is depressed.

The patient is brought into approximate posi-

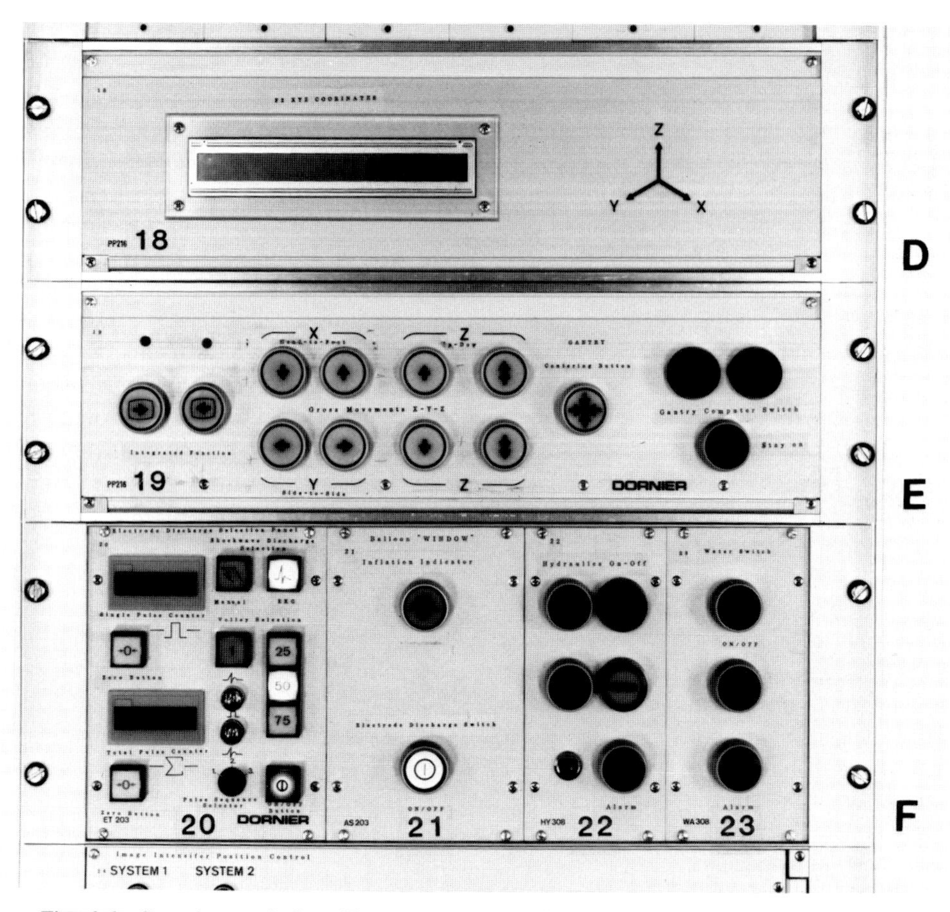


Fig. 6-6. Console: panels D to F; gantry movement is controlled using modules 18 and 19.

Fig. 6-7. The X, Y, and Z axes have been drawn on the video monitors to depict the direction of movement (on the screen) which the operator can anticipate when the respective buttons are depressed.

tion through communication between staff stationed at the console and tubside. A porthole at the head of the tub allows one to see when the area to be treated is above the ellipsoid/electrode. In many units, this part of the positioning is done by the nursing and technical staff, although the physician or anyone else trained in the use of the lithotripter can certainly do it.

## PREPARATION

### Preparatory Stone Localization

In the period immediately before ESWL, it is advisable to obtain a KUB film. Oblique, cone-down views may also be useful if the stone(s) are small or difficult to see. As with any other type of stone procedure, the exact location of the calculus must be known prior to the procedure in order to approach it properly. An upper ureteral stone may have passed into the distal ureter or a midcaliceal stone may have moved into the midureter since the last KUB film was obtained. Most ESWL units are equipped with a standard x-ray table or cystofluoroscopy table in the suite to facilitate this.

### Catheter Placement

If the stone lies in the ureter, we find it desirable to place a ureteral catheter just distal to the stone. Alternatively, the operator may choose to move the stone back up into the kidney by injecting normal saline, $CO_2$, or viscous lidocaine, or by dislodging the stone with a catheter or ureteroscope.

If a cystofluoroscopy table is not available in the immediate vicinity of the lithotripter, the patient may be cystoscoped on a litter. To accomplish this, the patient is frog-legged and a rolled sheet is placed under the sacrum to facilitate visualization of the trigone. Following a standard prep and draping, the urologist stands at the patient's side and inserts a rigid or flexible cystoscope.[1] The trigone and ureteral orifices are generally easily visualized. Experience gained in performing retrograde nephrostomy indicates that ureteral catheters may be inserted using this technique about 95 percent of the time.[2] A catheter with a single end hole and no side openings is best as this allows direct injection underneath the stone and maximizes the likelihood of repelling the stone into the kidney. A 65-cm 6 F angiographic catheter (Cook Urological, Spencer, IN) or a 7 F balloon occlusion catheter (Medi-Tech, Watertown, MA) may be used for this purpose.

Many centers favor placement of a urethral catheter before ESWL, particularly for large stones. The catheter can be placed prior to anesthesia induction or, in the interest of patient comfort, may be inserted in men after the patient is on the gantry and anesthetized. Stone manipulation should not be attempted unless fluoroscopy is available. With a small stone and a brief tub time, urethral catheterization may be omitted.

### Changing the Water in the Tub

The degassed water in the lithotripter tub should be kept at a temperature range of 35.8 to 37.5°C. Thermometers which measure the temperature of the water in the tub and the temperature of the incoming water are located near the foot of the tub. Water can be added or drained at any time if the water temperature becomes too cold.

It is not necessary to routinely change the water in the tub between each case. None of the six American F.D.A. sites reported problems associated with this practice even when epidural anesthesia was used. It is important to cover the epidural catheter with an adherent, waterproof, plastic covering. The tub water should be completely changed if any of the following are present: (1) menses, (2) ostomies, (3) patient defecation during lithotripsy, (4) escape of urine into the tub during ESWL, and (5) nephrostomy tubes placed during the previous 24 hours.

### Simulation

If questions arise regarding visualization and focusing of the stone with the patient in the tub, a simulation should be performed. Conditions which may mandate simulation include the following: (1) obesity, with large amounts of anterior abdominal or flank fat; (2) axial skeletal abnormalities resulting in decreased flexibility (osteoarthritis, Strümpell Marie disease, parkinsonism), acquired deformity (Pott's disease, fractured pelvis), or congenital deformity (spina bifida, scoliosis, or hemihypertrophy); (3) horseshoe kidneys; (4) transplant kidneys; and (5) the patient being a child. The simulation can be dry or wet. An awake patient may easily be placed in the tub and generally tolerates it quite well. If the calculus cannot be placed in focus or cannot be seen with the underwater fluoroscopy during simulation, an unnecessary anesthetic treatment can be avoided.

### Vaseline Application

It has been recommended that Vaseline or a similar material be applied to the patient's back in the area overlying the kidney.[3] It is felt that Vaseline prevents air bubble formation at the site of shock wave entry which could decrease the efficacy of the treatment.

At the University of Florida, more than 800 patients have been treated without applying anything to the patient's back and the investigators have not been impressed that stone breakup is diminished. Furthermore, Vaseline accelerates deterioration of the water displacement balloons (balloon windows). However, it is important to brush the bubbles, as seen through the viewing port at the head of the tub, away from the back after the patient has been immersed.

## RADIOLOGY

### General Concepts

The two underwater fluoroscopes are controlled by the upper three panels on the console (Fig. 6-8, panels A-C). The radiographic system consists of two integrated, but independent systems, which are oriented so that their beam paths intersect at F2 above the ellipsoidal reflector (Fig. 6-9). The stone is centered at F2, or in the "blast path" (see below), to provide optimal conditions for stone breakup.

Each x-ray system (I and II) consists of several components including: (1) an x-ray tube located beneath the patient in the floor of the tub; (2) an image intensifier with adjustable collimator and an inflatable cushion positioned next to the patient; (3) inflatable balloon "windows" positioned beneath the patient allowing the x-ray beam to pass through air before entering the patient, thus improving resolution; and (4) dual video monitors with a magnification factor of about 2.

The image intensifiers must be pivoted into the tub in order for the x-ray system to work. The patient contact balloons on the image intensifiers and the balloon windows underneath the patient should be inflated so that they are near, but not touching, the patient. Image intensifier movement may be controlled by buttons on either the console (Fig. 6-10, panel G, module 24) or the intensifiers (Fig. 6-11). A button at the top of module 21, panel F (Fig. 6-6) lights when the balloon windows are inflated. The balloons deflate and the light goes out when this button is depressed. The balloon windows are inflated during positioning to improve resolution, and deflated during shock wave administration to allow the shock wave to achieve maximum efficiency. If the balloon windows are overinflated, the balloons can actually move the patient and displace the stone from F2 when deflation occurs.

The balloon windows are long lived, but occasionally one will burst. Should this occur, one

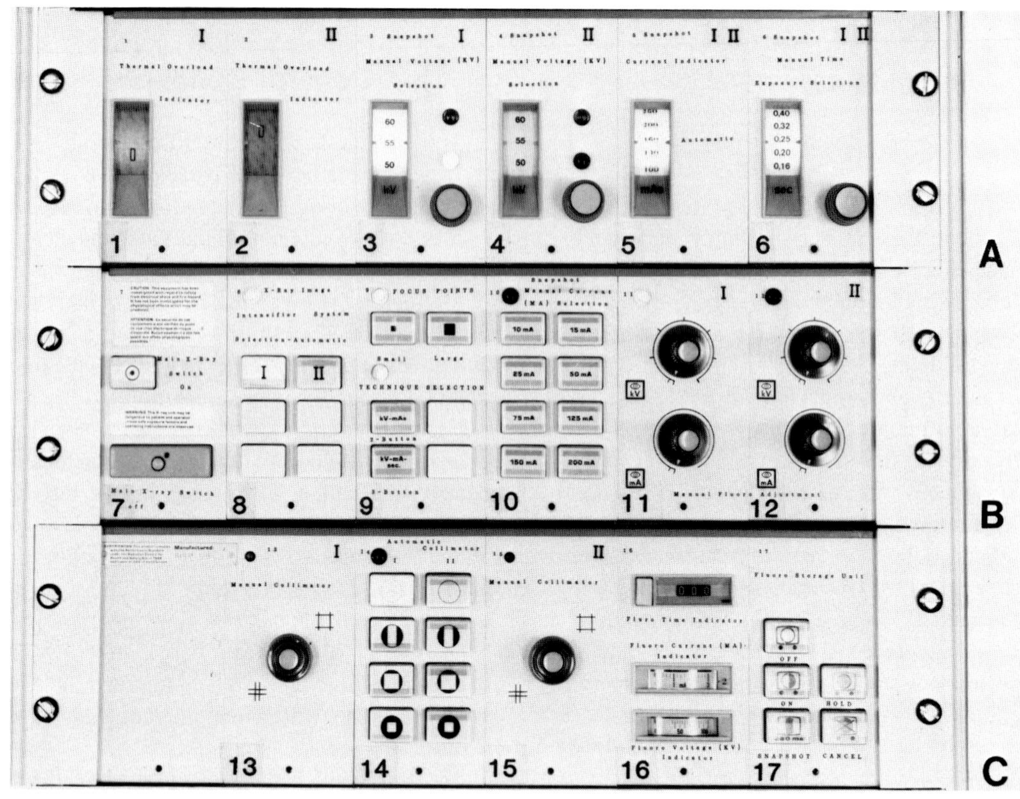


Fig. 6-8. Console, panels A to C. The lithotripter x-ray functions are controlled using these three panels (modules 1 to 17). (See text for further discussion.)

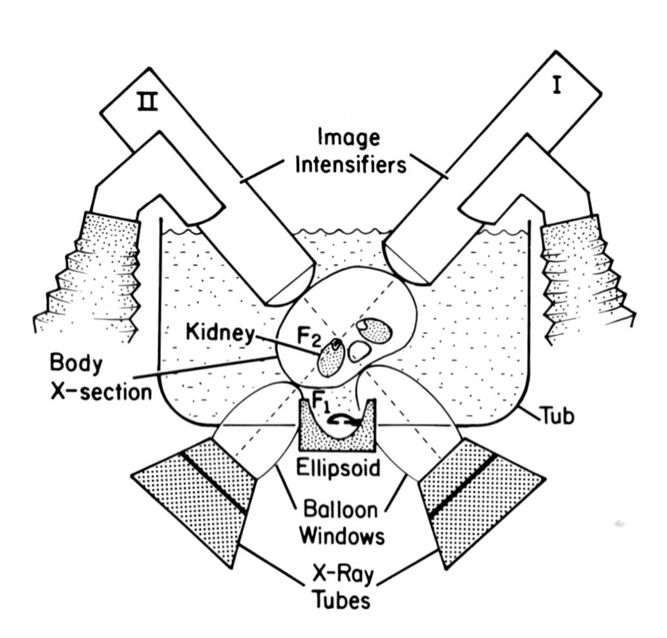


Fig. 6-9. Cross section of lithotripter tub. The x-ray tubes are oriented such that the beams intersect at F2. F2 corresponds to the intersection of the electronically produced crosshairs seen on the video monitors.

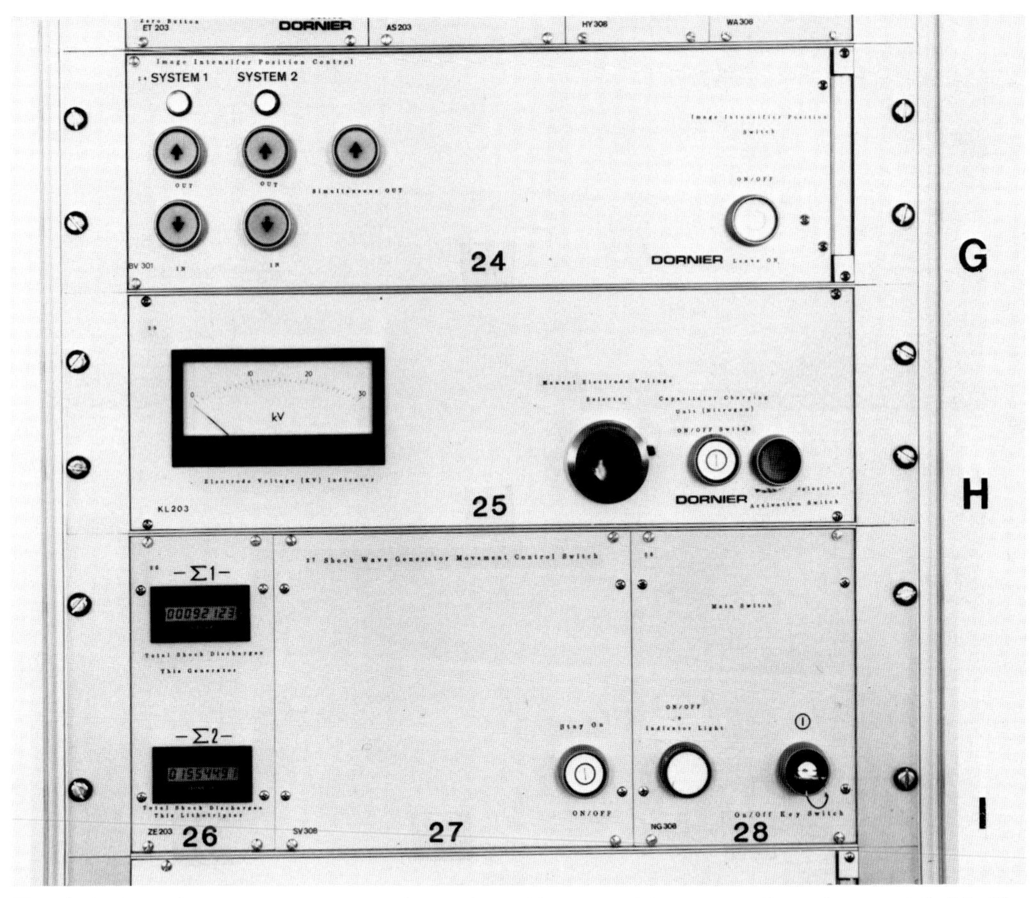


Fig. 6-10. Console, panels G to I. The image intensifier control buttons are located on module 24. (See text for further discussion.)

may substitute a toy balloon or inflated surgical glove to displace the water during x-ray exposure. To replace a damaged balloon, the water level must be below the housing apparatus to prevent damage to the x-ray source during balloon changing.

Using the controls on the console, the operator must make certain exposure selections before taking the radiograph. Kilovoltage (kV) determines the x-ray wavelength and penetration power; penetration power increases with increasing kV. Milliamperes (mA) or current is the quantity of energy released by the x-ray tube. The amount of time the patient is exposed to the x-ray beam helps to determine the quality of the film. The overall exposure is determined by the mAs (milliamperes × seconds). The mA and kV factors used for ESWL are generally higher than the settings used for conventional radiographs because exposure must be made through the water.

Once generated, the radiographic images seen on video monitors are positive images with the stones appearing dark rather than white. Magnification is about twofold. For example, if a stone lies 4 cm from the intersection of the crosshairs on the monitor, it is actually about 2 cm from F2. TV monitors are labeled I and II to correspond to the right and left image intensifiers. Brighteners and contrast adjustment knobs are positioned below the screen. The image intensifiers electronically produce crosshairs visible

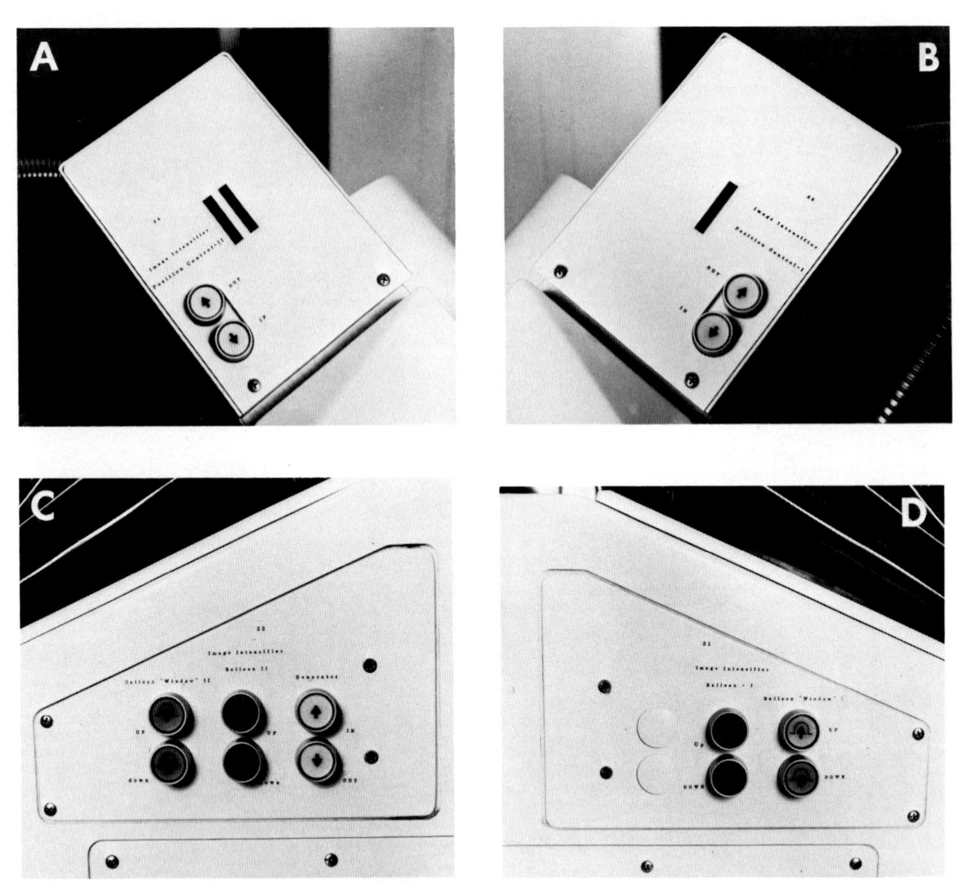


Fig. 6-11. **(A)** Control buttons located on image intensifier move the intensifier up and down (unit II). **(B)** Similar buttons control unit I. **(C)** Control panel at head of tub (left); middle buttons inflate or deflate balloon window on image intensifier II. Buttons on left inflate or deflate balloon window lying beneath patient in tub. Buttons on the right move the generator in and out making it possible to change the electrode without draining the tub. **(D)** Control panel at head of tub (right); buttons on the left control the balloon window on image intensifier I while those on the right inflate or deflate the balloon window lying beneath the patient on the right.

on the monitors. The crosshairs are superimposed on the projection of the point where the x-ray beams intersect at F2.

Seven modules (1, 2, 7, 8, 13, 14, and 15) in the x-ray portion of the control console (Fig. 6-8, panels A-C) allow the operator to select the required technique and imaging. The machine is activated by pressing the small ON button on module 7 or deactivated by pressing the large OFF button. Modules 1 and 2 are thermal overload indicators of their respective x-ray systems. As long as the indicator is green, fluoroscopy can proceed without difficulty. If one of the systems overheats, the indicator rotates into white and indicates how long the operator must wait before continuing. When the tube has cooled sufficiently, the green color will again become visible. It should be noted that use of the x-ray system when the thermal load indicators are in the white area is monitored and recorded internally, and "may result in voiding of the warranty."[4]

The image intensifier system in use is indicated automatically by a lighted button on module 8. The collimators are controlled in modules 13, 14, and 15. Pushing the button from side to side or up to down on module 13 (image intensifier I) or panel 15 (image intensifier II) allows one either to view the entire screen (a circular area) or "cone-down" the area in view. Module 14 has four pushbuttons; the top buttons give the maximum viewing area and the bottom buttons cone down on a small sector in the area of the crosshairs. The large circular viewing area should be used to localize and focus the stone. Coning down the collimators decreases radiation being delivered to the patient and also makes the image sharper and clearer. Every attempt should be made to limit the radiation dose; therefore, after initial localization, the smallest collimated image should be used to follow stone disintegration.

## The X-Ray Hand Control

By using the toggle switch on the handheld control (Fig. 6-12A), one can switch back and forth between x-ray systems I and II. A fluoroscopic (dynamic) image is seen on the selected screen when the dime-sized button in the middle of the handheld switch is depressed. When this button is released, the x-ray system is deactivated. A snapshot ("Quick-pic" or static image) is obtained by depressing the dime-sized black button on the top of the hand control. A two-tone chime indicates that a radiograph has been obtained. The static image generated is transmitted to the video monitor. The snapshot image tends to be sharper and clearer than that seen using fluoroscopy. The buttons (2) and the lights (2) near the top of the hand control activate and indicate usage of the optional video cassette recorder and camera.

## Snapshot Mode

The controls for the snapshot made are shown in Fig. 6-8 (panels A and B, modules 3, 4, 5, 6, 9, and 10). The "three-button" technique is used most frequently and is activated by depressing the kV-mA-sec button on module 9 (Fig. 6-8, panel B). When using this technique, one can manually select the voltage (kV, modules 3 and 4), current (mA, module 10), and

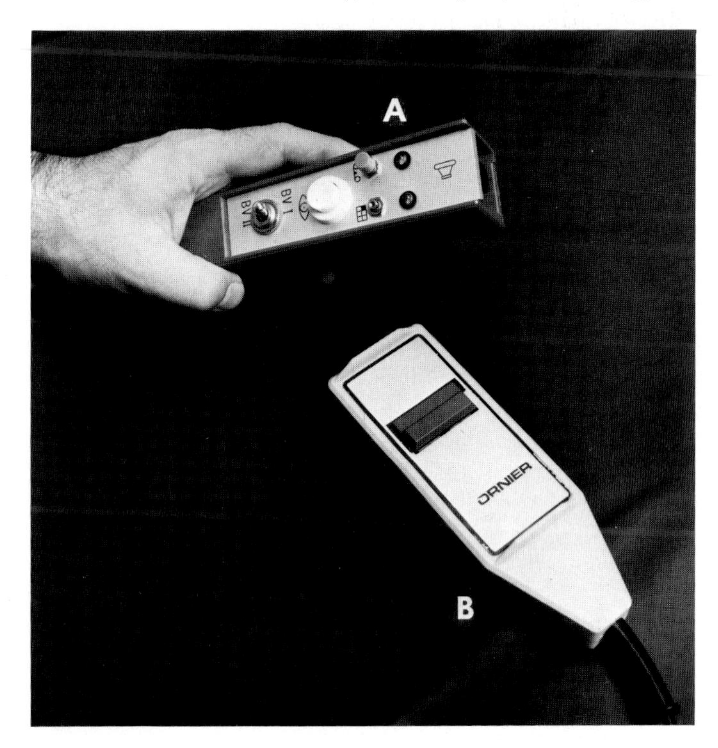


Fig. 6-12. Hand-held switches: **(A)** x-ray controls; **(B)** shock wave discharge button.

exposure time (sec, module 6); hence the name "three-button technique." When using the three button mode, the focus point (module 9) of the electronic x-ray beam is automatically selected. Though the voltage may be varied, 50 to 70 kV are often suitable for a 70 kg patient. A slightly higher setting (5 to 10 kV) is used for the x-ray system traversing the largest tissue mass (e.g., system II when treating a right-sided stone). An optimal image can usually be obtained with a current of 150 to 250 mA. In certain circumstances, lowering the current to 10 to 200 mA may be appropriate. X-ray exposure time can be varied from 3 msec to 5 sec, but most exposures are made at 160 to 250 msec. If the green indicator light in modules 3 or 4 goes out, one of the values—kV, mA, or time—has been set too high and must be lowered before proceeding.

The "two-button" snapshot technique is selected by pushing the kV-mAs button on module 9. This mode allows one to use the full energy of the generator, but increases the radiation exposure of the patient. When using this technique, a large or small focus (module 9) may be selected. Generally, sharper focus and better resolution may be expected when the small focus button is selected. The kV and exposure time are again set manually using modules 3, 4, and 6 while mAs is set simultaneously with the time exposure selection and is indicated on module 5. Because only the kV and exposure time are set manually, this mode is called the "two-button" technique. In practice, we almost never use it.

Recently, Eusek and associates at the University of Washington developed a technique for in-bath filming during ESWL.[5] The "in-bath" technique consists of the following: A plastic film holder (Eusek-Bush in-bath filming apparatus, patent pending) is attached to one of the image intensifiers; the inflatable rubber balloon from a large sphygmomanometer cuff is placed between the film holder and the patient's abdomen to create an air space through which the x-ray beam will pass; and the "two-button" snapshot technique is used to obtain a film. The kV-mAs button (Fig. 6-8, panel B, module 9) is pushed and kV and mAs selections are made; for an average sized patient, 65 kV and 50 mAs are used. The authors state that in 81 percent of 103 patients in which the technique was used, good or excellent quality images were obtained. They further stated that the technique is useful for localization of stones difficult to visualize and for determination of adequacy of fragmentation.

## Fluoroscopy Mode

The controls for the fluoroscopy mode are shown in Fig. 6-8 (panels B and C, modules 11, 12, and 16). Voltage (kV) and current (mA) selections used during fluoroscopy are made by turning the dials on modules 11 and 12. If the knobs are turned maximally in the counterclockwise direction to the "off" position, a current of 0.3 to 6.0 mA will be selected automatically, depending on the patient's body mass. Generally, this current provides an image with adequate contrast and resolution at minimal radiation exposure. However, a better fluoroscopic image may be obtained by turning the mA knobs fully clockwise and the kV knobs fully counterclockwise or "off" position. The current and voltage used are registered on the indicators in the lower half of module 16. The amount of fluoroscopy time is registered on a counter at the top of this same module. A buzzer sounds when 300 sec (5 min) of fluoroscopy time has ticked off and a button must be pressed to reset the device. It should be remembered that this timer records only fluoroscopy time, and therefore, no record of static picture or snapshot mode exposure time is generated. An indicator registering total radiation exposure is in the interior of the lithotripter, but is not accessible to the operator of the HM-3 machine.

## The Storage Module

Module 17, panel C can be set to function during fluoroscopy or when taking snapshots. To be effective, the desired button must be pushed before an exposure is made. When the "off" button is selected, x-rays can be made using either mode, however, no images will be stored. The "on" button activates the unit and automatically stores the last image seen on the screen, be it snapshot or a fluoroscopic image.

The "hold" button prevents erasing of the last image taken. Fluoroscopic or snapshot modes can be used once this button is activated. However, when the fluoroscopy button is released or a snapshot exposure is made, the previously made image will be visible on the video monitor.

If the "snapshot" (280 ms) button in the storage module is pushed and then the snapshot button on the hand control is pressed, the shortest possible x-ray pulse that will produce a stored image is released. When this button has been selected, a reproduction of a dynamic fluoroscopic image cannot be obtained on the monitor. In fact, what occurs is that the machine is preset to make a fluoroscopic exposure for 280 ms and that static image is then displayed on the screen.

Pressing the cancel button erases the image on the screen and it cannot be recalled. Switching from system I to system II has no effect on storage function, regardless of the button which has been selected.

## Radiation Exposure

Variables affecting the radiation exposure received by the patient during ESWL include patient weight, abdominal girth, distance of the patient from the x-ray tube, tube collimation, exposure settings (kV and mAs), and individual operator technique. These variables, including the fact that machine settings must sometimes be changed during a given case, make it difficult to estimate actual radiation dose in rads to the patient during treatment.

At New York Hospital–Cornell Medical Center, spot films performed at 150 mA, with an average kV(p) of 60 for 280 msec result in an approximate calculated surface exposure of 0.483 roentgens (R) for each spot film. During fluoroscopy, the tube output (kV(p)/mA) is automatically controlled and the setting of 80 kV(p)/1.5 mA results in approximately 2 R/min surface exposure.[6]

Carter and Riehle used these values to estimate patient entrance exposure. Fluoroscopy time recorded in seconds for 298 patients with an average stone burden of 19.3 mm (3 to 64 mm) averaged 160 sec (3 to 509 sec). The total number of spot films taken per patient was 5 to 68, averaging 26. Estimated surface radiation exposure in roentgens for each procedure, based on fluoroscopy time (2 R/min) and number of spot films (0.483 R/film), averaged 17.8 R.

Radiation exposure increased with increased stone burden (sum of stone diameters) and higher patient weights; it also varied according to stone position. Average radiation exposure during ESWL for patients in the 1 to 10 mm stone size group was 15.5 R compared to 22.5 R for patients with stones measuring $>50$ mm (Figs. 6-13 and 6-14).

Stone location also influenced radiation exposure independent of stone burden (Fig. 6-15). A comparison of renal pelvic, caliceal, and ureteral stones revealed that average radiation exposure was greatest for the patient group with ureteral calculi, even though size of these stones was significantly less than the mean. An average skin level exposure of 19 R was noted for ureteral calculi, compared to 16 R and 16.4 R for renal pelvic and caliceal stones, respectively.

Preliminary data from two other institutions indicate that the mean patient exposure lies between 6[7] and 18 rads per case (personal communication, David McCullough, Bowman Gray School of Medicine, Winston-Salem, NC). Of note, early investigators in Munich reported an average patient exposure of 8.9 rads.[8] In general, levels as detected by dosimeters are lower than calculated surface radiation exposure.

It should be noted, however, that radiation doses to whole body and specific organs cannot be equated to entrance skin doses, the former being dependent on the radiation energy and field size. Entrance skin doses are measurements of the maximum dose that could possibly be

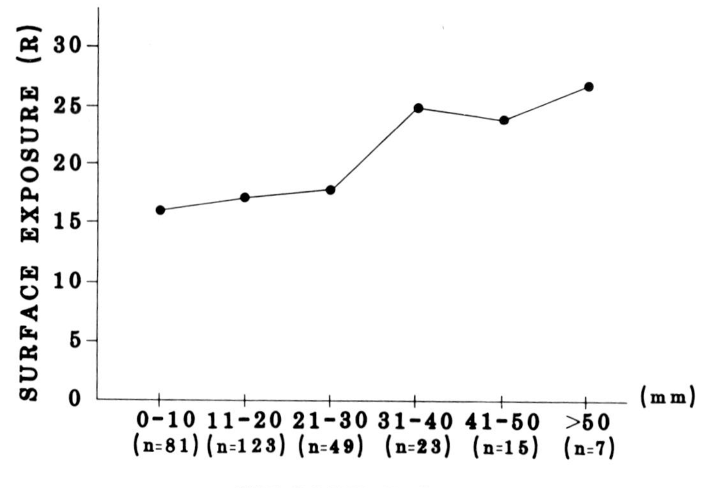


Fig. 6-13. Calculated radiation exposure versus stone load.

received, since x-rays are attenuated by tissue. For example, during a barium enema examination using fluoroscopy, a surface dose of 480 mrads/min results in a gonadal dose of 11 mrads/min.[9] Previous studies have shown that surface exposure rates for various fluoroscopic procedures such as upper gastrointestinal series, small bowel series, and barium enemas are between 0.1 and 0.5 rads/min using image intensification.[10] Because of long fluoroscopy times, the highest doses are found during angiographic procedures.[11,12] Bush et al studied radiation exposure during percutaneous nephrostolithotomy (tract dilation and stone extraction) and found an average skin surface exposure of 25 rads per procedure during an average fluoroscopy time of 24 min.[13]

Although precise quantification of patient radiation exposure during ESWL is not yet available, an estimate of skin surface exposure using

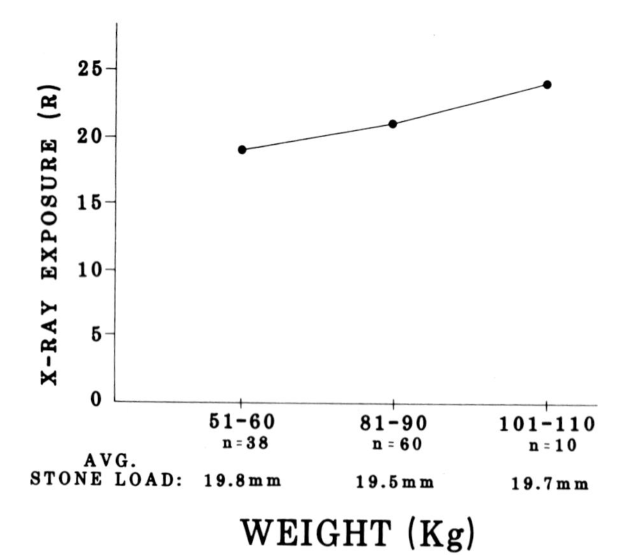


Fig. 6-14. Calculated radiation exposure according to patient weight.

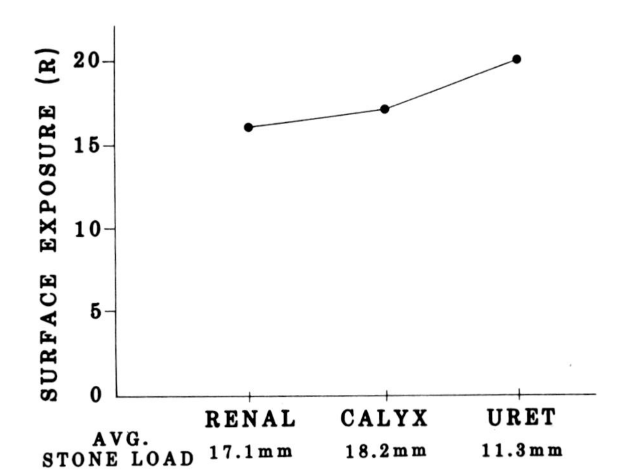


Fig. 6-15. Calculated radiation exposure depending on stone location.

measurements of the x-ray tube output can be employed to examine the factors important in determining patient radiation exposure.

In summary, it appears that the most important factors influencing radiation exposure during ESWL are stone size, stone location, degree of calcification, and patient weight.

Radiation scatter outside the tube should be measured locally by the institution's radiation physicist. Opinions vary from center to center about the need to wear lead protective shielding when staff are close to the tub. For those within 5 ft of the lithotripter on a daily basis (e.g., nurses, technicians, or anesthesiologists), wearing a lead apron seems prudent.

It is best for all concerned to minimize the amount of radiation administered. The collimators should be used to reduce the size of the viewing field. Frequently, it is necessary for the collimators to be fully open when initially localizing the stone. However, after the calculus is moved to the center of the viewing screen, the field can be collimated down. Obviously, limiting the number of snapshots and minimizing fluoroscopy time will decrease overall patient radiation exposure.

In general, it has been recommended that the fluoroscopic mode be used whenever possible to diminish overall radiation exposure. The relationship between the amount of radiation emitted during fluoroscopy and during a snapshot is currently being studied.[7]

## Stone Positioning

### CONVENTIONAL POSITIONING

Before stone treatment can begin, the patient on the support system must be maneuvered hydraulically so that the calculus is brought into the focus of the shock wave (F2). As described earlier, the stone is first localized in the vicinity of F2 when positioning the gantry. The x-ray imaging systems are then used to bring the largest stone into the precise focus area. Picture resolution is usually best—particularly in fluoroscopy—when using the equipment on the side being treated. System II is used first when focusing a left-sided stone. When beginning the stone localization process on a right-sided stone, we usually use x-ray system I; using this system results in less radiation exposure since the body mass through which the beam must pass is less than when using x-ray system II.

Once the stone is seen on the monitor, it should be brought to the horizontal axis of the crosshairs using the X and/or Z controls. Once the stone is on the horizontal video axis, movements can be restricted to use of the "integrated" function. If the stone is moved to the right or left using this function on screen I, the calculus will remain on the horizontal video crosshair axis on screen II. In contrast, if the X and Z controls are used to move the concretion previously placed on the horizontal video axis on screen I, when one looks at screen II, the stone will no longer be on the horizontal crosshair axis. In this circumstance, one could not use a single button—the integrated function—to bring the stone into focus. It should be realized, however, that repositioning during the procedure to direct the shock waves at a different portion of the stone may require use of the X or Z functions.

Positioning a calculus in the electronic crosshairs is the exact opposite of aiming a rifle. The operator must realize that the stone is moved into the crosshairs rather than the crosshairs being moved to the stone. Put another way, one might imagine that the calculus is a car and the ESWL operator is the driver. One must "drive" the stone to the crosshairs; the crosshairs will not come to the stone. As an example, assume that an object is centered on screen II, but lies to the left of center on screen I (Fig. 6-16A). To bring the object into focus on screen I, depress the integrated function button with an arrow pointing to the right. Once accomplished, the object will be in focus on both screens (Fig. 6-16B). This becomes even clearer when one realizes that the patient and stone are being moved, rather than the image intensifier; in other words, the stone moves beneath the intensifier. The operator should pretend he is inside the intensifier and is using his x-ray vision to look through the patient.

When positioning the stone, the x-ray system (fluoroscopy or snapshot) does not need to be activated at all times. As one gains experience, it is possible to anticipate the amount of movement which will occur when a button is depressed for a given amount of time. Using the equipment in this fashion lessens the radiation exposure to all concerned.

## The Blast Path Theory

An alternative method for focusing and treating stones is based on the "blast path" theory described by Madorsky and Finlayson.[14] These investigators proposed that stones can be effectively disintegrated at $\leq$8 cm from F2 in the upper medial quadrant on the video monitors. They note that if the Dornier F2 aligning device is placed on the ellipsoid, the device points at F2 from the lower lateral quadrants of the fluoroscopic monitors and lies in the central axis of the shock wave (Fig. 6-17). Based on an analysis of the decomposition of the shock wave pressure vectors, they note that the approaching wave is compressive and the departing wave has a large tensile component. Kaneko et al showed that stones are more susceptible to tensile stress than to compressive stress by a factor of about seven.[15] It follows that stones placed on a line running through F2 into the video monitor's upper medial quadrants at an angle of 25° from the horizontal (i.e., on the blast path) will be effectively pulverized at modest shock numbers. A hierarchy of therapeutic effectiveness of the quadrants around F2 (i.e., $A > C \gg B, D$) may be deduced based on the preceding arguments (Fig. 6-17).

Graphically, the blast path may be drawn on the video monitors by making a dashed line through F2 and a point 10 cm medial to F2 and 4.66 cm above the horizontal axis (Fig. 6-17). The stone is placed as close as is convenient to F2 on one monitor and then brought to a similar position, in the blast path, on the second monitor using the integrated function. Madorsky and Finlayson proposed that the blast path be used specifically for horseshoe and transplant kidneys, for stones in the distal ureter, for particularly hard stones (i.e., cystine) and, perhaps, with any upper tract stones. In addition, it is very helpful in the patient who is too thick to allow positioning of the stone at F2.[14] Problems arise in the latter circumstances because the distance between F1 and F2 is fixed by the configuration of the ellipsoid. The investigators' initial results are promising, but it should

Fig. 6-16. Stone positioning: **(A)** Monitor I, ball on metal rod to left of center; monitor II, ball on metal rod centered. **(B)** Monitor I, ball on metal rod centered; monitor II, ball on metal rod centered. When a stone lies on the horizontal axis but is not centered, pressing the integrated function (Fig. 6-6, panel E, Module 19) brings the calculus into the center of the crosshairs.

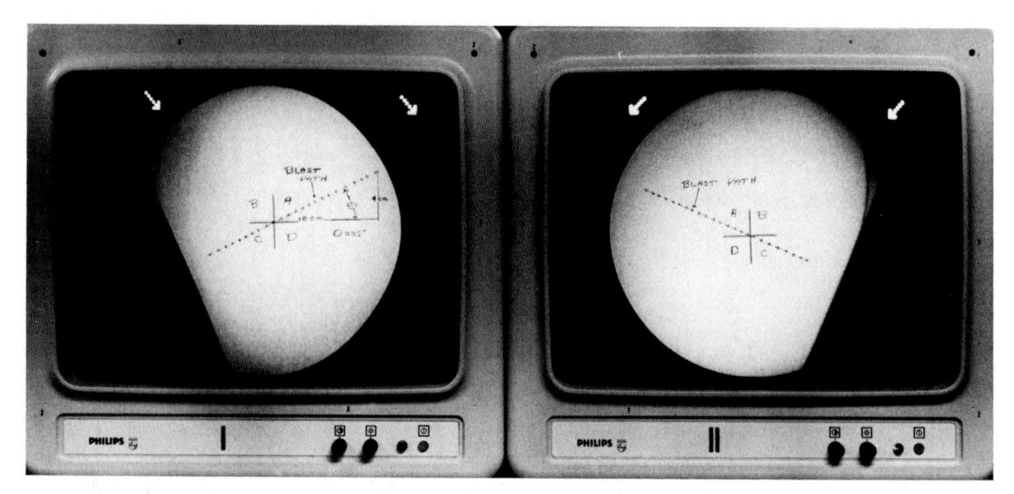


Fig. 6-17. Madorsky–Finlayson blast path. Stones can be effectively disrupted if positional at F2 or along the blast path in the upper medial quadrant on both monitors. (See text for further discussion.)

be noted that limited clinical data are available to support the theory at this time.

## SHOCK WAVE ADMINISTRATION

With the stone in position, the operator must select the voltage and sequence for shock wave delivery. The energy level of the shock is determined by the kilovolt (kV) setting on the capacitor charging unit. A setting of 18 to 20 kV corresponds to a pressure of approximately 900 to 1100 bar at F2 (1 bar = 14.7 psi). As the voltage is increased, the wave energy produced increases. The higher the voltage, the shorter the lifespan of the electrode.

Many stones break up at the 18 to 20 kV level, though some calculi (e.g., cystine) are particularly resistant and may require energy levels of 24 to 28 kV. When higher energy levels are used, the stone tends to fragment into larger pieces rather than disintegrate into fine, sand-like particles. If a hard stone is fractured into several large pieces, lesser voltages may be adequate to break up the pieces.

Electrode lifetime is dependent on the voltage selected and the number of shocks administered at a given voltage. It has been stated that the electrode must be changed after 700 to 800 shocks.[16] Measurements of pressure at F2 discussed in Chapter 2 and published elsewhere[17] indicate that the pressure at F2 can be maintained with a single electrode for up to 4,000 shocks if the initial voltage is 18 kV and is increased by 1,500 V per 1,000 shocks. As a practical matter, at the University of Florida, treatment is frequently begun at 18 to 20 kV and increased in 1 to 2 kV increments when misfiring is audible up to a maximum of 2,000 shocks. There is an inverse relationship between electrode life and the kV setting; thus, the electrode will not last as long at 28 kV as it will at an 18 kV setting.

Next, the pulse sequence knob (panel F, module 20 in Fig. 6-6) is turned to select whether the electrode will be discharged after each R wave (1), after every other R wave (2), or after every third R wave (3). In practice, the shock wave is almost always set to discharge after each R wave. In the event the patient is tachycardic, every second or third R wave may be more appropriate. Also, if an arrhythmia occurs during treatment, a change to a 2:1 ratio may eliminate the irregularity. Both the single pulse and total pulse counters should be zeroed by pressing the appropriate "0" (zero) button at the outset of the case (panel F, module 20 in Fig. 6-6).

The shock wave may be discharged manually (by depressing the "hand" button and the button marked "1") or triggered off the R wave (by depressing the button "QRS complex" button and the "1" button) for single shock wave discharge. Alternatively, the shocks may be fired in volleys of 25, 50, or 75 by depressing the "QRS complex" button and the appropriately numbered button. The shock waves are actually discharged by depressing the handheld switch (Fig. 6-12B). This is a "dead-man switch"; if the red button is released, the shock wave discharge ceases. Discharge also ceases when the volley (e.g., 50) of shocks has been completed. The single pulse counter must then be zeroed in order to fire another volley. In general, the patient should be fluoroscoped after the first 100 shocks to make sure that neither the patient nor the stone has moved. As discussed, the balloon windows must be inflated in order to obtain reasonable resolution during x-ray exposure. During lithotripsy, 100 to 200 shock waves can be delivered between x-ray exposures depending on size or the position of stone being treated. If the operator wishes, shock wave discharge may be continued while x-rays are being taken. The balloon windows are deflated thereafter.

Hematuria, generally noted at some point during the procedure, is probably due to urothelial trauma. It usually clears in 24 to 48 hours, often by the evening of the treatment day. Should the urine be particularly bloody, administration of a diuretic (e.g., Lasix, 40 mg intravenously) may help prevent colic secondary to clot formation.

### Sound Exposure

The delivery of focused shock waves is accompanied by an auditory sonic wave component easily heard within the lithotripter unit and adjacent areas. Enhanced sounds reverberate from the tub and radiate against sheet steel and other structural lithotripter components, such as the overhead track for the gantry. Soundproofing to reduce vibrations and reverberations in the room has been installed in some units, and newer models of the immersion tub are insulated. Noise measurements at New York Hospital–Cornell Medical Center using a precision impulse sound level meter revealed that impulses originating from the lithotripter were approximately 105 dB peak sound pressure level.

With regard to damage/risk criteria for hearing loss, the number of impulses emanating from the instrument on a daily basis does not appear to be hazardous by OSHA's 1974 Proposed Hearing Conservation rule. In this case, impulses of 140 dB were not permitted more than 100 times daily, with an additional tenfold increase in occurrence permissible with each decrease of 10 dB. However, the guidelines for impulse noise are much more complicated than those for continuous noise (constant level) (personal communication, Marc Kramer, Ph.D., New York, NY).

Investigators at the University of Virginia agree that the impulse sound created by the lithotripter should not result in "damage to the auditory system in most individuals."[18]

## TREATMENT STRATEGY

### Size

Treatment strategy varies depending on several factors. A solitary stone measuring up to about 1.5 cm may be treated by centering the stone in the crosshairs. For smaller stones, high-frequency jet ventilation (HFJV) has proven to be particularly useful. Using this ventilation technique, the concretion or portion thereof being treated remains in the desired location at all times as opposed to moving in and out of focus. Presumably, because the stone is in focus, the treatment is more efficient.

Larger stones are treated in sections beginning with the inferior portion of the stone nearest the ureteropelvic junction (UPJ). If the portion of the stone lying closest to the UPJ is not well broken up, fragments will not be able to flow out the ureter even if other portions of the stone are well disintegrated. As the medial portion of the stone breaks up, the operator can move to a more lateral and then a more superior position.

## Location

Generally, stones in lower calyces or in other relatively dependent portions of the kidney should be treated first. With fragmentation, the small pieces tend to obey the law of gravity and often fall into the lower pole calyces. When stone fragments admix with untreated concretions, it becomes more difficult to differentiate treated and untreated fragments. Thus, calculi in the upper calyces should be treated last to avoid this problem.

Most centers are now treating bilateral calculi during the same session. The most symptomatic stones are treated first in case there is reason to terminate treatment. If all goes well with the first side, the contralateral stones are subjected to shock wave therapy.

## Opacity

Dense calcium oxalate stones generally present no problem as far as treatment. Treatment of faintly opaque calculi is often facilitated by pretreatment placement of a ureteral catheter or nephrostomy tube. Such a catheter serves as a useful landmark and allows the operator to inject contrast into the pelvocaliceal system when the stone's position is in doubt.

Lucent stones, on the other hand, can challenge the operator. In this situation, dilute contrast should be injected through a ureteral catheter or nephrostomy tube. The filling defect is then focused and treated.

If stones cannot be seen with or without contrast injection, and one can ascertain their position by studying the pretreatment intravenous pyelogram (IVP), one can proceed by treating the calyx in which the concretions are presumed to lie. Obviously, this technique is less than desirable; nevertheless, it has been effective in our hands on several occasions.

## End Point Determination

The end point of treatment for a given stone is subjective; however, a number of indicators are used by the experienced operator. Generally, the stone becomes less dense and tends to spread out as the treatment progresses (Fig. 6-18A–E). In some cases, it will virtually disappear. Particles tend to fall laterally, inferiorly, and in some cases, superiorly, depending on the intrarenal anatomy of the calyces. Fragments (dark images) visualized on both screens must be treated. Fragments seen only on one screen, however, may actually be disintegrated and visualized "on end" (en face); these do not need further treatment.

If the stone is in a tightly enclosed space, it may change very little as the treatment progresses. Examples would include medullary sponge kidney stones in renal tubules, calculi proximal to an unrecognized obstructed calyx, and stones completely filling all or a portion of the collecting system. In such circumstances, shocks should be delivered to all portions of the concretion; the operator must arbitrarily determine the total number of shocks to be delivered.

If the stone being shocked is lucent, the treatment endpoint is even more vague than usual, because one usually cannot see when and if the stone fractures. In addition, an increase in the number of filling defects seen may be due to the presence of clots in the collecting system. Here again, one must arbitrarily select a number of shocks to administer and work toward that goal. If there is any question as to the adequacy of stone fragmentation, it is much less expensive for the patient to administer a few hundred more shocks as opposed to the individual returning days to weeks later for a subsequent treatment.

## Length of Treatment

The average treatment time is $54 \pm 25$ min[19] with a range of 10 to 185 min. As with other procedures, the amount of time varies according to the operator's experience. The patient's body

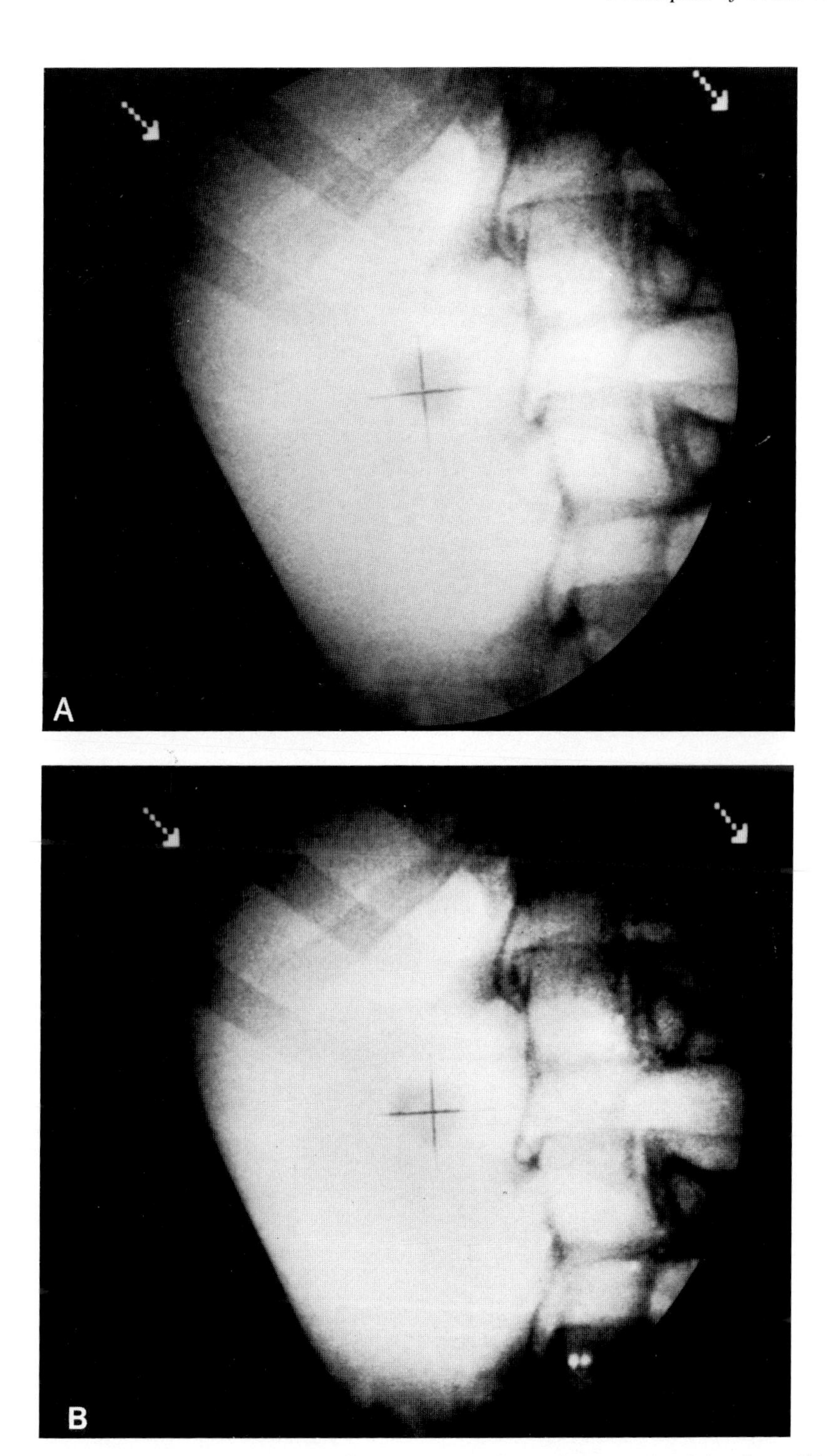


Fig. 6-18. Fluoroscopic sequence demonstrating treatment of a solitary renal pelvic stone. **(A)** Focused stone prior to treatment; **(B)** 200 shocks. (*Figure continues.*)

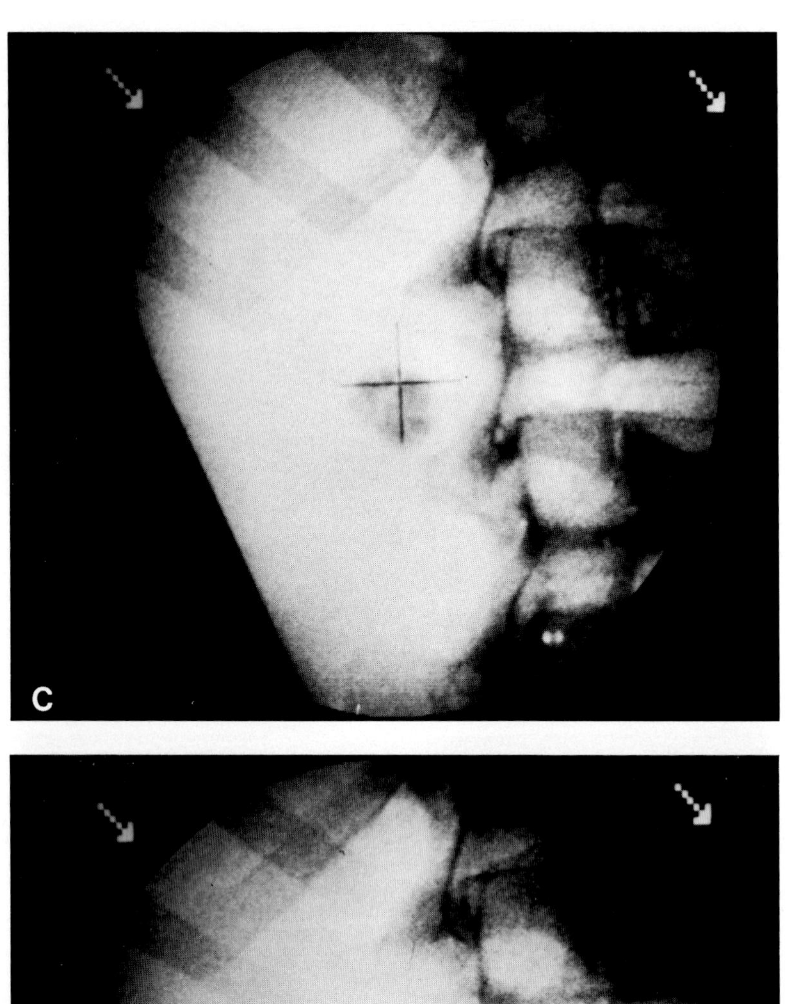


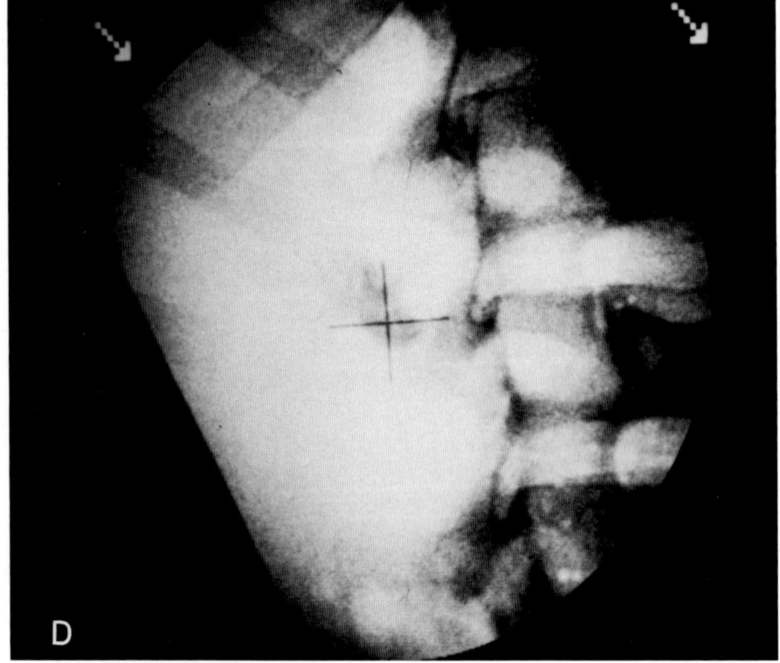


Fig. 6-18 (*continued*). **(C)** 400 shocks; **(D)** 600 shocks. (*Figure continues.*)

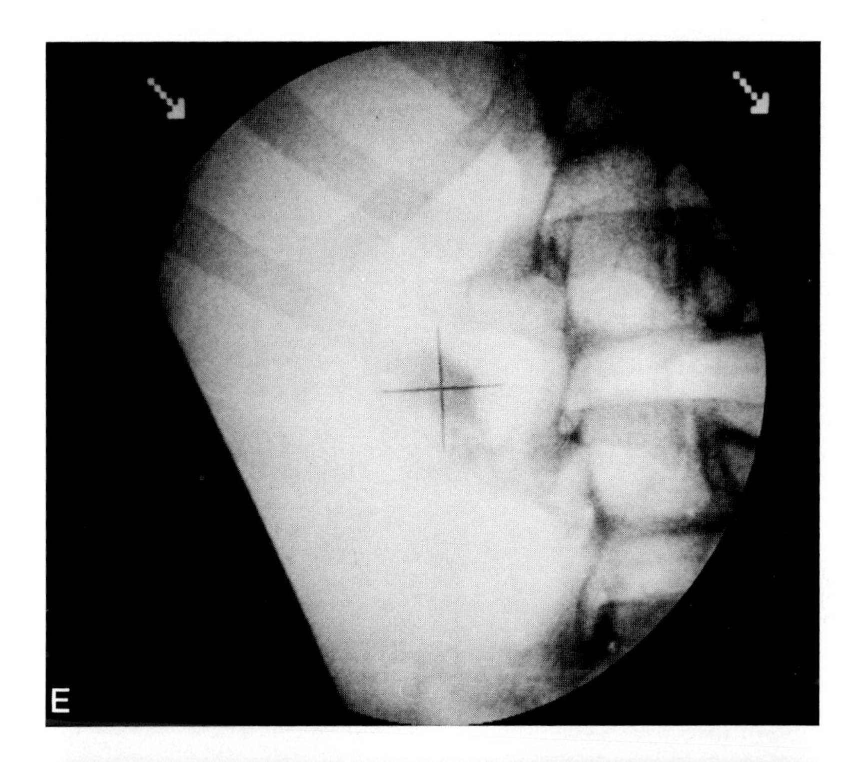


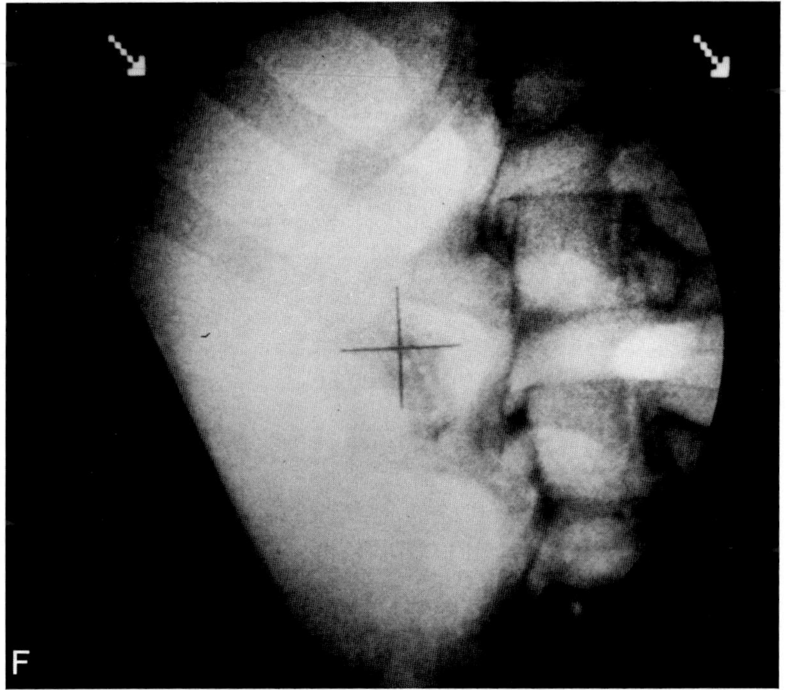


Fig. 6-18 (*continued*). **(E)** 800 shocks; **(F)** 1,000 shocks. (*Figure continues.*)

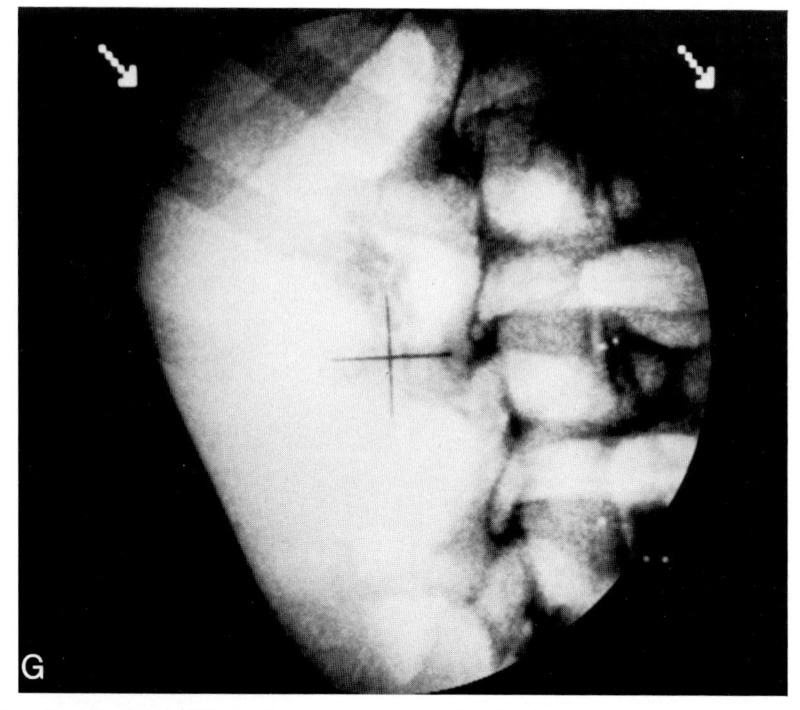


Fig. 6-18 (*continued*). **(G)** 1,200 shocks—treatment completed. Note that stone gradually spreads out and becomes less dense as it is broken up.

habitus and the size, location, radiographic density, and composition of the stone all play major roles as well.

### Determination of the Total Number of Shocks to be Administered

From series to series, the average number of shocks administered in one treatment varies from 978[19] to 1,382.[20] This too depends on the stone size, the number of the stones being treated, and the hardness of the calculus material. As few as 150 shocks[20] and as many as 4,600 shocks[21] have been given to a single kidney in one treatment. The maximum number of shocks one can safely give remains unknown; however, the FDA recommends a maximum of 2,000 shocks per kidney.

## IMMEDIATE POST-ESWL ASSESSMENT

Radiographic resolution may be less than desired when using the fluoroscopy or snapshot modes. Better definition is almost always obtained when a conventional kidney–ureter–bladder (KUB) film is taken. If the operator is uncertain about stone breakup, it may be useful to remove the patient from the tub while anesthetized, take a KUB film, and resume treatment if stone breakup seems inadequate. This is rarely necessary if in-bath filming technique is available.

A film should be taken during the 24 hours after treatment. This allows the operator to assess the efficacy of treatment and advise the patient about what to expect over the coming days and weeks.

## ACKNOWLEDGMENT

This work was supported in part by a grant from OmniMed, Inc. and by NIH Grant #20586.


## REFERENCES


1. Clayman RC, Reddy P, Savage PA: Flexible fiberoptic and rigid rod lens endoscopy of the lower urinary tract: A prospective controlled comparison. J Urol 131:715, 1984
2. Hunter PT, Newman RC, Hawkins I, Finlayson B: Endourology VI: Retrograde nephrostomy and percutaneous stone removal. World Urology Update Series. 2:33, 1985
3. Dornier Kidney Lithotripter Operating Manual HV-954–300 BH. Dornier Medical GmbH, Marietta, GA, 1984
4. Dornier Kidney Lithotripter—Interactive Learning System, II-31. Dornier Medical GmbH, Marietta, GA, 1985
5. Eusek JF, Bush WH, Burnett LL, Gibbons RP: In-bath filming during ESWL. Radiology 158:850, 1986
6. Carter H, Riehle R: Variables influencing radiation exposure during extracorporeal shock wave lithotripsy: A review of 298 treatments. J Urol, in press
7. Newman R, Mauderli W, Finlayson BF: Radiation exposure during ESWL. Unpublished data.
8. Jocham D, Brandl H, Chaussy C, Schmidt E: Treatment of nephrolithiasis. p. 35. In Gravenstein JS, Peter K (eds): Extracorporeal Shock Wave Lithotripsy. Technical and Clinical Aspects. Butterworths, Stoneham, MA, 1986
9. Sinclair W: Effects of low level radiation and comparative risk. Radiology 138:1, 1981
10. Shigetoshi A, Walter JR: Dose to the active bone marrow, gonads, and skin from roentgenograph and fluoroscopy. Radiology 101:669, 1971
11. Gustafsson M, Lunderquist A: Personnel exposure to radiation at some angiographic procedures. Radiology 140:807, 1981
12. Whalen JP, Balter S: Radiation risks associated with diagnostic radiology. In Disease-a-Month, vol. 28. Year Book, Chicago
13. Bush W, Brannen G, Gibbons R et al: Radiation exposure to patient and urologist during percutaneous nephrostolithotomy. J Urol 132:1148, 1984
14. Madorsky M, Finlayson B: ESWL blast path considerations in patient treatment strategies. Presented at the Second Symposium on ESWL, Indianapolis, IN, March 1986
15. Kaneko H, Watanabe H, Takahashi T et al: Studies on the application of micro explosion to medicine and biology. IV. Strength of wet and dry urinary calculi. Nippon Hinyokika Gakkai Zasshi 70(1):61, 1979
16. Kidney lithotripter instructions for electrode change. Dornier Medical Systems, Marietta, GA
17. Hunter P, Finlayson BF, Newman R, Drylie D: Geometry of the Dornier Extracorporeal Lithotripter and Pressures Around F2. p. 19. In Gravenstein JS, Peter K (eds): Extracorporeal Shock Wave Lithotripsy. Technical and Clinical Aspects. Butterworths, Stoneham, MA, 1986
18. Arnold W, Ruth R, Ross W, Jenkins A: The Lithotripter as a sound hazard. Anesthesia, 63:A17, 1985
19. Chaussy C, Schmeidt E, Jocham D et al: Extracorporeal shock wave lithotripsy (ESWL) for treatment of urolithiasis. Urology (special issue) 23:59, 1984.
20. Riehle RA, Fair WR, Vaughn ED: Extracorporeal shock wave lithotripsy for upper urinary tract calculi: experience at a single center. JAMA 255:2043, 1986
21. Finlayson BF: A favorable comment on the practice of outpatient ESWL. Endourology Newsletter 1:1, 1986

# 7

# Techniques of Anesthesia

Robert C. Newman
Robert A. Riehle, Jr.

Extracorporeal shock wave lithotripsy (ESWL) has presented a new challenge to the anesthesiologist. Before the advent of this new modality, there were no guidelines for administering anesthesia to a patient immersed in water. However, the need for patient analgesia during treatment has been served by general, regional, and local techniques adapted and modified at lithotripsy centers throughout the world. Immersion anesthesia for ESWL offers a new frontier for investigating the effects of immersion of the anesthetized patient, and the anesthesiologist has become a crucial member of the ESWL team.

## PREOPERATIVE EVALUATION

### Risk Assessment

Preoperative assessment of the patient is an essential element in preparation for an anesthetic. The American Society of Anesthosiologists (ASA) adopted a classification of physical status to be used in evaluating the patient prior to surgery. These ASA classes (1974)[1] are as follows:

**Class 1.** The patient is in good condition with a localized rather than systemic disease process. Example: acute appendicitis in an otherwise healthy adolescent male.

**Class 2.** Mild to moderate systemic illness is evident which may or may not be related to the surgical illness. Example: mild diabetes or extreme obesity.

**Class 3.** Severe systemic disease is present. Example: severely limiting organic heart disease.

**Class 4.** A severe systemic illness is present which may or may not be surgically correctable. The disease may be life threatening. Example: advanced hepatic, pulmonary, or endocrine insufficiency.

**Class 5.** Moribund patients with little chance for survival, regardless of measures which might be undertaken. Example: shock associated with a ruptured abdominal aortic aneurysm.

**Emergency (E).** A patient in any of the aforementioned classes who must undergo surgical intervention on an emergent basis. Example: a previously straightforward inguinal hernia that has become incarcerated and is associated with nausea and vomiting.

This classification is used as a means of assessing the patient's preoperative condition, rather than a measure of anesthetic risk. This is an important distinction since many significant factors that may affect the clinical outcome for the patient are not taken into consideration. Such parameters as length of anesthesia, position of the patient, type of surgery, transfused blood products, and the competence and technical ability of the anesthesiologist and surgeon are not addressed.

Thus, the anesthesia risk to the patient derives in part from the disease, the actual surgical procedure, and the medical staff caring for the patient.[2] It is often difficult to identify adverse factors attributed solely to anesthesia. By the same token, it is not possible to assume that a given method of anesthesia (e.g., general or epidural anesthesia) carries more or less risk to the patient than another techinque.

One can say that sicker patients tend to have more intraoperative and postoperative complications, regardless of the type or length of the procedure. More complex situations can be approached with maximum effectiveness when a good knowledge base is attained. Patients in categories 1 through 4 have, however, been treated successfully at a number of ESWL centers.

## Premedication

The administration of a preanesthetic medication can help to allay the anxiety and fear of a patient prior to surgery. A discussion of the planned procedure, the type of anesthesia to be used, and what the patient can expect before, during, and after ESWL often helps to assuage anxiety. Administration of a hypnotic (e.g., Dalmane 30 mg) at bedtime the night before the procedure is useful, particularly if the patient is used to taking a sedative.

Preanesthetic medications should be individualized. Barbiturates (e.g., secobarbital 75 to 200 mg), tranquilizers (diazepam 5 to 10 mg), opiates (e.g., morphine sulfate 10 mg or meperidine 100 mg) and/or anticholinergics (e.g., atropine 0.4 mg) can be administered in the immediate preoperative period. In most cases, the administration of a tranquilizer alone is adequate. The medication selected depends on the type of anesthetic selected, the patient's condition, and the anesthesiologist's preference. In some circumstances, it is advisable to omit preoperative drugs. For a detailed discussion of these and other agents, the reader is referred to one of the major anesthetic texts.[3,4]

## Monitoring Techniques

Blood pressure, pulse, temperature, respirations, and EKG are monitored throughout ESWL. Lead II is used to display the cardiac cycle, as dysrhythmias can be detected easily using this lead. In addition, the peak of the R wave configuration in this lead is usually optimal for triggering electrode discharge. To insure good contact of the EKG monitor electrodes with the patient, the EKG pads should be covered with adherent plastic drapes (Fig. 7-1).

When a general anesthetic is chosen, a neuromuscular blockade monitor attached to the wrist allows the operator to monitor muscle relaxation. During high frequency jet ventilation (HFJV), a pulse oximeter to measure peripheral oxygen saturation, an inspired $O_2$ monitor, and an in-line infrared capnograph (Hewlett-Packard Model 47210A) should be used.[5] A Swan-Ganz catheter and arterial line are helpful in the poor-risk patient with significant cardiopulmonary disease.

## Positioning

The gantry (patient support apparatus) of the Dornier machine provides an adjustable support device for the upper extremities (see Fig. 6–1). The upper arm is positioned at an angle of approximately 90° from the torso while the forearms are strapped onto cushioned supports. Abduction of the elbows/arms is important when using this support structure to minimize the risk of upper extremity nerve damage.

Alternatively, one may use inner-tube-like flotation devices (Swimmies, Kransco, San Francisco, CA) to support the arms (Figure 7-2) (personal communication, D. Griffith, M.D., Baylor University College of Medicine, Houston, TX). The Swimmies allow the arms to

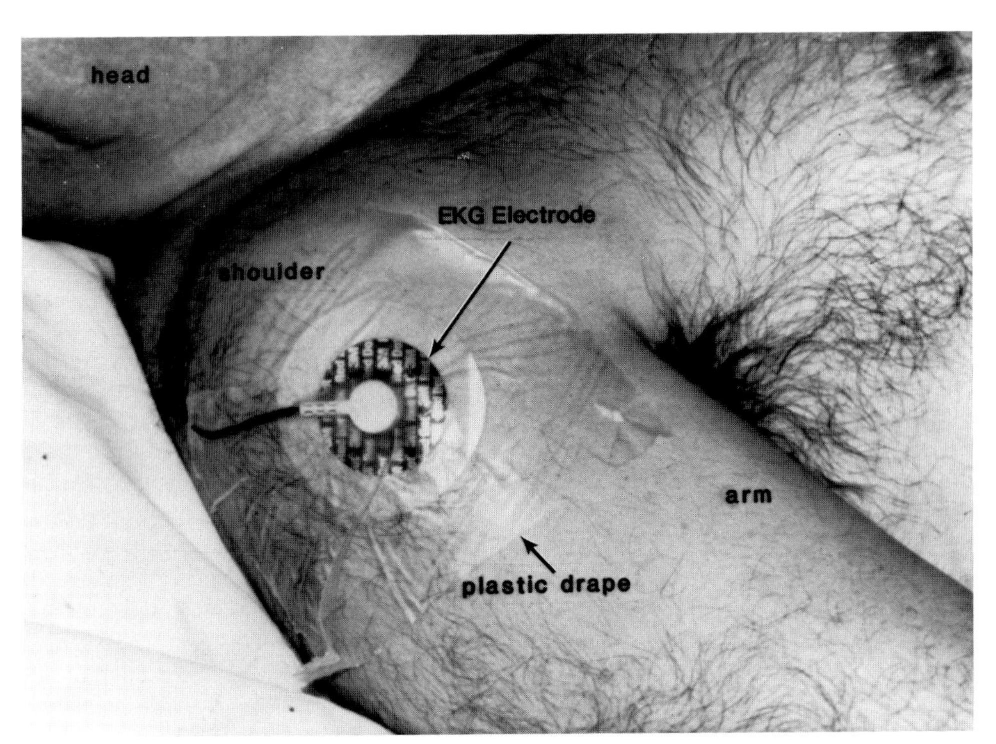


Fig. 7-1. EKG electrode covered by plastic occlusive drape which serves as a barrier to water.

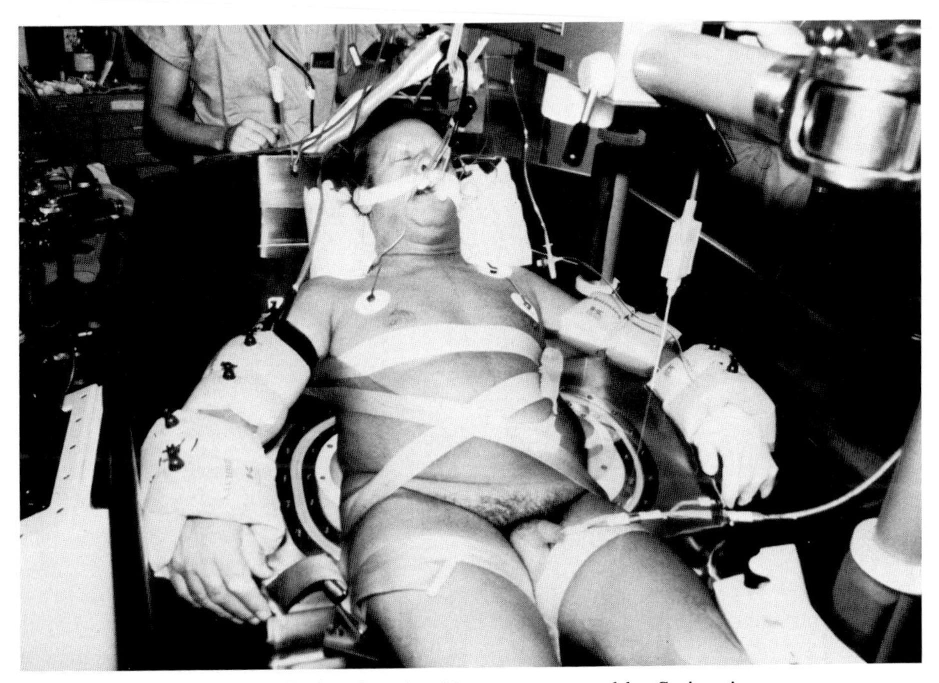

Fig. 7-2. Patient in tub with arms supported by Swimmies.

lie in a more natural position on top of the water and help to prevent interference with monitoring devices due to immersion. If monitors are not used on the arms or hands, the arms can float unsupported in the water. If cervical ribs or vascular disorders are present, one of the latter techinques should diminish the probability of the occurrence of neurovascular disorders.

The body is borne by adjustable back and leg supports which are a part of the gantry (see Fig. 6-1). Leg supports are adjusted according to the patient's height, and the patient is placed in a semireclining sitting position. Webbed straps are used across the chest, abdomen, and legs to counteract buoyancy. Straps may be placed underneath the lumbar area if additional support is needed. Generally, the patient is rotated in the support system toward the side on which the stones lie. When the stone to be treated lies low (e.g., at or below the bony pelvis), flattening the body may allow the operator to bring the stone into focus.

### No Anesthesia

Because of the pain associated with shock wave administration, ESWL can be performed without anesthesia in only a very limited patient group. In general, paraplegics with a T6 or higher sensory level can be treated without anesthesia. At least five patients with prior spinal cord injury have been treated in this manner with anesthesia standby only.[10]

## PERTINENT ANATOMY

The kidney, adrenal, and upper ureter are innervated from the renal plexus, which receives branches from the dorsal roots of the eleventh and twelfth spinal nerves, the sympathetic fibers of the celiac axis, the semilunar ganglion, the splanchnics, and the vagus nerve. These nerves innervate the intrarenal arterial and caliceal systems.[6] Renal afferent pain fibers travel along sympathetic ganglia to reach the spinal cord.[7]

The great, lesser, and least splanchnic nerves arise from the fifth through twelfth thoracic segments. The exact anatomy may vary, but in general, these nerves pass downward in the thorax, through the crus of the diaphragm, and give off fibers to the celiac and renal plexuses. The splanchnic nerves are strictly autonomic rather than somatic, and supply many intraabdominal organs in addition to the kidney. These nerves can be blocked using a posterior approach such as that described by Kappis or Labat.[8]

Beneath the skin of the flank, the latissimus dorsi, posterior serratus, external and internal oblique, and transversus abdominis muscle are innervated by the intercostal nerves. The intercostal neurovascular bundle lies within the subcostal groove, bounded externally by a bony lip of the rib which projects downward. The intercostal nerve lies inferior to the artery and vein in the erect position, and both lateral and superficial to the vessels in the flank position. The distal portion emerges from under the tip of the rib and courses between the lumbodorsal fascia and the internal oblique muscle to arborize into its muscular and cutaneous segments.

The intercostal muscles are three, thin, superimposed muscular layers (the external, internal, and intimi) which arise from the lower border of the rib and insert into the cranial border of the rib below. The internal intercostal becomes an aponeurotic membrane posterior to the angle of the ribs, and the intimi layer, which is more developed in the lower intercostal spaces, extends from the inner surface of one rib to the inner surface of the second or third ribs below. The intercostals intimi are external to the endothoracic fascia and internal to the external and internal intercostal muscles as well as the neurovascular bundle.[9]

## TYPES OF ANESTHESIA

### Local Anesthesia

At the Klinikum Grosshädern in Munich, anesthesiologists attempted to provide analgesia for ESWL treatment by infiltrating local anesthetics into the skin at the entry site of the shock wave. However, in their hands, this infiltration technique failed to provide adequate analgesia.[11] Recently, groups at the University of Iowa (personal communication) and New York Hospital[11] have reported the successful use of local anesthesia for ESWL.

At New York Hospital–Cornell Medical Center, Malhotra has introduced a regional block technique using intercostal nerve block and local infiltration. The patient is given intravenous diazepam (5 to 10 mg) and fentanyl (50 to 100 μg) and placed in the prone position. The position of the stone is then measured in relation to the lumbar spine and twelfth rib. These measurements are used to estimate the location of the stone and the point of shock wave entry on the patient's flank. A 15 cm × 15 cm area is then mapped out; this usually extends to the midline medially and to the iliac crest inferiorly (Fig. 7-3).

The ninth through the twelfth intercostal nerves are blocked by injection at a point between the posterior axillary line and the angle of the rib; a 22-gauge, 1½-inch needle attached to a 10 ml syringe is used to inject a 5 ml mixture of equal amounts of 1 percent lidocaine and 0.5 percent bupivacaine with 1:200,000 epinephrine. This mixture provides rapid onset and a prolonged effect. Epinephrine reduces the intravascular absorption of the anesthetic. The needle is then "walked off" the lower edge of the rib, advanced 2 mm, and 5 ml of the anesthetic is injected at each intercostal site. Care should be taken to avoid intravascular injection and pneumothorax.

The previously demarcated 15 cm × 15 cm area is then infiltrated subdermally with 40 to 60 ml of 0.25 percent bupivacaine. Subdermal injection is less painful than intradermal injection and allows wider spread of the anesthetic solution.

Intravenous sedation may be used to supplement local anesthesia. At New York Hospital, small incremental doses of fentanyl (50 to 100 μg) have proved useful. Significant bruising and ecchymosis at the skin infiltration site, over and above that which may accompany shock wave therapy alone, sometimes follows the use of a local anesthetic.

As always, proper selection of candidates for local anesthesia should include consideration of stone size and composition, stone location, patient personality and level of anxiety, and the operator's experience with awake lithotripter patients.

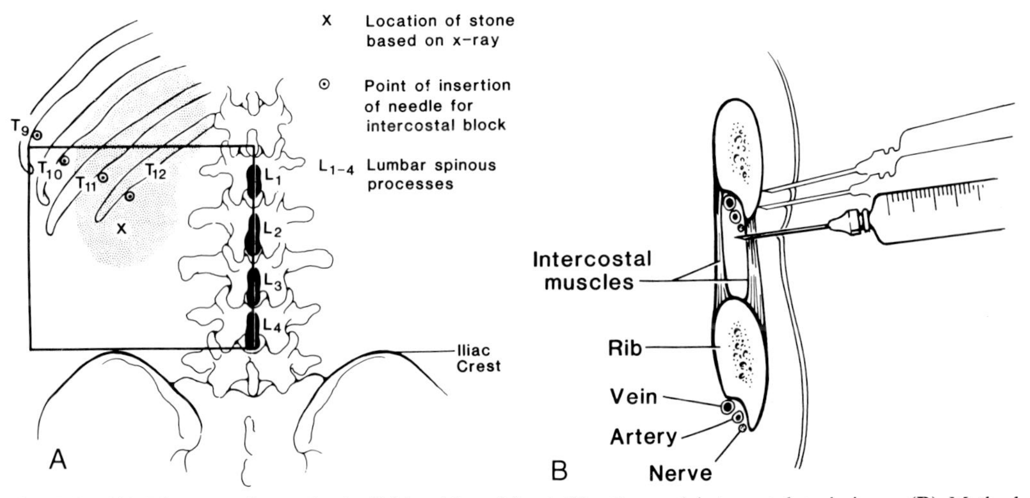


Fig. 7-3. **(A)** Diagram of anesthesia field achieved by infiltrative and intercostal technique. **(B)** Method for performing intercostal block.

## Epidural Anesthesia

Because of the autonomic effects of the drugs used, pretreatment by hydration is always advisable. Infusion of 500 to 700 ml of crystalloid should prevent hypotension. The L1-2, L2-3, or L3-4 interspace area is prepped using sterile technique; a needle puncture is made; and the infusion catheter is advanced 3 to 4 cm into the epidural space. The tubing is then taped to the back using a clear plastic, self-adherent drape. To increase adherence, sterile tincture of benzoin may be sprayed on the back before placing the plastic (Fig. 7-4). Bupivacaine 0.5 percent or lidocaine 2 percent may be used to obtain the desired T6 level. In a series of 1,000 cases, there were no reported incidences of infection or inflammation, despite the fact that the patient is immersed during treatment.[10]

Continuous infusion of anesthetic agents into the epidural space offers some advantages. The awake patient can communicate with the anesthesiologist and can use the arms to assist with positioning, thereby minimizing the possibilities for nerve damage due to arm positioning. Additional medication may be injected during the procedure or during the recovery period if needed; 60 to 70 percent of the initial dose is required.[10] On rare occasions, it is desirable to remove the patient from the tub, take a conventional KUB film to determine the adequacy of stone break up, and then re-immerse the patient. Because there is no need to be concerned with endotracheal tube dislodgment when an epidural is used, this technique offers an advantage over general anesthesia. At the Klinikum Grosshädern, investigators feel that these advantages are of such magnitude that the epidural technique is used almost exclusively. Headphones are often used to block out the sound from electrode discharge and to allow the patient to sleep or listen to music.

The most common complication of epidural is hypotension. In one series of 100 patients,[12] relative hypotension, defined as "blood pressure less than 80 percent of preanesthetic levels" was seen in 10 percent of patients. At New York Hospital, London and Riehle noted similar findings in 26.3 percent (5/19) of spinals and 18.7 percent (29/155) of patients having epidural anesthesia[13] (see Table 7-1). Vasopressors and hydration generally remedies this problem if it is not corrected by immersion alone.

Epidural anesthesia has potential disadvantages particularly in centers treating a high volume of patients. First, there is more "down time" between cases because of the time required to properly position the epidural catheter. This problem can be eliminated if two anesthesia teams are used so that one patient may be prepared while another treatment is in progress. Second, diaphragmatic excursion is more pronounced with epidural than with some general anesthetic techniques. This makes stone positioning and treatment more cumbersome. Because the patient is awake, unexpected move-

**Table 7-1. Percent Change in Mean Arterial Blood Pressure[a] in Comparison to Anesthetic Technique**

| Percent Change in MAP | Epidural ($N$ = 155) | Epidural to Spinal[b] ($N$ = 11) | Spinal ($N$ = 8) |
|---|---|---|---|
| (+) 1–20 | — | — | 1 |
| 0 | 121 | 9 | 4 |
| (−) 1–20 | 5 | — | — |
| **Incidence of Hypotension** | | | |
| (−) 21–30 | 12 | — | 3 |
| (−) 31–40 | 13 | 2 | — |
| (−) >40 | 4 | — | — |
| Total | 29/105 (18.7%) | 5/19 (26.3%) | |

[a] Mean arterial pressure (MAP) = diastolic BP + (1/3) (systolic − diastolic).

[b] Spinal anesthesia used when epidural technique failed.

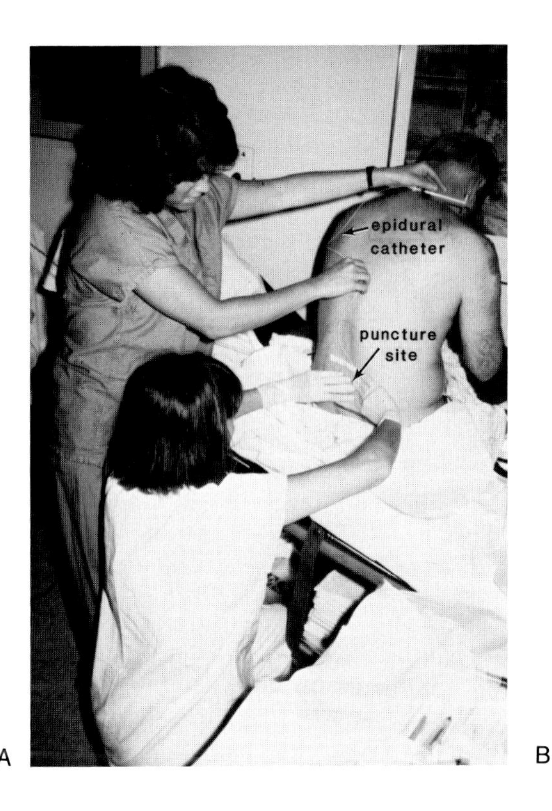


A

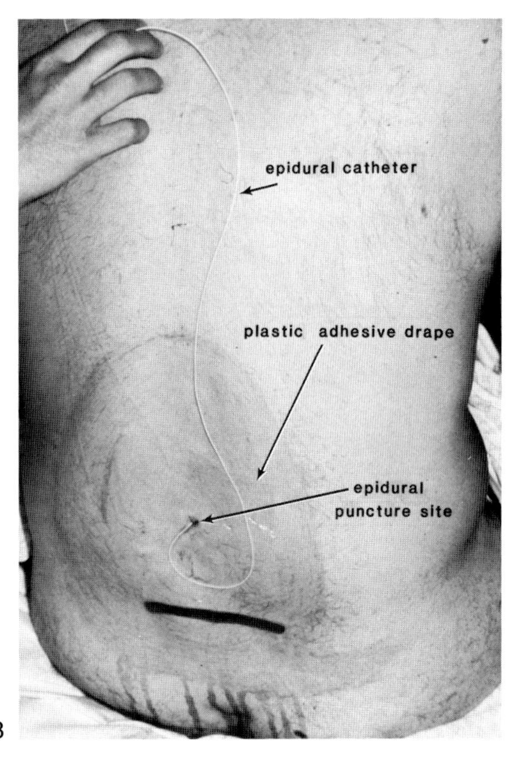


B

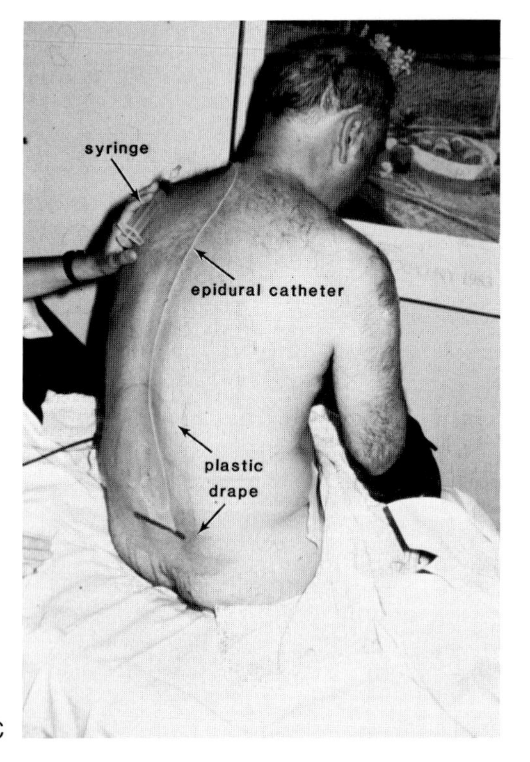


C

Fig 7-4. Placement of an epidural catheter. **(A)** Patient in sitting position for placement of epidural catheter. **(B)** Back of patient with epidural catheter; note puncture site at L3-4. **(C)** Patient ready for positioning in gantry; adherent plastic drapes have been placed over the catheter to protect it from the water. A syringe is used to inject incremental amounts of bupivacaine.

ment occurs from time to time. Seemingly small movements can place the stone completely out of the focus of the shock wave. Tranquilizers can be used judiciously, but with recognition of the potential for respiratory arrest.

## Spinal Anesthesia

For spinal anesthesia, hyperbaric bupivacaine, 0.5 percent, or lidocaine, 4 percent, is injected at the same site as for the epidural technique. Using the hyperbaric technique, the onset of action when using bupivacaine is 15 to 30 min.[10] A small 25-gauge needle should be used to avoid postspinal headache. A T6 level is usually satisfactory though the level depends on the stone's location. In general, the higher the stone, the higher the sensory level required.

Initiation of spinal anesthesia may be quicker than epidural depending on the skills of the anesthesiologist. However, re-injection is not possible if the patient begins to experience pain. Anesthesia may be "patchy," requiring institution of general anesthesia. The propensity for patient and/or physician anxiety may be higher depending on the temperament of those involved. As a result, use of spinal anesthesia is limited, but useful if the epidural space cannot be entered, and general anesthesia is contraindicated.

## General Anesthesia

Conventional general anesthetic techniques are used in many centers. Etomedate or sodium thiopental, curare (3 to 4.5 mg), oxygen (100 percent), fentanyl (3 to 5 mg/kg), and succinylcholine (1 to 2 mg/kg) may be used.[14] Generally, the medication dosages are similar to those for other general procedures. Intubation is accomplished in the dorsal recumbent position with the patient lying on a litter or in the gantry (Fig. 7-5). The latter eliminates the need for

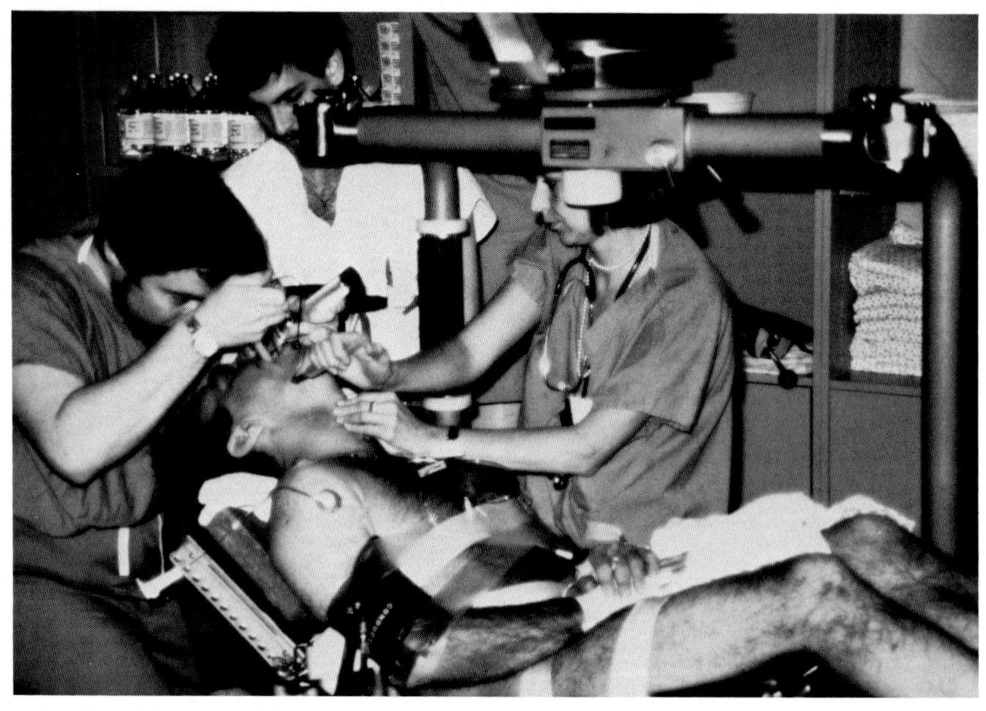

Fig. 7-5. Patient being intubated on the gantry.

lifting the anesthetized patient. Anesthesia is maintained using inhalational agents such as halothane or Enflurane or a balanced technique using intravenous fentanyl infusion (1 to 2 mg/kg/hr), nitrous oxide (50 to 70 percent), oxygen, and muscle relaxants. Continuous mechanical ventilation (CMV) with a tidal volume of 12 to 15 ml/kg and rate of 8 breaths/min to keep the oxygen saturation greater than 95 percent and end-tidal $PCO_2$ ($P_{ET}CO_2$) between 30 and 37 mmHg is suitable.[14]

In a healthy patient with a heart rate in the 50 to 70 bpm range, atropine (0.4 mg) may be administered intravenously to increase the heart rate and speed up the treatment. Atropine should not be used for this purpose in the patient with significant cardiac problems because the increase in heart rate may increase the susceptibility to ventricular ectopic activity.[15]

Because of the potential for endotracheal tube dislodgment, airway management may be a problem with general anesthesia if it proves necessary to remove the patient from the tub for radiographs during ESWL. Fortunately, this is seldom necessary. General anesthesia does make it more difficult to assess possible upper extremity malposition. Allowing the arms to float or using Swimmies seems desirable, though the arm supports on the gantry can be safely used if caution is exercised. General anesthesia eliminates patient anxiety during the treatment and allows more precise control of diaphragmatic movement through manipulation of the volume of gas inspired. It also allows one to use high-frequency jet ventilation.

## High-Frequency Jet Ventilation

High-frequency ventilation (Fig. 7-6) is a method of mechanical ventilation in which compressed gas is delivered through a small-bore catheter placed in the trachea. Standard general anesthesia induction techniques are used. There are three types of high frequency ventilation: high frequency, positive pressure ventilation

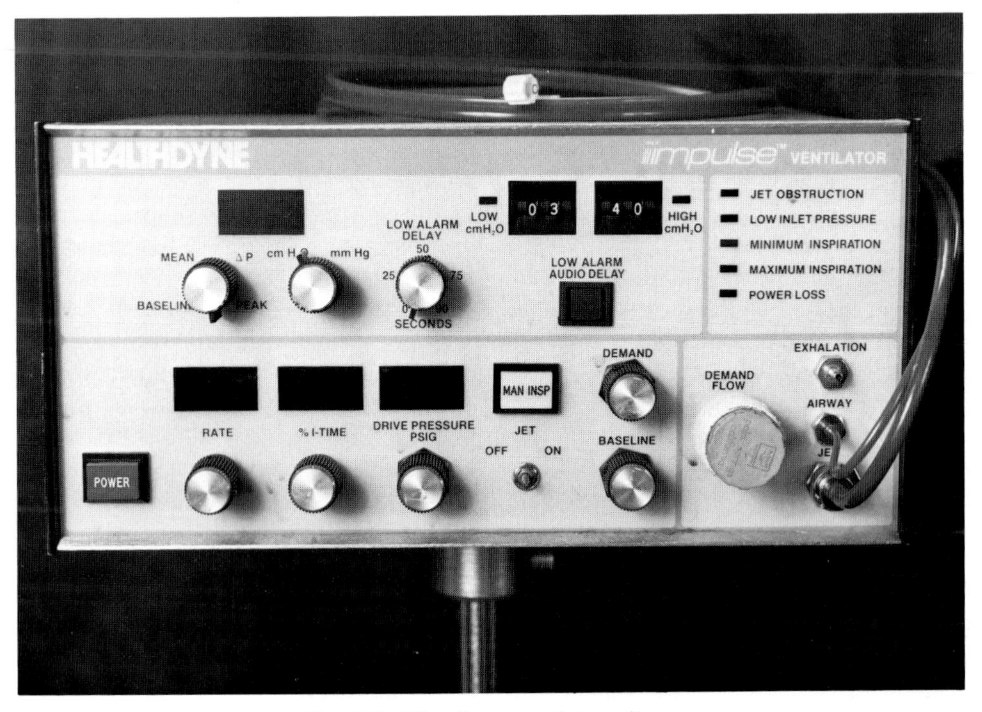


Fig. 7-6. High frequency jet ventilator.

(HFPPV), in which gas in 150 to 200 ml volumes at a rate of 60 to 90 breaths per minute is used; high-frequency jet ventilation (HFJV), using rates of 100 to 200 breaths per minute and 50 to 100 ml volumes; and high-frequency oscillation (HFO), using rates of 900 to 3,000 breaths per minute and even smaller volumes.[16,17] With all three techniques, gas exchange occurs at ventilatory frequencies higher than those used during conventional CMV, peak airway pressures are lower than with CMV, and tidal volumes are less than or equal to anatomic dead space volumes.[17]

At the University of Florida, HFJV was selected for use with ESWL.[18,19] Using this technique, gas is delivered via a Hi-Low jet endotracheal tube[18] by a "fluidic" controlled ventilator or solenoid valve and electric tuner. After each jet is delivered, gas is passively exhaled. The system also results in entrainment of gas from the surrounding airway by the Venturi principle.[20] The mechanism of action remains elusive.

The HFJV is incorporated into the anesthetic circle, and $N_2O$ and $O_2$ controlled by flow meters on the anesthesia machine are entrained by the jet ventilator. A fentanyl (1 to 2 mg/kg/hr) drip and muscle relaxant are required. A ventilatory frequency of 100 breaths per minute and 20 percent inspiratory time (TI) are used. $P_{ET}CO_2$ may be determined during shock wave administration and should be maintained at 30 to 37 mmHg. Peripheral oxygen saturation may be kept at greater than or equal to 95 percent by adjusting the $F_IO_2$.[14] Inhalation agents can be added as needed. A capnograph and pulse oximeter are essential when using this technique. CMV is used until the patient is submerged; then a switch is made to HFJV.

A decrease in $O_2$ saturation and an increase in end-tidal $CO_2$ signal that drive pressure should be increased. Patients with bronchospasm should not be treated with HFJV. Experience at the University of Florida using HFJV in over 600 patients indicates that it is a safe and effective technique. No long-term sequelae directly attributable to HFJV when used with ESWL are known to have occurred.

### CLINICAL IMPLICATIONS OF HFJV

In the only known study available at the time of this writing, 50 patients treated using CMV were compared to 66 patients undergoing ESWL who were ventilated using HFJV. No clinically significant changes were noted when the pre- and immediate post-ESWL hematocrit, serum amylase, and quantitative renogram were compared.[21] In both groups, 20 percent required parenteral analgesia. There are no data indicating that HFJV is harmful to ESWL patients if used properly.

A distinct advantage of HFJV is that it decreases diaphragmatic excursion significantly. As a result, kidney and stone movement are minimal. In a series of 30 patients, mean renal stone movement was 30.6 mm with CMV and 2.17 mm with HFJV. This difference was highly significant by the paired t-test ($P < 0.0001$).[14] When trainees at the University of Florida training center were offered an opportunity to use CMV or HFJV, the latter was chosen unanimously.

## Effects of Immersion

Research efforts investigating the effects of immersion have been limited. It is known that with awake subjects immersed to the level of the diaphragm, the heart volume increases by approximately 130 ml. The heart becomes distended with the increased volume and central venous pressure (CVP) rises from 2.5 to 12.8 mmHg when the water level rises from the diaphragm to the neck. Engorgement of the pulmonary vascular system is accompanied by a decrease in vital capacity.[22]

The relative central hypervolemia brought about by immersion induces marked changes in fluid and electrolyte balance including natriuresis, kaliuresis, diuresis, suppression of the renal aldosterone system and of antidiuretic hormone release. The changes are rapidly reversible in the normal subject once the patient is taken out of the water.[23]

Other investigators have studied healthy, immersed, unanesthetized volunteers under differing experimental conditions and have demonstrated increases in cardiac output to 62 percent, increased stroke volume to 79 percent, marked increase in pulmonary blood flow, no change in pulmonary capillary volume, and a shift in blood volume of about 700 ml into the central vascular compartment.[24,25,26,27,28]

Weber and his associates in Munich conducted some investigational work on the anesthetized patient during ESWL.[29] They studied ASA status I and II patients with calculi being treated by lithotripsy using general anesthesia. The water level in the tub was gradually increased from trochanter level, to pelvic crest, to navel, upper abdomen, xiphoid, and lower and upper third of the sternum and clavicle; CVP was measured at each water level. In a second group, pressures were measured with the patient supine before positioning in the gantry, after positioning in the gantry, repeatedly during immersion to the level of the clavicle, after removal from the bath, and with the patient supine after transfer from the gantry.

A decrease in CVP was noted in the semireclining position and was thought to be due to venous pooling. As the water level was increased, a direct correlation with CVP was noted. Esophageal pressure increased in a similar fashion.

The studies indicated that the decrease in CVP resulting from the semireclining position was countered by immersion to the umbilical level. Immersion to xiphoid level increased only about half as much as immersion to the clavicle. Systemic arterial pressure sometimes decreased with immersion. The hypothesis was that this may be due to the vasodilatory effects of the warm bath water (35.5 to 36.1°C).

Blood pressure and central venous pressure changes seemed to have little significant clinical effect in healthy patients. However, Weber suggests that if the patient is at high risk from the cardiac standpoint, it may be best to raise the water level no higher than the umbilicus to avert increase of CCVP to dangerous levels. Water temperature should be kept in a thermoneutral range (e.g., 35.5 to 36°C) to diminish the probability of significant hypotension which could adversely effect coronary perfusion.[29]

## COMPLICATIONS

### Cardiovascular

In the early studies, dysrhythmias associated with shock wave lithotripsy were observed in 80 percent of the patients.[30] At that time, the delivery of shock waves was hand-triggered without synchronization. To prevent rhythm disturbances, electrode discharge timing was changed so that shock wave generation was triggered by the R wave in the cardiac cycle. Electrode discharge occurs 20 msec after the R wave of the QRS complex. Following electrode discharge, a refractory period of at least 300 msec eliminates the possibility of subsequent discharge on the T wave.[31] The latter could induce severe dysrhythmias.

Despite these changes, ventricular tachycardia has spontaneously occurred at the University of Florida when shock wave discharge was being triggered by automatic ECG sensing. The patient was a 63-year-old man who had had an inferior myocardial infarction 5 years before treatment. The patient was promptly removed from the tub, 100 percent oxygen was administered, and a 100 mg lidocaine bolus was given. Sinus rhythm and a normal blood pressure ensued 30 sec later and the treatment was discontinued. Serial cardiac enzymes and ECGs revealed no evidence of infarction and the patient recovered uneventfully.[31]

On another occasion, at the University of Florida unit, the lithotripter failed to sense the R wave and the electrode would not discharge. Hand triggering was initiated and followed immediately by premature atrial contractions, premature ventricular contractions, and ventricular tachycardia. The medical history of this 42-year-old white woman was negative, except for diastolic blood pressures in the 85 to 90 mmHg range. ESWL was discontinued, and the patient was removed from the tank, ventilated with 100 percent oxygen, and a precordial thump was followed by two 100 mg boluses of lidocaine. Conversion did not occur, and closed chest massage was begun. A short while later, a sinus rhythm ensued, the ventricular tachycardia being of less than 2 min duration. A lidocaine infusion (4 mg/min) was initiated and the patient was transferred to ICU. Serial cardiac enzymes and ECGs were normal and there were no sequelae.[31]

Carlson et al point out that ill timed shocks may also be triggered by electrode movement, defective wiring, or possibly signals from other electrical equipment, such as a nerve stimulator. Ventricular tachycardia is generally treated in the manner described above. Although defibrillation may be necessary, it was not used in these two cases.[32,33] If defibrillation is required, the patient should be placed on a litter away from the water.

Transient atrial fibrillation and acute bradycardia have also occurred during ESWL.[10] Some centers have treated patients with pre-existing atrial fibrillation,[10] while others feel that pre-ESWL conversion to a normal sinus rhythm is desirable to avoid the possibility of triggering shock waves at an inappropriate point in the cardiac cycle (personal communication, J.S. Gravenstein, M.D., Gainesville, FL). Signs of congestive heart failure may also become evident during lithotripsy.

## Upper Extremity Paresis/Paralysis

Neurologic damage can result from positioning of the arms. In one instance,[10] at the Klinikum Grosshädern in Munich, the damage was thought to be due to pressure on the arm by the image intensifier. The patient regained full use of the arm.

Improper arm positioning may cause a stretch injury of the brachial plexus. For example, at the University of Florida, a previously neurologically normal patient could move neither arm immediately after ESWL, presumably due to brachial nerve plexus stretch. A mild motor defect remained bilaterally 6 months following

**Table 7-2. Anesthetic Techniques for ESWL**

| Anesthetic Parameter Considered | None | Local | Spinal | Epidural | General: Conventional | General: HFJV |
|---|---|---|---|---|---|---|
| Paraplegia/quadraplegia | ++ | + | + | + | + | ++ |
| Ease of emersion for x-rays | + | + | + | + | − | − |
| Ease of emergency removal from tub | + | + | + | + | + | + |
| Preparation time required | + | −− | − | −−− | + | + |
| Unexpected patient movement | +/− | −− | −− | −− | + | + |
| "Patchy" or inadequate anesthesia | +/− | − | +/− | +/− | + | + |
| Ability to supplement anesthesia | − | − | − | + | + | + |
| Pain associated with anesthesia injection or induction | + | −− | +/− | + | + | + |
| Patient anxiety potential | − | − | +/− | + | + | + |
| Stone movement | − | − | − | − | − | +++ |
| Potential for upper extremity positioning complications | + | + | + | + | − | − |
| Contraindication to general anesthesia | + | + | + | + | − | − |
| Contraindication to regional anesthesia | + | − | − | − | + | + |

− = disadvantage; + = advantage

this incident. A similar case at New York Hospital resulted when adjustment of the arms supports during treatment caused an extreme hyperextension of the upper arm. Injuries of this nature have not occurred when Swimmies have been used.

## SUMMARY

General and regional anesthesia have been successfully used for ESWL worldwide.[12–14] Each method has inherent advantages, disadvantages, and the potential for complications (Table 7-2). In general, the complication rate of immersion anesthesia has been quite low.

The preliminary results using local infiltration with intercostal nerve block seem indicative for future widespread use, especially in outpatient treatment centers. However, as always, selection of anesthesia techniques at each center depends on stone parameters, the patient's mental and medical condition, daily caseload, and the skills of the anesthesiologist.

## ACKNOWLEDGMENT

This work was supported in part by a grant from OmniMed, Inc. and by NIH Grant #20586.

## REFERENCES

1. Gravenstein JS, Paulus DA (eds): Monitoring Practice in Clinical Anesthesia. JB Lippincott, Philadelphia, 1982
2. Goldstein A, Keats AS: The risk of anesthesia. Anesthesiology, 33(2):130, 1970
3. Dripps RD, Eikenhoff JE, Van Damm LD (eds): Introduction to Anesthesia. The Principles of Safe Practice. WB Saunders, Philadelphia, 1972
4. Wylie WD, Churchill-Davidson HC (eds): A Practice of Anesthesia. Year Book Chicago, 1972
5. Carlson CA, Gravenstein JS, Banner MJ, Boysen P: Monitoring techniques during anesthesia and HFJV for extracorporeal shock wave lithotripsy. Anesthesiology 63:A178, 1985
6. Lich R, Howerton L, Amin M: Anatomy and Surgical Approach to the Urogenital Tract in the Male. P. 8. In Harrison J, Permutter A et al (eds): Cambell's Urology. Vol. 1, 4th Ed. WB Saunders, Philadelphia, 1978
7. DeWolf W, Fraley E: Renal pain. Urology 6:403, 1975
8. Adriani J: Labat's Regional Anesthesia—Techniques and Clinic Applications. 3rd Ed. WB Saunders, Philadelphia, 1967
9. Riehle R, Lavengood R: The eleventh rib transcostal incision: Technique for an extrapleural approach, J Urol 132:1089, 1984
10. Lehmann P, Weber W, Madler C et al: Anesthesia and ESWL: 5 years experience. p. 61. In Gravenstein JS, Peter K (eds): Extracorporeal Shock Wave Lithotripsy—Technical and Clinical Aspects. Butterworths, Stoneham, MA, 1986
11. Malhotra V: Local infiltration technique for ESWL. Endourology Newsletter (in press)
12. Duvall TO, Griffith D: Epidural anesthesia for extracorporeal shock wave lithotripsy. Anesth Analg 64:544, 1985
13. London R, Riehle, R: Immersion anesthesia for extracorporeal shock wave lithotripsy. Urology 28:86, 1986
14. Carlson CA, Boysen PG, Banner MJ, Gravenstein JS: Conventional versus high frequency jet ventilation for extracorporeal shock wave lithotripsy. Anesthesiology 63:A530, 1985
15. Dauchot P, Gravenstein JS: Bradycardia after myocardial ischemia and its treatment with atropine. Anesthesiology, 44:501, 1976
16. Gallagher TJ: High frequency ventilation. Med Clin N Am 63:633, 1983
17. Quan SF, Calkins JM, Waterson CR, Otto CW: High frequency ventilation—current concepts. Arizona Medicine XL(5):319, 1983
18. Boysen PG, Carlson CA, Banner MJ, Gravenstein JS: Ventilation during anesthesia for ESWL. p. 69. In Gravenstein JS, Peter K (eds): Extracorporeal Shock Wave Lithotripsy—Technical and Clinical Aspects. Butterworths, Stoneham, MA, 1986
19. Schulte AM, Esch J, Koches E, Meyer WH: High frequency jet ventilation during extracorporeal shock wave lithotripsy. Anaesthetist 34:294, 1985
20. Gillespie J: High frequency ventilation—a new concept in mechanical ventilation. Mayo Clin Proc 58:187, 1983
21. Finlayson BF, Newman R, Hunter P, Graven-

stein JS, Wilkes R: Efficacy of ESWL for stone fracture. p. 87. In Gravenstein JS, Peter K (eds): Extracorporeal Shock Wave Lithotripsy—Technical and Clinical Aspects. Butterworths, Stoneham, MA, 1986

22. Risch WD, Koubenec H, Beckmann U, Lange S, Gauer OH: The effect of graded immersion on heart volume, central venous pressure, pulmonary blood distribution, and heart rate in man. Pflügers Arch 374:115, 1978
23. Epstein M: Cardiovascular and renal effects of head-out water immersion in man. Circ Res 39:619, 1976
24. Arborelius M, Jr., Balldin UI, Lilja B, Lundgren CEG: Hemodynamic changes in man during immersion with the head above water. Aerospace Med 43:592, 1972
25. Begin R, Epstein M, Sackner MA et al: Effects of water immersion to the neck on pulmonary circulation and tissue volume in man. J Appl Physiol 40:293, 1976
26. Farhi LE, Linnarsson D: Cardiopulmonary readjustments during graded immersion in water at 35°C. Respir Physiol 30:35, 1977
27. Levinson R, Epstein M, Sackner MA, Begin R: Comparison of the effects of water immersion and saline infusion on central haemodynamics in man. Clin Sci Molec Med 52:343, 1977
28. Loellgen H, von Nieding G, Horres: Respiratory and hemodynamic adjustment during head out water immersion. Int J Sports Med 1:25, 1980
29. Weber W, Madler C, Keil B, Pollwein B, Laubenthal H: Cardiovascular effects of ESWL. p. 101. In Gravestein JS, Peter K (eds): Extracorporeal Shock Wave Lithotripsy—Technical and Clinical Aspects. Butterworths, Stoneham, MA, 1986
30. Chaussy C, Schmeidt E, Joacham D, Schüller J et al: Extracorporeal shock wave lithotripsy (ESWL) for treatment of urolithiasis. Urology (Special issue), 23:59, 1984
31. Carlson C, Gravenstein JS, Gravenstein N: Ventricular tachycardia during extracorporeal shock wave lithotripsy—Etiology, treatment, and prevention. p. 119. In Gravenstein JS, Peter K (eds): Extracorporeal Shock Wave Lithotripsy—Technical and Clinical Aspects. Butterworths, Stoneham, MA, 1986
32. Zoll PM, Leventhal AJ, Gibson W et al: Termination of ventricular fibrillation in man by externally applied electric countershock. N Engl J Med 254:727, 1956
33. Alexander S, Kleiger R, Lown B: Use of external electric countershocks in the treatment of ventricular tachycardia. JAMA 177:916, 1961

# 8

# Patient Management and Results after ESWL

Robert A. Riehle, Jr.
Erik Näslund

The management of the patient after extracorporeal shock wave lithotripsy (ESWL) often requires more wisdom, experience, and surgical judgment than the treatment itself. Although the ESWL procedure requires only 40 minutes, the urologist is responsible for monitoring and advising the patient during the ensuing days and weeks of colic, fragment passage, and possible endourologic manipulation. Important decisions must be made for a patient who is usually ambulatory and quite able to view his own x-rays, monitor particle passage, and appreciate the possible significance of residual fragments. The newness of the technology increases the operator's responsibility to reassure both patient and family members as well as closely monitor the sequelae and results of the shock wave lithotripsy.

As with any new therapeutic modality, it is only through actual clinical trials that parameters for management and follow-up are derived. Treatment philosophies regarding patients after ESWL are continually evolving, and management principles have changed rather markedly since the introduction of the ESWL procedure in Munich in 1980. As confidence has been increased regarding the safety and efficacy of ESWL, more streamlined and efficient management protocols have been developed, and fewer x-rays, lab tests, and office visits are required.

During the earlier phases of clinical research in ESWL therapy, renal scans, sonograms, intravenous pyelograms, magnetic resonance imaging, and a battery of laboratory studies were performed. After performing lithotripsy on more than 100,000 patients worldwide, the safety and efficiency of ESWL therapy have been established, and post-treatment management can now be tailored to the individual patient situation.[1-5] Nonetheless, close follow-up of the post-ESWL patient is mandatory, and patients must be encouraged to remain in touch with their urologist and the lithotripsy unit.

## THE TREATMENT PERIOD

In general, complications during ESWL are related to anesthesia.[6] Hypotension, arrhythmias, chest pain, nausea, tachypnea with rapid renal excursion, abdominal and flank pain, and patient movement occur during general or regional (epidural, spinal, or local) anesthesia.

At completion of treatment in the lithotripsy suite, the patient is lifted from the bath and immediately covered with warm blankets. Under the same anesthesia, the patient may be brought to the x-ray suite for final kidney–ureter–bladder (KUB) films to verify adequate disintegration of the stone. This step can be omitted if underwater spot kidney films are available, or if definite and effective disintegration has been displayed on the fluoroscopic monitor. If the KUB shows that inadequate disintegration has occurred, and less than 2,400 shocks have been given to the kidney, the patient may be reimmersed, and further treatment may be administered. The patient is then brought to a recovery room, and subsequently discharged through the outpatient unit, or returned to the hospital bed for observation.

## THE POST-TREATMENT PERIOD

It is exceptionally difficult to objectively quantify the incidence of immediate ESWL complications unless strict definitions of fever, sepsis, renal hematoma, renal colic, or cardiopulmonary events are used. The US Cooperative Study reports the following:[7] urinary tract infection (5 percent), colic (depends on stone size), hematoma (less than 1 percent), and nonurologic complications (7 percent). Other reports have produced similar estimates.[1]

It is imperative for the patient to be well hydrated during the first 24 hours after treatment. A large volume of urine allows an increased rate of ureteral peristaltic bolus activity and a good flow of fragments into the distal ureter with dilution of fragment density. Lasix (furosemide) may be given during the treatment or in the first twelve hours after treatment; however, at the New York Hospital–Cornell Medical Center it has not been noted to increase rapidity or thoroughness of fragment passage, and often increases colic and symptoms, necessitating parenteral analgesia.

Discomfort and pain as reported by patients after shock wave lithotripsy are difficult to quantify and localize. After emergence from anesthesia, some report a generalized flank ache, suggesting muscle and renal irritation. This is especially prevalent in thin patients and usually is more noticeable after longer treatments with higher voltage. Patients with superior caliceal stones or caliceal stones in general seem to have more discomfort, probably because of the concentrated area of shock wave entry. After regional anesthesia or local infiltration anesthesia, numerous needle penetrations often precipitate patient discomfort.

It is unusual to see renal colic in the first 6 to 8 hours after treatment. If a patient reports severe pain in the recovery room or during the first several hours, this may suggest intrarenal or perirenal hematoma, and a physical examination of the anterior abdomen and flank should be performed to detect any large retroperitoneal hemorrhage. For persistent pain without fragment passage or steinstrasse, a sonogram or, if necessary, a CT can confirm the diagnosis of intrarenal or perirenal hemorrhage.

Certainly, quantification of post-procedure pain is subjective and individualized. At the New York Hospital, during the initial phases, nurses treated any patient discomfort by giving parenteral analgesics, the same as if a surgical operation had been performed. After re-education of the staff, use of narcotics was greatly reduced. In general, parenteral analgesics are required for only 15 to 20 percent of patients during the first 12 hours,[7,8] and are rarely necessary after that period. It is exceptionally unusual for a patient, once discharged, to be readmitted to the hospital for pain control.

Certainly, as fragments progress to the ureterovesical junction, patients tend to have more abdominal pain with bladder spasms and urinary urgency. Subjectively, it seems that patients with large, well disintegrated stone volumes have more discomfort if the volume rapidly appears in the supravesical ureter within 12 hours. Perhaps tightly packed sand (just before passage) proximal to a spastic ureterovesical junction can be more obstructive than less compact 2 to 4 mm fragments. Of note, in the U.S. Cooperative Study, patients with stones larger than 10 mm were much more likely to require intramuscular narcotics prior to discharge.[7]

At the Atlanta Stone Center, a study was undertaken in which patients with renal pelvic stones were randomly selected for insertion of 6 F ureteral catheters 24 hours prior to ESWL.

Investigators wished to determine if there was any reduction in the number of intramuscular injections of narcotics required after ESWL; however, no such reduction in parenteral requirements was found (personal communication, T. W. Schoborg, M.D., Atlanta, GA, 1985).

Certain outpatient centers have significantly reduced the rate of intramuscular injection of narcotic analgesia. Proper selection of patients with small stone burdens and the judicious insertion of double pigtail ureteral catheters before shock wave lithotripsy have decreased the incidence and severity of immediate renal colic. In one series, more than 75 percent of patients admitted for pain control were discharged within 12 to 24 hours.[8]

Operators and investigators are continually amazed that patients pass assorted fragments up to 6 mm in size relatively asymptomatically. Several theoretic explanations can be advanced: (1) whereas a single ureteral stone causes ureteral spasm (smooth muscle contraction), multiple irregular stones in the ureter prevent effective ureteral wall apposition and gripping of the stone; (2) although a single stone with ureteral spasm can cause high-grade obstruction and pain, the multiple stones of the steinstrasse continue to allow antegrade percolation of urine through and past the fragments; and (3) any ureteral manipulation with a stent, regardless of the time indwelling, will dilate and relax the narrow points of possible stone impaction.

Most patients begin to pass some fragments within 12 hours of treatment. The higher the level of activity (i.e., in most cases, the younger the patient), the more rapid and complete the stone passage. Rates of fragment passage are variable, and patients with large stone burdens report intermittent fragment passage for 4 to 8 weeks after treatment.[20]

## Ileus

Gastrointestinal ileus after treatment can occur, especially in symptomatic patients who rapidly pass large amounts of stone material into their ureters. In part, this ileus is probably reflex, secondary to hydronephrosis and generalized retroperitoneal edema and irritation. Usually, hydration by vein and analgesia allows for prompt resolution, and unless anxiety prevails, persistent ileus is rare. Nausea can be controlled by antiemetic suppositories. In addition, many patients report an overall achiness, as if they had influenza, and many will report feeling weak and tired, even if they did not have general anesthesia.

## Infection

Temperature elevation on the first evening and within the first 24 hours is usually of pulmonary (atelectasis) or of urologic etiology. Showers of bacteria liberated during stone disintegration can cause a lower or upper urinary tract infection. Of course, the combination of increased bacteria within the kidney and partial temporary hydronephrosis can rapidly cause a patient to become symptomatic and septic; so patients with temperatures, even on antibiotics, should be monitored carefully to determine the need for decompressive nephrostomy. Other sources of fever, such as atelectasis, pancreatitis, phlebitis secondary to intravenous infiltration, drug allergy, and possible retroperitoneal hemorrhage should also be considered.

### ANTIBIOTICS

Antibiotics must be individualized for each patient; however, in general, guidelines at the New York Hospital include the following: (1) for stones of infection history or well documented urinary tract infections with urease producers, patients are given parenteral antibiotics with hydration for 24 hours before treatment in an attempt to dilute and diminish bacterial counts, even if preadmission culture is negative; (2) for patients with small stone burdens, usually calcium oxalate in composition, and prior history of positive culture, perioperative antibiotics are administered 1 hour prior to treatment and continued for one or two doses after treatment; (3) if a patient is manipulated with ureteral stent, and has the potential for more than temporary

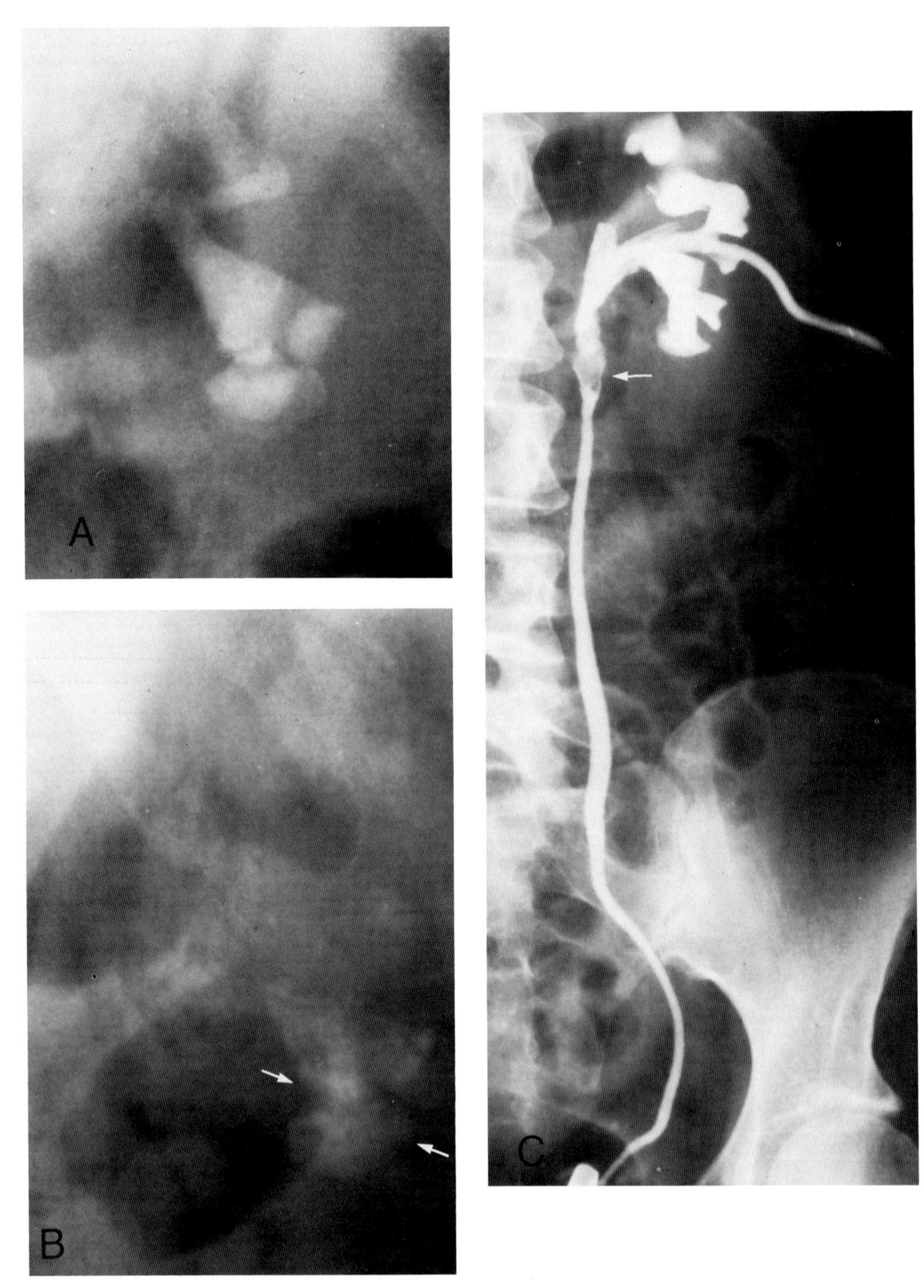


Fig. 8-1. **(A,B)** Lower partial staghorn calculus requiring emergency decompressive nephrostomy for fever and ileus, despite parenteral antibiotics. **(C)** Antegrade pyelogram reveals large proximal ureteral fragment requiring retreatment.

upper tract obstruction, periprocedural antibiotics are administered.

Prophylactic antibiotics for prosthetic heart valves, vascular grafts, and orthopedic hardware are prescribed as would be the case for any operation.[9] Patients with diabetes or indwelling nephrostomy tubes placed more than 48 hours prior to ESWL are also given periprocedural antibiotics. Except for the above instances, antibiotics are not routinely used at the New York Hospital.

To determine the value of prophylactic antibiotics in the prevention of post-ESWL fever, a prospective study of 225 patients was performed at the Atlanta Stone Center of Georgia Baptist Medical Center. The study excluded patients with an indwelling catheter, a positive urine culture, or a history of urinary tract infection during the two weeks prior to treatment. The patients were divided into two groups; one group received antibiotic prophylaxis; the other group received no antibiotics. After comparing the two groups, the study showed no significant difference in the occurrence of fever after lithotripsy (personal communication, T.W. Schoborg, M.D., Atlanta, GA, 1985).

It seems reasonable, therefore, to administer perioperative antibiotics only to those patients with (1) suspected or documented urinary tract infection, (2) partial or complete staghorn calculi of infection etiology (Fig. 8-1), or (3) prosthetic heart valves, vascular grafts, diabetes, etc. Low-grade fever can be managed on an outpatient basis unless signs of progressive obstruction or sepsis are noted. Patients passing disintegrated struvite stones should be maintained on appropriate oral antibiotics until they are stone-free.

## Renal Hematoma

Detectable subcapsular renal hematoma, secondary to the direct effects of the shock waves, occurred after less than 1 percent of treatments as reported by Chaussy.[4] At the New York Hospital, one symptomatic subcapsular hematoma necessitating transfusion was reported in a series of 518 treatments,[5] and two additional renal hematomas were documented during the next 1,000 treatments (Fig. 8-2). Although some investigators believe there is a direct relationship between the number of shock waves delivered and the development of hematoma, to date, no data substantiate this. Whether patients with indwelling stents, coagulation defects, or caliceal stones are more likely to develop such perirenal or intrarenal hematomas has not yet been confirmed.

Certainly, patients receiving anticoagulant therapy should be off such medication for an appropriate period of time to ensure relatively normal PT and APTT values. At Grosshädern Clinic in Munich, hematoma of the spinal cord resulting in paralysis occurred in one anticoagulated patient (personal communication, D. Jocham, Munich, F.R.G., 1985).

## Effect on Adjacent Organs

Effect of shock waves on adjacent organs is often suspected, but rarely demonstrated. During the FDA monitored trials, occasional transient elevations of liver function tests or amylase were reported, depending on the side treated.[7] At the Atlanta Stone Center, no significant elevation of CPK (MB band) was noted (personal communication, T.W. Schoborg, M.D., Atlanta, GA, 1985). Thus, at present, it seems unnecessary to continue monitoring these enzymes on a routine basis, as so few clinically significant elevations occur.

Since there is a significant difference in acoustic impedance between lung parenchyma and body tissue or water, pulmonary damage to patients theoretically may occur if such tissue is exposed to the path of the focused shock waves. This situation is most likely to occur in children because of the small distances between the lung parenchyma and kidney. Prevention of lung contusion in children is best accomplished by proper shielding of the lower thorax. The Methodist Institute for Kidney Stone Disease in Indianapolis has incorporated lead-lined plywood as lung shields for their pediatric patients.[10]

Adults with upper-pole caliceal stones occasionally may have a post-procedural chest x-ray compatible with lower lobe atelectasis or mild contusion, but associated clinical symp-

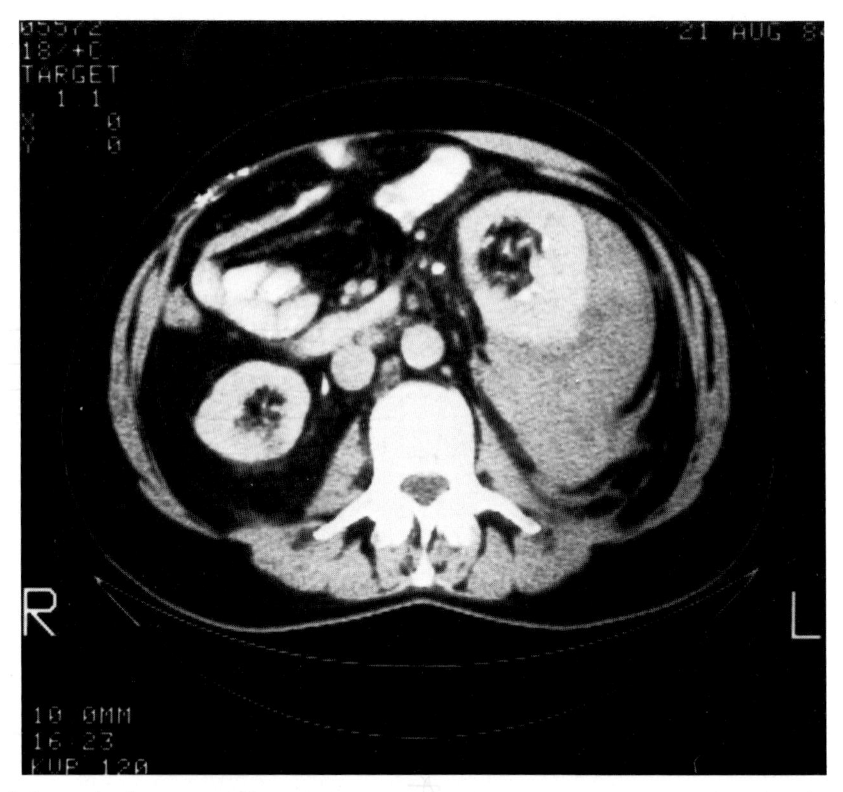


Fig. 8-2. Subcapsular hematoma after 1,200 waves at 20 kV for ureteral stone displaced to lower calyx. Follow-up scan at 3 months revealed no change in size or shape of hematoma.

toms (hemoptysis) are rare, and radiologic changes are transient.

## Roentgenograms

Follow-up roentgenograms must also be individualized, and depend on the particular case treated. Obviously, radiolucent stones do not need standard KUB films. Flat plates taken the morning after treatment are often obscured by bowel gas and minor ileus, and cannot always be fully interpreted. A film taken directly after treatment is sometimes more informative in terms of effective disintegration because the fragments can be seen better (Fig. 8-3). In other cases, however, the film taken the next morning will detect dispersion of disintegrated fragments, and give a better representation of the effectiveness of treatment. A film taken the morning after treatment also allows the operator to assure the patient that his stone is disintegrated. The presence or absence of a steinstrasse may explain any nausea and vomiting reported by the patient. Although 15 percent of discharge KUB films show no residual fragments, it is difficult to predict a stone-free state from the post-procedural KUB, and this determination, therefore, should be left to the 3-month follow-up.

In general, intravenous pyelograms are not necessary in the immediate period after shock wave lithotripsy. Especially during the first few days, the diuretic element associated with the osmotic load of dye causes increased ureteral peristalsis, increased intrarenal pressure, and increased patient discomfort which may delay patient discharge. If a patient reports persistent

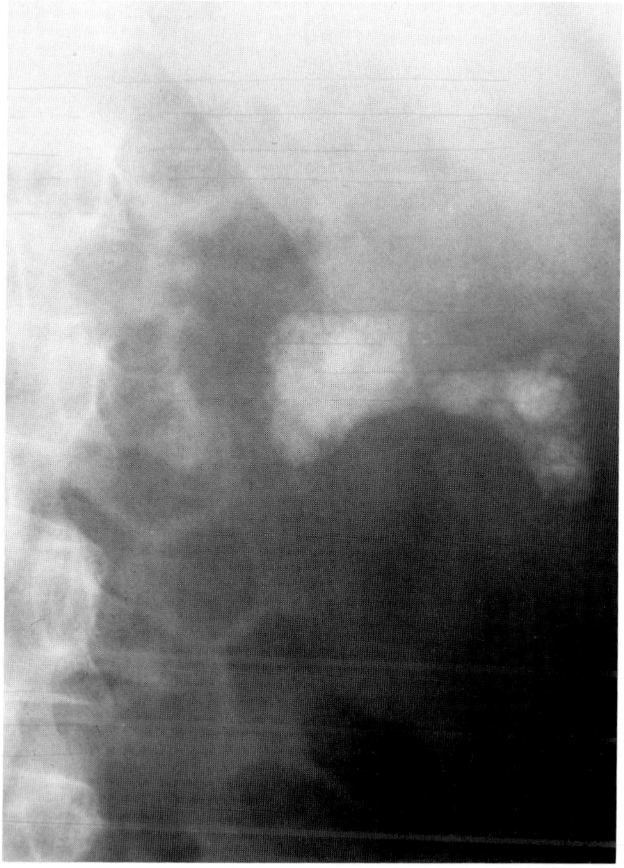

Fig. 8-3. Immediate post-treatment film.

pain out of proportion to the KUB stone pattern, a sonogram will detect any intrarenal hematoma, as well as indicate the degree of hydronephrosis.

## Stent Management

Ureteral stents can be managed in several ways after treatment. First, if the stent has been inserted to enhance fluoroscopic visualization, the stent should be removed immediately after the treatment. Second, if a ureteral stone is manipulated to the kidney and successfully disintegrated, the stent can be removed immediately after treatment. Third, if a ureteral stone has been manipulated, but remains within the ureter, the stent should remain in place for 24 hours, since ureteral edema occurs after treatment with high number of shock waves in one area of the ureter. Although distintegrated, these stones may be held by this edema in one area, thereby increasing the degree of obstruction, possible pain, or obstructive pyelonephritis. Fourth, if manipulation of the ureteral stone was unsuccessful, and the stent was left below the stone as a marker during treatment, this stent should then be removed since it will impede fragment passage.

If the treated renal stone was large, and possibly of infection etiology, the operator may choose to allow the external stent to remain for 24 hours of decompression during parenteral antibiotic therapy. With finely disintegrated stones, the sand will pass around the stent into the distal ureter, and sometimes into the bladder. The day after treatment, the stent is removed, and the patient discharged; often, the patient will pass huge volumes of particles within hours. If an indwelling double-J catheter was placed before treatment, it usually remains indwelling for a period of 5 to 7 days, allowing decompression, and dilating of the ureterovesical junction. Small particles may pass around the stent into the distal ureter (Fig. 8-4). However, large fragments (5 to 6 mm) will probably be held in the renal pelvis by the stent, possibly creating a scenario for late onset of colic. Patients with long distances to travel home after lithotripsy are ideal candidates for indwelling double-J catheters.

With indwelling external ureteral stents, often a stentogram (ureterogram) allows dye flow into the kidney and down the ureter, allowing an appraisal of the degree and success of treatment. This is especially useful for large, poorly calcified, or uric acid calculi.

Stent placement can be viewed as a double edged sword. While initial colics and hospitalizations are minimized, often fragments are prevented from passage. Since stone-free status is one treatment objective, the rate of complete fragment passage *after* stent removal must be considered. Stents must be judiciously, but not indiscriminately, used to maximize successes for all types for the patient.

## Discharge Medication

Patients discharged from the ESWL unit must be able to hydrate themselves well and to ingest oral medications. Discharge prescriptions to be considered by the urologist operator include: (1) analgesics—Tylenol with codeine, Percodan, if necessary; (2) antibiotics—as outlined in above section; (3) antiemetics—usually in suppository form; (4) antispasmodics—for bladder spasms; and (5) alkalinizers—K-lyte or Urocit, for uric acid or cystine calculi.

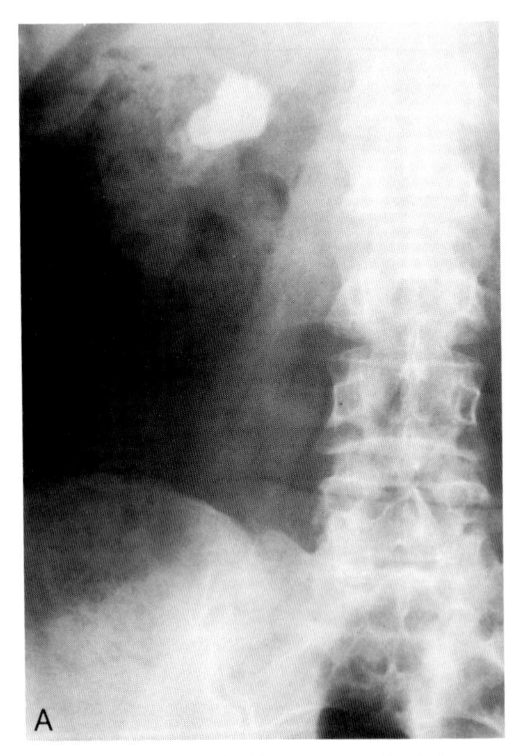


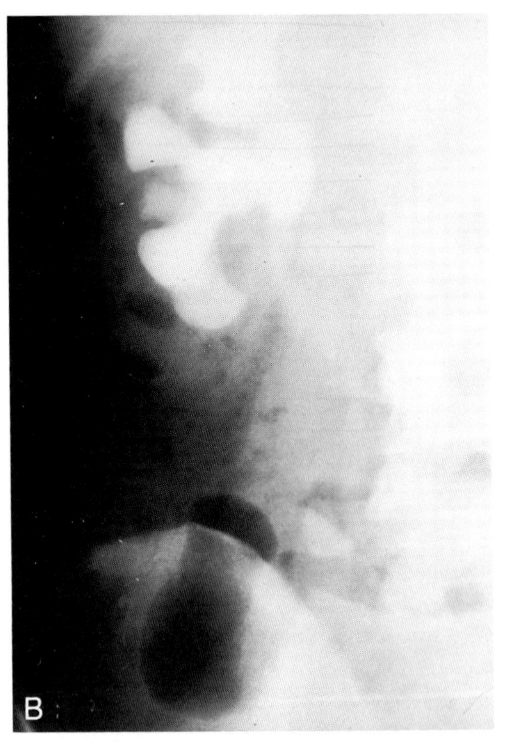


Fig. 8-4. **(A,B)** Large renal pelvic stone with hydronephrosis. (*Figure continues.*)

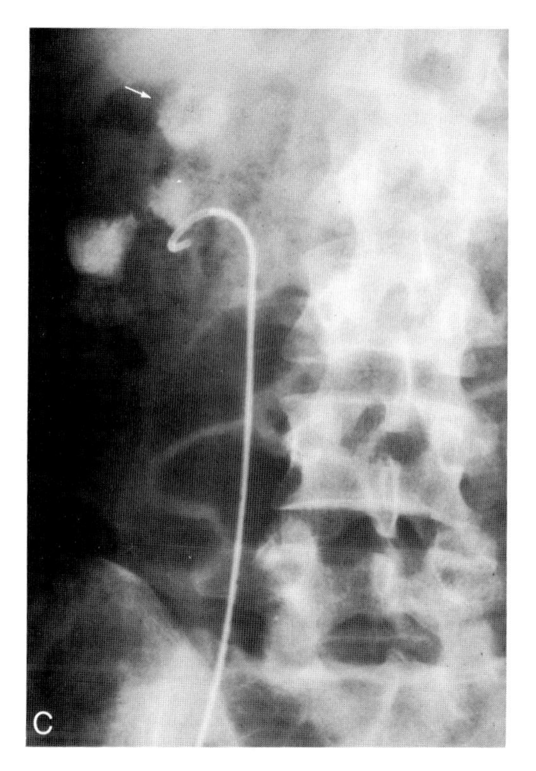


Fig. 8-4 (*continued*). **(C)** before treatment, a double-J ureteral stent was passed. **(D)** Passage of stone fragments around stent to a point of anatomic inertia.

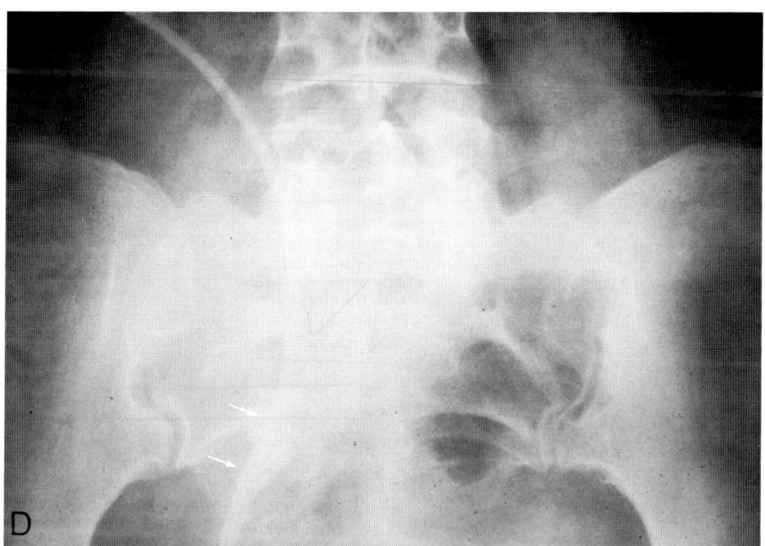


## Hospital Stay

During 1985, most ESWL patients in the US were treated as inpatients. The patient was generally hospitalized the day before or early on the morning of treatment and discharged the following or subsequent days. The U.S. experience represents a significant change from an average hospital stay of 7.5 days during the first two years of clinical trials in Munich.

The minimum hospital stay for ESWL patients during the FDA trials in 1984 was 2 days (48 hours). At the New York Hospital, patients were admitted the afternoon before the treatment and dischargd by noon the day following ESWL (2 days, or 48 hours). In general, patients with stones less than 20 mm diameter had a lower average stay per treatment than those with large or multiple stones (2.6 days versus 3.5 days). Treatment for multiple, large stones and partial or full staghorns required the longest hospital management before and after ESWL. Hospital stays averaged greater than 6 days for treatment of stone burdens greater than 40 mm (Fig. 8-5).[5]

With increased use of double-J stents, better patient screening, and more percutaneous nephrostolithotomies before ESWL, the hospital stay per patient has markedly decreased. Yet, it is still impossible to predict which patients with small stones will require additional hospital days for hydration and analgesia. Patients referred from distances outside the geographic area of the unit often cannot be treated as outpatients, and patient and family concerns about this new treatment modality require constant medical staff support and reassurance, whether ESWL is performed as an inpatient or outpatient procedure.

## Outpatient

During initial FDA trials at the first six centers, less than 3 percent of all ESWL treatments were performed as outpatient therapy.[7] With increasing numbers of ESWL facilities, outpatient treatment units emerged. The immediate post-ESWL management of outpatients must be tailored to prevent complications which would require hospitalization. More aggressive techniques to prevent obstruction, particularly the frequent use of double-J internal ureteral catheters, are employed in the outpatient setting, based on the following assumptions: (1) indwelling ureteral catheters allow continued passage of urine via and around the ureteral lumen, while allowing some spontaneous passage of particles (sometimes up to 5 mm diameter); (2) the ureter initially responds to the insertion of a catheter by constriction around the foreign object, but secondarily it becomes fatigued and dilates. Thus, upon removal of the ureteral stent several days later, larger particles subsequently may pass unimpeded through a relaxed ureterovesical junction.

## Steinstrasse

Ureteral steinstrasse is common after ESWL (Fig. 8-6). Proximal or distal ureteral sand or fragments are seen days to weeks after treatment. Common sites of temporary inertia are the proximal ureter as it crosses anteriorly over the psoas muscle, the midureter at the level of the iliac vessels, and the prevesical ureterovesical junction. Patients may pass these granules daily in small amounts, or sporadically after a few hours of urinary symptoms. Fortunately, these fragments usually pass spontaneously if given enough time (Fig. 8-7). Both patient and doctor must be patient. If the patient is asymptomatic, large amounts of ureteral sand can simply be observed until passage (Fig. 8-8).

However, ureteral fragments can cause temporary hydronephrosis with symptomatic colic, decreased ipsilateral renal function, and gastrointestinal ileus; occasionally, obstructed pyelonephritis occurs (Fig. 8-9).

Since most fragments progress to spontaneous passage, the indications for endourologic intervention are (1) prolonged fever or ileus with symptomatic hydronephrosis; (2) severe colic without fragment passage; (3) ureteral fragment inertia; (4) premature cessation of fragment passage; and (5) large residual renal fragments refractory to ESWL.

Prolonged proximal ureteral fragment inertia should be treated with stent or ureteroscopic repositioning of fragments retrograde to the kidney, and retreatment with ESWL (Fig. 8-10).

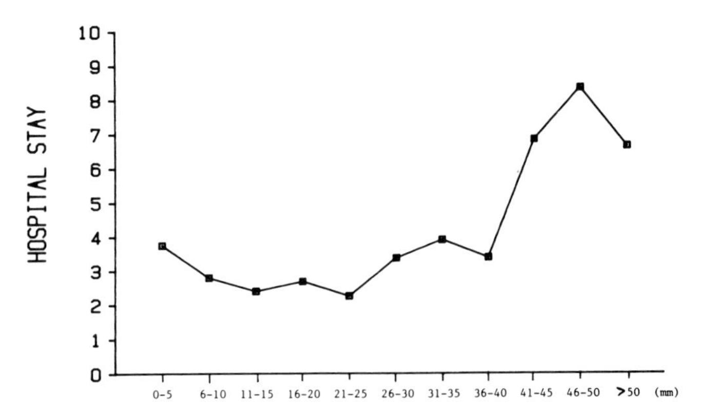


Fig. 8-5. Average hospital stay in days versus stone burden (300 treatments).

In the febrile patient with obstructed pyelonephritis, a percutaneous nephrostomy tube may need to be placed prior to manipulation. Impacted distal ureteral particles are best removed with ureteroscopic ultrasonic lithotripsy, unless urosepsis makes decompression nephrostomy the better clinical choice. Frequently, nephrostomy drainage allows subsequent spontaneous passage (Fig. 8-11).

As expected, ureteral stents rarely can be passed through these impactions. If the stone is disintegrated finely to sand, and the steinstrasse is short and distal, a ureteral stent with irrigation and rotation may be effective (Fig. 8-12).

Secondary endourologic procedures after ESWL require modification of previously described techniques.[5, 11] Ureteroscopy of a distal steinstrasse is particularly difficult because (1) a guide wire cannot be passed through the impaction; (2) conventional baskets, loops, and forceps cannot be used; (3) multiple ureteral fragments require multiple reintroductions of the ureteroscope with accompanying ureteral edema and hemorrhage; and (4) the direct vision ureteroscope with offset lens has a small, delicate ultrasonic probe which clogs easily (Table 8-1).

**Table 8-1. Number of Treatments Requiring Secondary Procedures (New York Hospital–Cornell Medical Center, 1984 to 1985)**

| | |
|---|---|
| Cystoscopy | 7 |
| Percutaneous nephrostomy | 17 (6 acute) |
| Percutaneous nephrostolithotomy | 5 |
| Ureteroscopy | 8 |
| Ureterolithotomy | 3 |
| Rate 8% (40/518 treatments) | 5 |

Riehle R, Fair W, Vaughan E: Extracorporeal shock wave lithotripsy for upper urinary tract calculi. JAMA 255:2043, 1986. Copyright 1986, American Medical Association.

## Follow-Up Visits

Patients are usually seen 7 to 10 days post ESWL in the office, and a KUB film is obtained at that time. The film is viewed with the patient, and unless it shows the patient to be stone free, further appointments are made for continued follow-up. At two to three weeks, sonography may be indicated if ureteral fragments suggest a high-grade obstruction. Determining the degree of associated hydronephrosis helps determine how soon a secondary procedure should be performed.

Laboratory studies during the first post-ESWL follow-up visit may include urine culture, complete blood count, serum chemistries, and creatinine, depending on the clinical progress of each patient. However, patients who have passed all

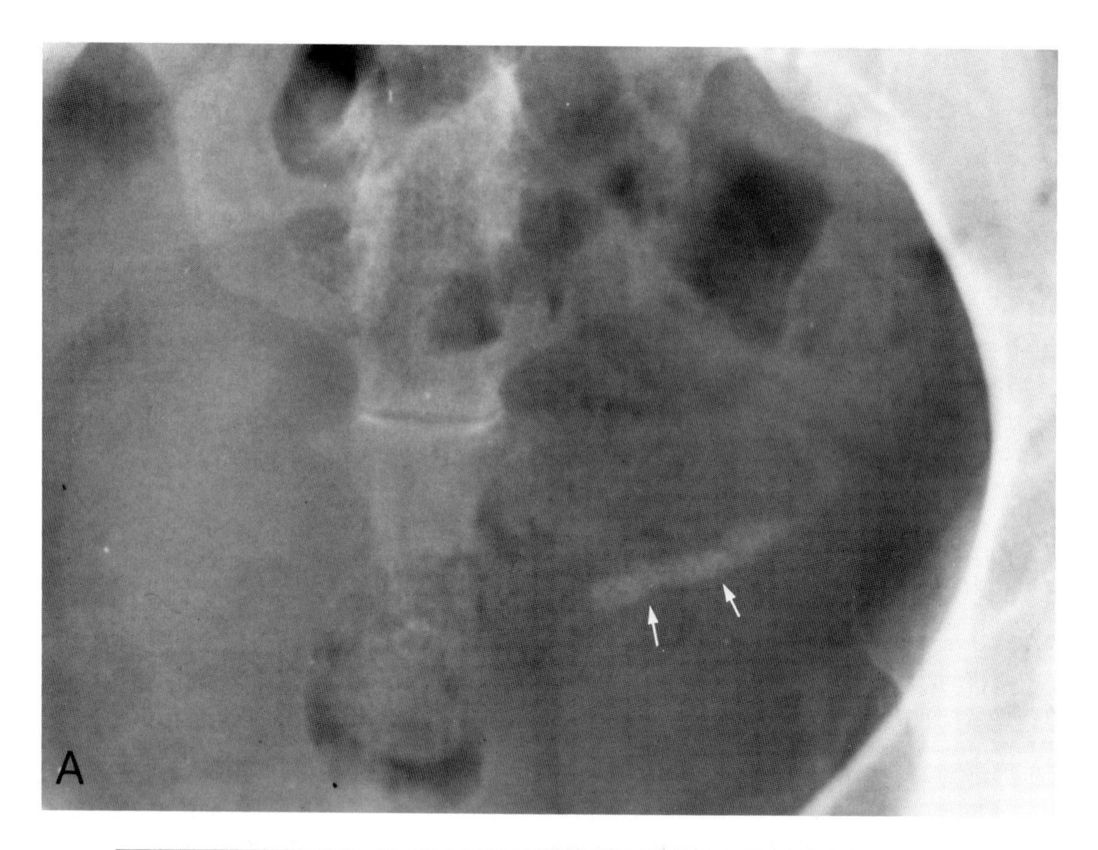


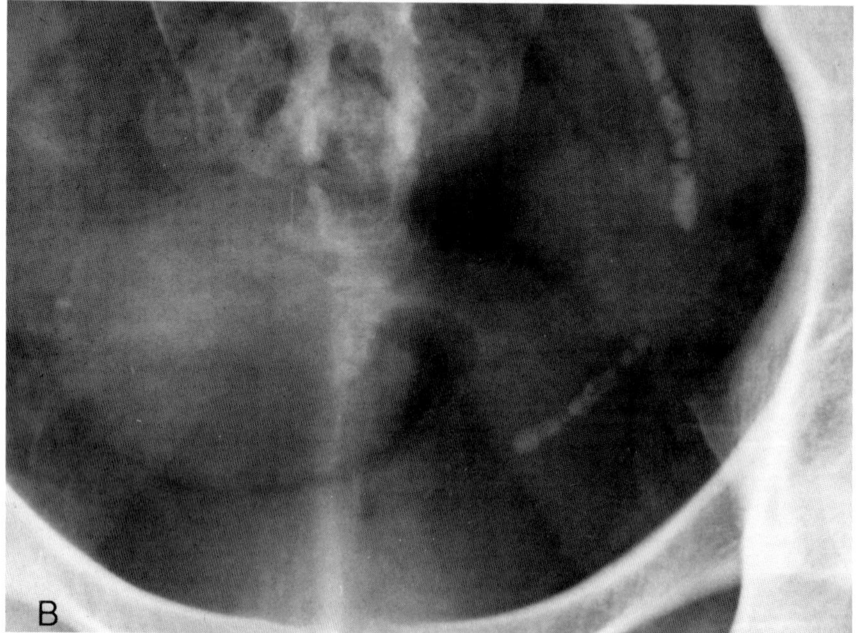


Fig. 8-6. **(A)** Distal ureteral sand (arrows). **(B)** Larger ureteral fragments.

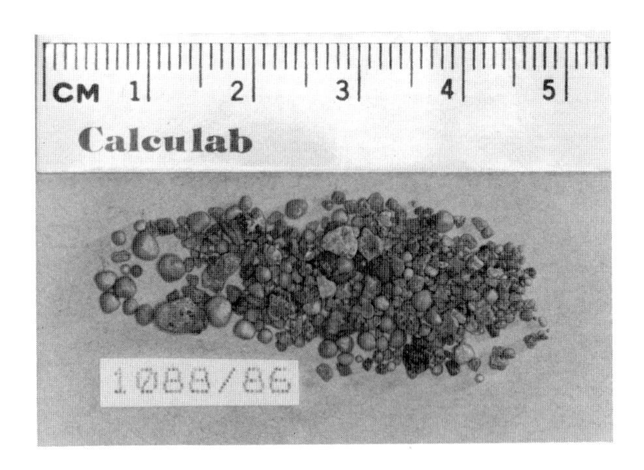


Fig. 8-7. Typical sample of spontaneously voided fragments. Analysis reveals calcium oxalate dihydrate (90%), monohydrate (10%). (Riehle R, Carter B, Vaughan E: Quantitative and crystallographic analysis of voided stone fragments after ESWL. J Urol, in press. © The Williams & Wilkins Co., Baltimore.)

fragments, who are symptom free, and who have normal KUBs, need no further laboratory studies. Determination of creatinine levels is indicated in patients with solitary kidneys and retained ureteral fragments.

A 3-month visit with KUB film is recommended for any patient with a prior KUB film showing residual calculi in the kidney or ureter. Any retained fragments in the ureter at this point must be carefully monitored and, depending on the patient's clinical status and renal function, will probably have to be removed endoscopically.

During the follow-up period, one cannot minimize the importance of the metabolic stone evaluation to detect an underlying metabolic abnormality and the need to institute medical therapy for the prevention of future stone recurrence.[11-13]

## RESULTS

Successful ESWL treatment requires both complete disintegration of the targeted calculus and complete spontaneous discharge of all fragments as monitored by a KUB 3 months after treatment. Since 1982, the experience of Chaussy with ESWL in West Germany has predicted that 80 percent of symptomatic upper urinary tract stones could be treated with noncontact disintegration by shock waves. In a series of 1,000 carefully selected cases reported by Chaussy, a stone-free rate of 90 percent was achieved with minimal complications.[3, 5, 16] Other European centers have confirmed the efficacy of shock wave lithotripsy, but specific stone-free rates between 60 to 90 percent have been reported depending on stone size, location, composition, and prior percutaneous extraction procedures.[14, 15]

At the New York Hospital, our overall stone-free rate of 75 percent was achieved in 300 treatments with an average stone burden of 17.8 mm. While lower than initially expected, the rate parallels the F.D.A.-monitored U.S. clinical trial results. The 87 percent success with category A stones contrasts with the 65 percent stone free rate for category B stones (Fig. 8-13). A decrease in the rate of successful spontaneous passage can be expected as the total stone burden in the kidney increases. Two percent of the treatments were failures, confirming a 98 percent disintegrative effect of ESWL on all treated stones (Table 8-2).

**Table 8-2. Results**

| | Category A | Category B | Overall (A + B) |
|---|---|---|---|
| Success | 87% (111/127) | 65% (113/173) | 75% (224/300) |
| Fragments remain | 12% (15/127) | 32% (55/173) | 23% (70/300) |
| Failure | 1% (1/127) | 3% (5/173) | 2% (6/300) |
| Stone burden (avg.) | 12.1 mm | 21.9 mm | 17.8 mm |

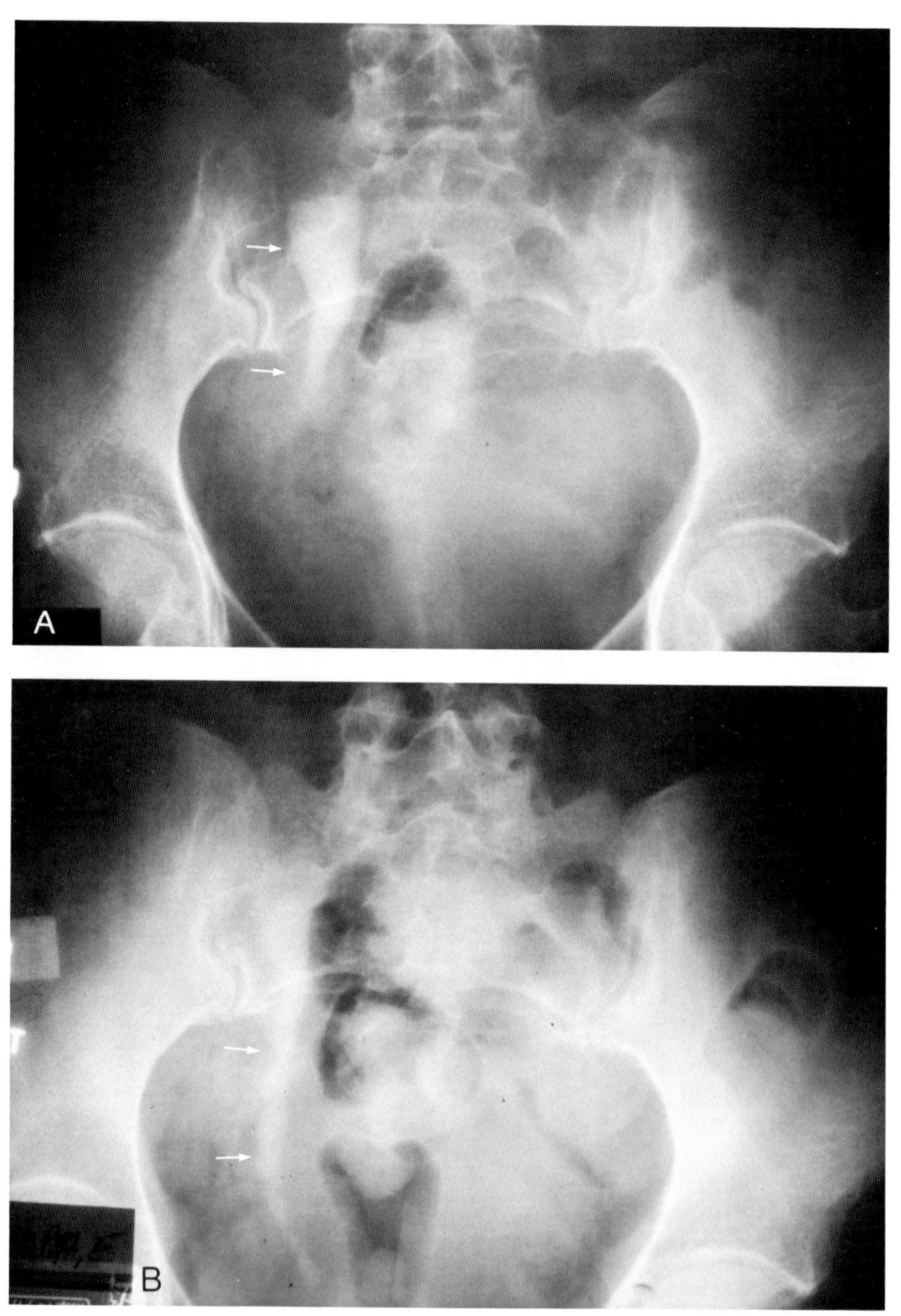


Fig. 8-8. **(A)** Distal ureteral sand 2 weeks after removal of double-J catheter (see Figure 8-4). **(B)** Slow passage 6 weeks after treatment. (*Figure continues.*)

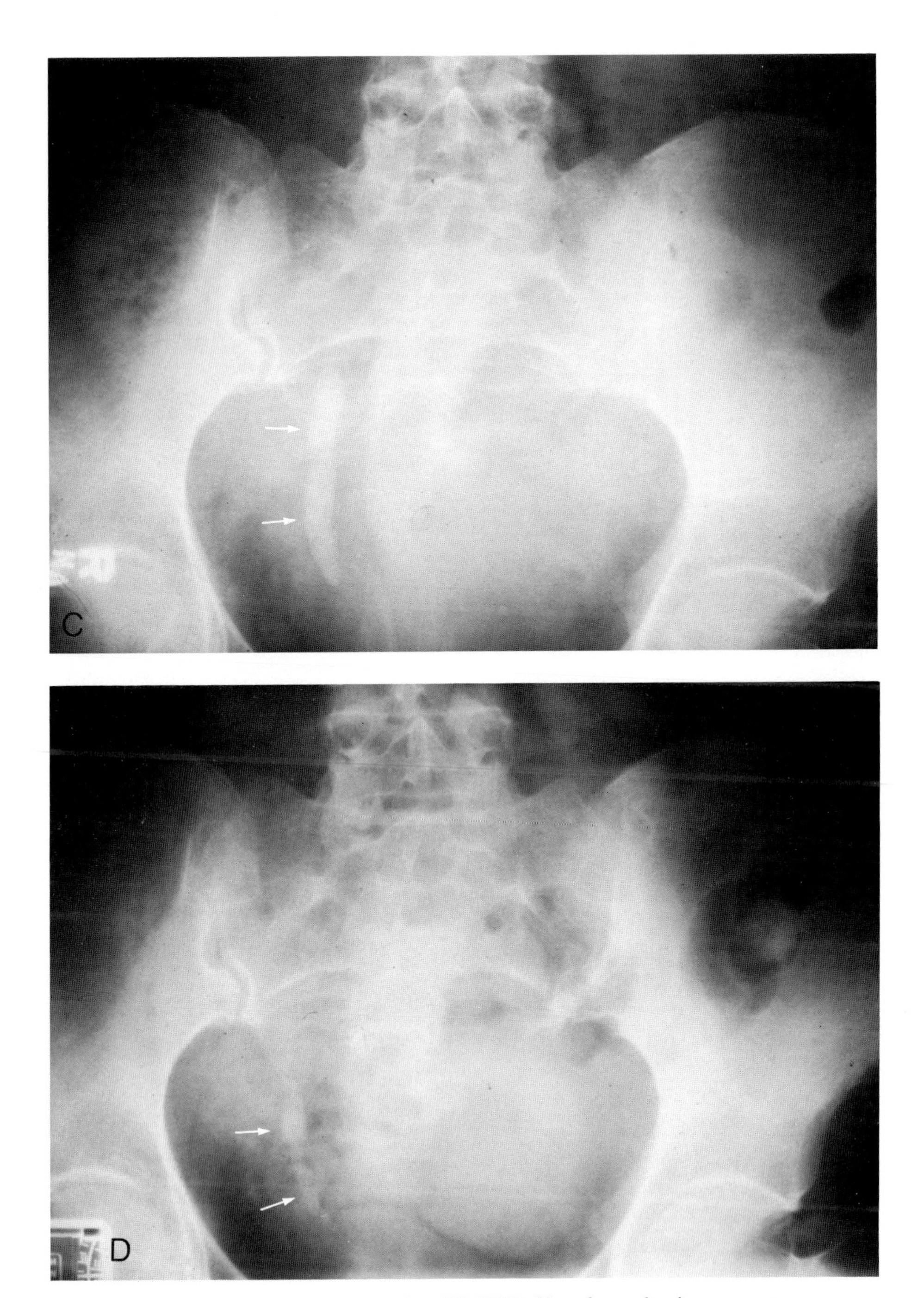


Fig. 8-8 (*continued*). **(C)** KUB film at 3 months. **(D)** KUB film after voluminous spontaneous passage at 14 weeks.

In detail, from May 1, 1984 to June 1, 1985, 467 patients underwent shock wave lithotripsy at the New York Hospital. Twenty-five patients received bilateral treatments; these were sometimes performed sequentially under the same anesthesia. In total, 518 treatments were performed; of these, 5 percent (26/518) were retreatments of the same kidney. Ninety-five percent of stones were completely treated with one ESWL session. Sixty-five percent of the treatments were performed on males, and 57 percent of treated stones were on the left side.

Twenty-three percent (119/518) of treatments were preceded by auxiliary cystoscopy and retrograde pyelogram or stent placement. The indication for endoscopy was (1) delineation of distal ureteral anatomy, (2) manipulation of ureteral stones retrograde, (3) better localization of poorly calcified or radiolucent renal calculi, and (4) injection of contrast during ESWL to visualize ureteral or renal fragments. Six treatments were performed on patients with indwelling nephrostomy tubes.

The average number of shock waves delivered per treatment was 1,382 (150 to 2,300). Category B stones received more shock waves per treatment than did category A stones (1,507 average versus 1,212 average). Kidneys with a stone burden greater than 20 mm received, on the average, more than the mean number of shock waves per treatment. Despite the smaller size of ureteral stones, the number of shock waves delivered per treatment was close to the overall average for all stones. In most cases, stone loads greater than 35 mm received the maximum number of waves allowable per treatment (Fig. 8-14).

As of August 31, 1985, a retrospective analysis of 277 patients (300 treatments) had been performed. Follow-up data were compiled from our center's referring physician letters, documented x-ray reports, and telephone reports solicited from physicians. Information on treatment effect was available on 58 percent (300/518) of ESWL procedures performed. The age distribution of these patients is similar to the total group treated (Fig. 8-15).

Forty-two percent (127/300) of the treatments were performed on category A stones, and 58 percent (173/300) were performed on category B stones. The average stone burden for all treatments was 17.8 mm.

Treatment time averaged 44.5 minutes for category A stones and 45.6 minutes for category B stones. The average fluoroscopy time per patient treatment was 186 sec with mean of 175.3 sec for category A and 193.4 sec for category B.

The overall stone free rate (success) as deter-

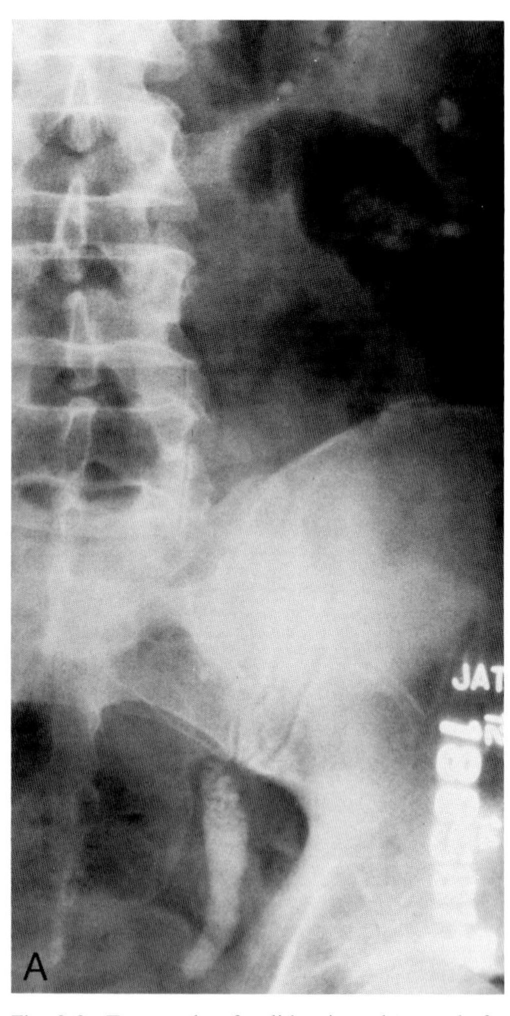


Fig. 8-9. Two weeks after lithotripsy, low-grade fever with ileus and flank pain; ureteroscopic decompression and stent passage performed. (*Figure continues.*)

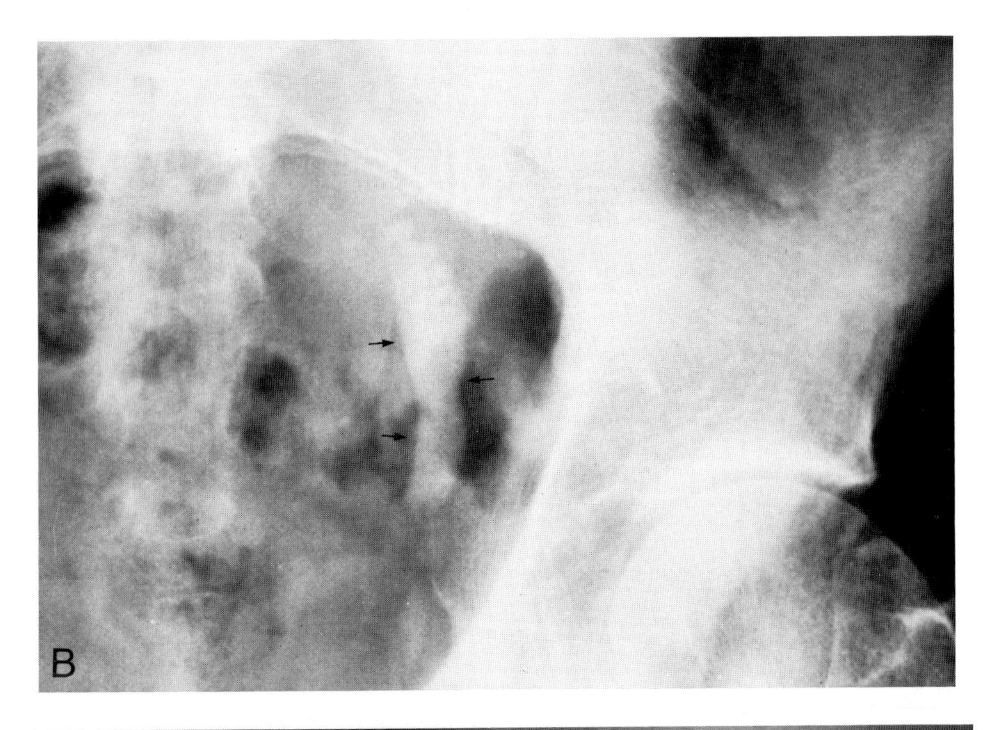


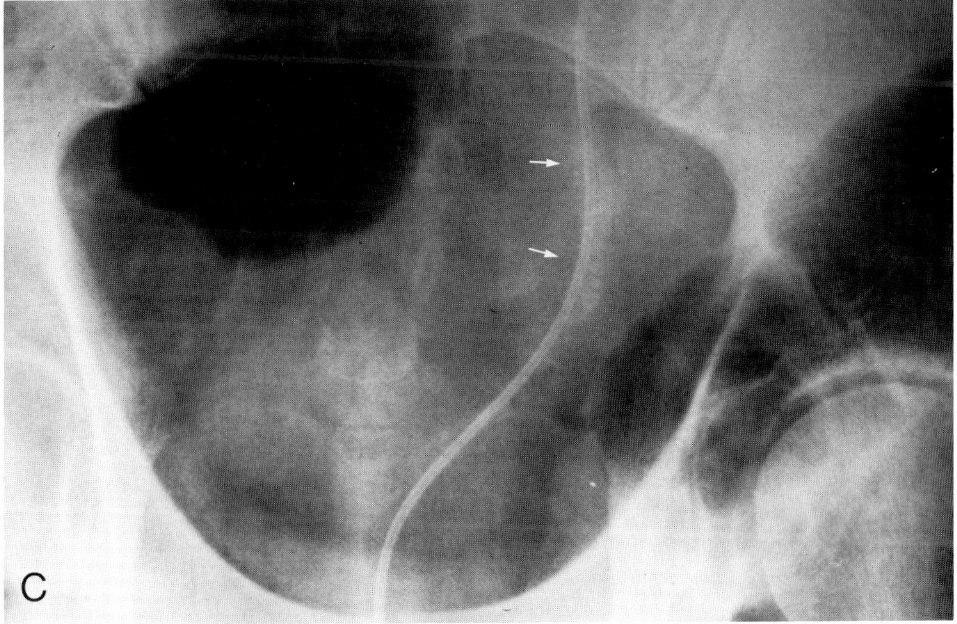


Fig. 8-9 (*continued*).

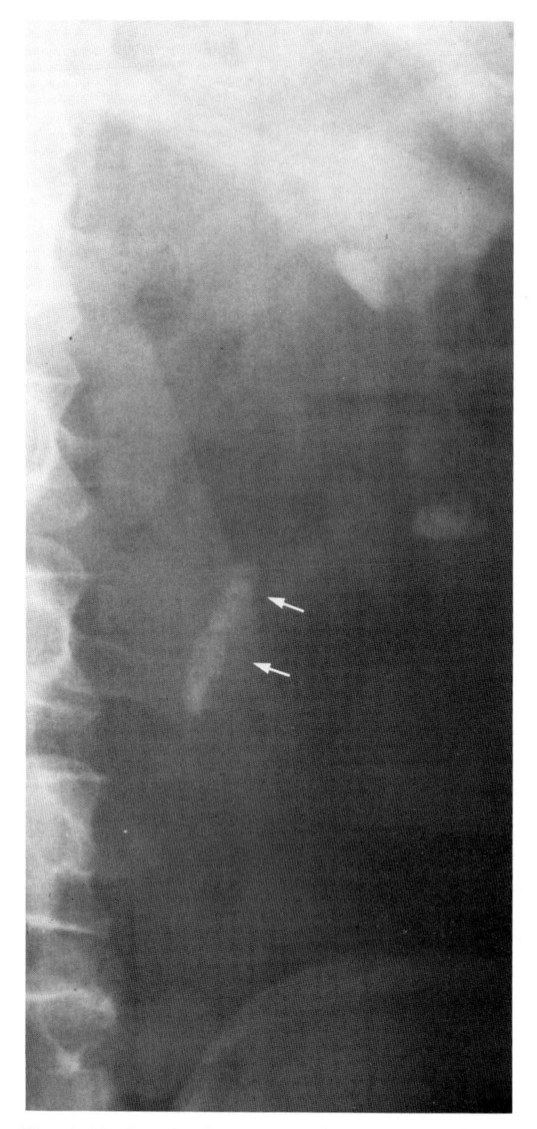

Fig. 8-10. Proximal ureteral steinstrasse for retreatment.

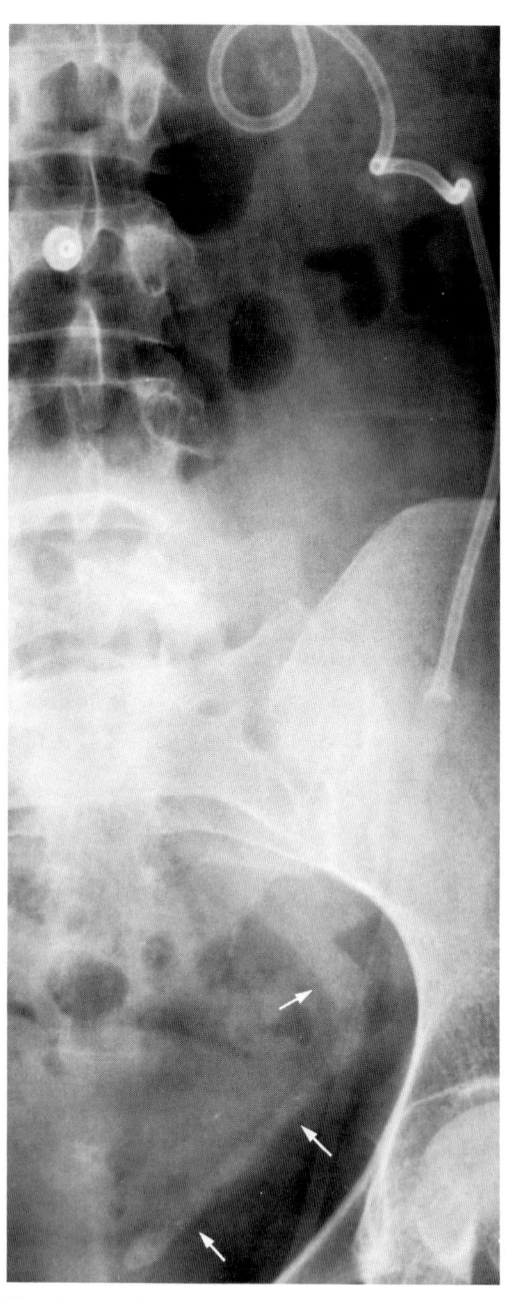

Fig. 8-11. After nephrostomy for fever, proximal particles dropped into distal ureter.

mined by a single KUB film obtained at a 3-month interval after treatment was 75 percent (224/300). Twenty-three percent (70/300) had some remaining detectable fragments, and 2 percent (6/300) revealed no significant disintegration or effect of ESWL. The average stone burdens were as follows: 15.6 mm (stone free), 25.2 mm (remaining fragments), and 17.0 mm (failure).

The stone-free success rate for category A stones was 87 percent (111/127) and for category B, 65 percent (113/173). The smaller average stone burden for category A patients (12.1 mm) contrasts with the 21.9 mm average stone burden

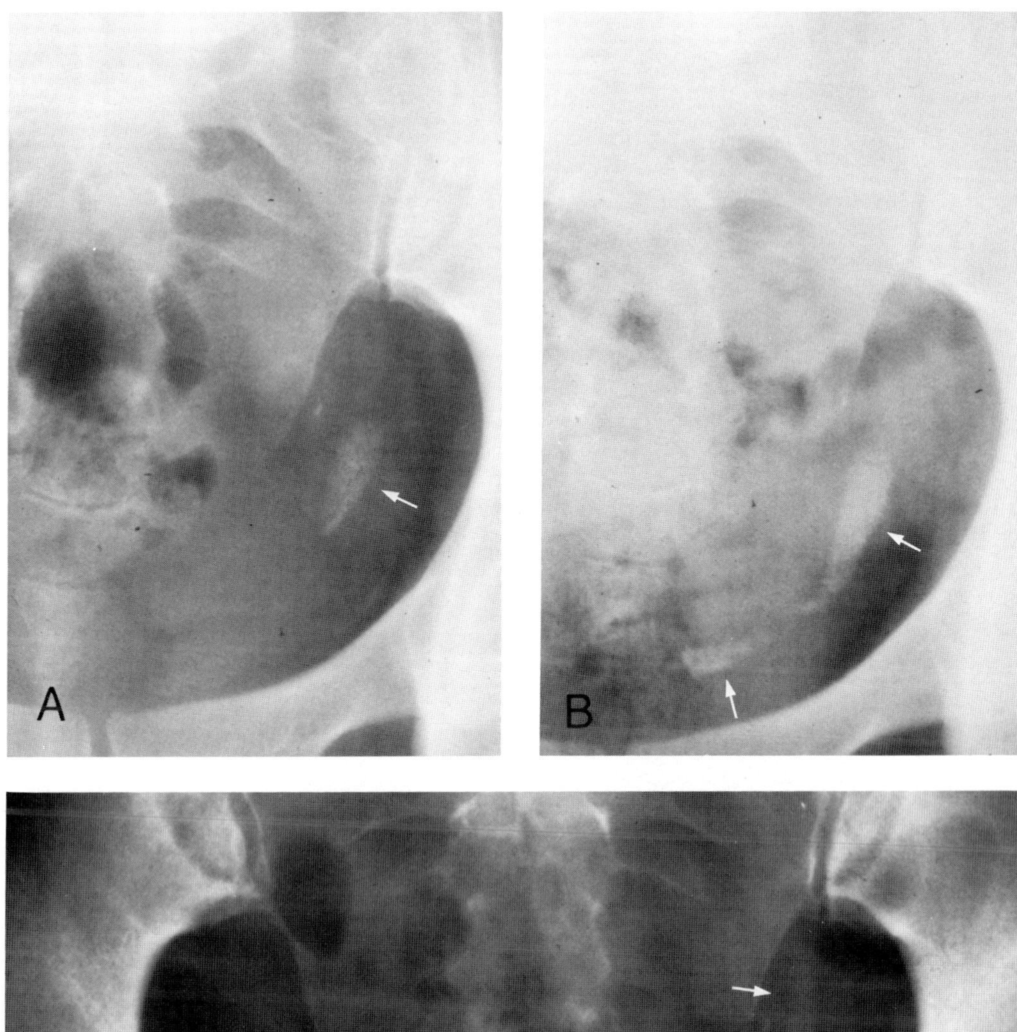


Fig. 8-12. Staghorn stone (infection type) with progressive distal steinstrasse, disimpacted as outpatient with rat-tail catheter, irrigation, and stent rotation.

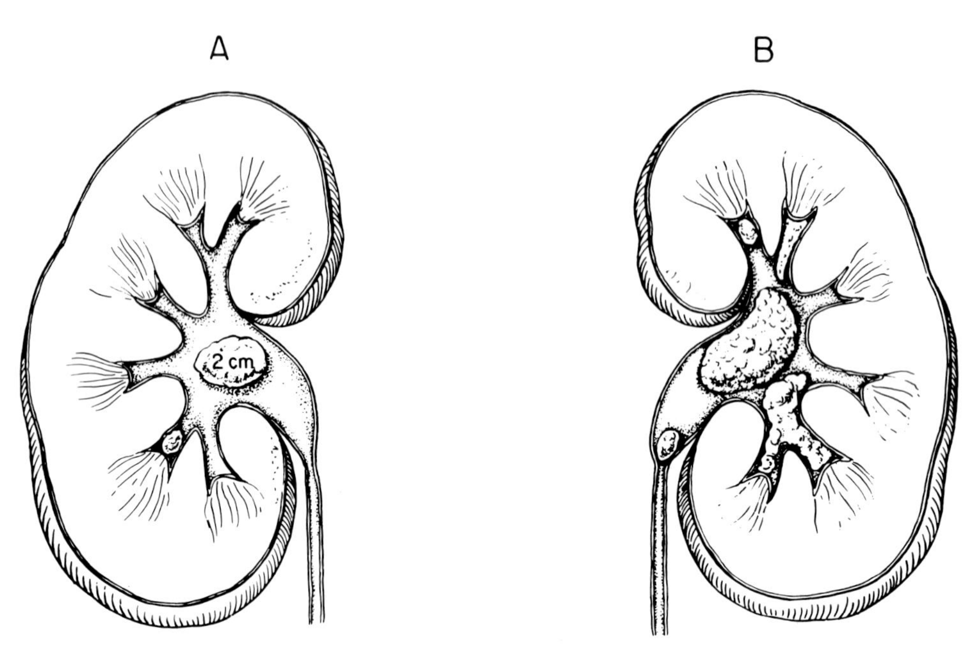


Fig. 8-13. FDA classification. Type A must be a solitary renal, pelvic, or caliceal stone.

for category B patients. Two percent (6/300) of treatments failed to disintegrate the target stone (Table 8-2).

For category A stones, 91 percent of renal pelvic stone treatments were stone-free at 3 months; yet only 78 percent of treatments for a solitary caliceal stone were completely successful.

For category B stones, the average stone load and success rate correlate inversely (Fig. 8-16). The stone-free rate diminished in proportion to increasing stone burden and also increasing age. Older patients were less likely to completely pass all fragments by the 3-month evaluation period.

The lowest stone-free rates were with multi-

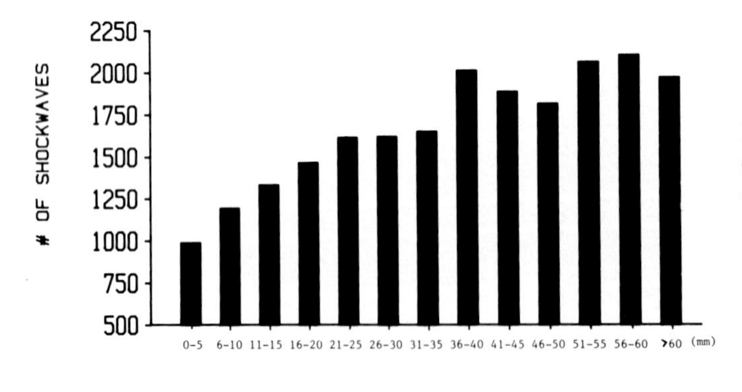


Fig. 8-14. Stone burden versus average number of shock waves delivered during treatment (518 treatments).

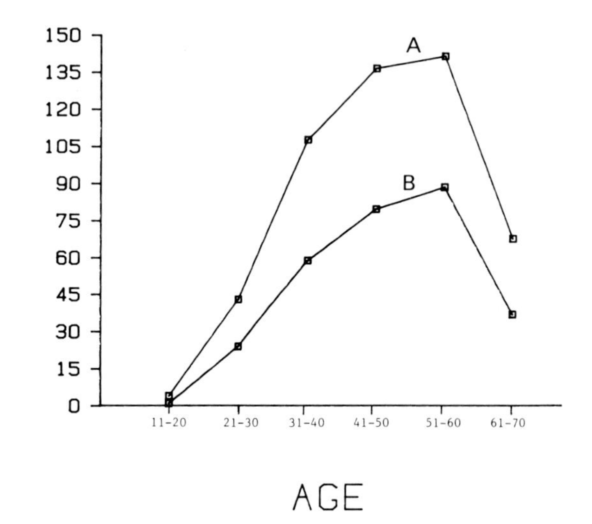


Fig. 8-15. Age distribution of patient treated (518 treatments, line A) and patients with 3-month follow-up (300 treatments, line B).

ple, large calculi (43%) and full staghorn calculi (50 percent); these two groups also had the highest pretreatment stone burdens.

In an analysis of 100 symptomatic primary upper ureteral stones (9.8 mm average diameter) treated using ESWL, 72 percent of the patients underwent preprocedural cystoscopy for stone manipulation, and 39 percent (28/72) of these stones (impacted) could not be dislodged proximally to the kidney or upper ureter. The stone-free success rate at 3 months for dislodged stones (9.1 mm average) was 93 percent. Of 56 ureteral stones treated in situ (impacted or not manipulated), 47 patients (84%) were rendered stone-free at 3 months. Four stones (13.6 mm average) treated in situ failed to disintegrate.[17]

Lithotripsy of small renal stones prior to ureteral migration and proximal ureteral stones early in the symptomatic course may significantly alter the incidence of distal ureteral calculi requiring hospitalization, cystoscopy, or ureteroscopy.

## Pediatric Patients

Management of the child after ESWL follows the same guidelines as for the adult, with special attention to shielding the lungs during treatment and preventing ureteral impactions which may require ureteroscopy. Newman and associates reported 36 treatments performed on 28 children

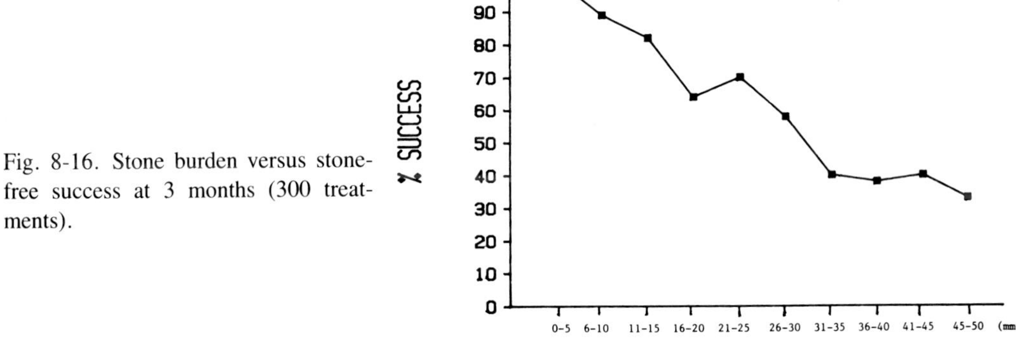


Fig. 8-16. Stone burden versus stone-free success at 3 months (300 treatments).

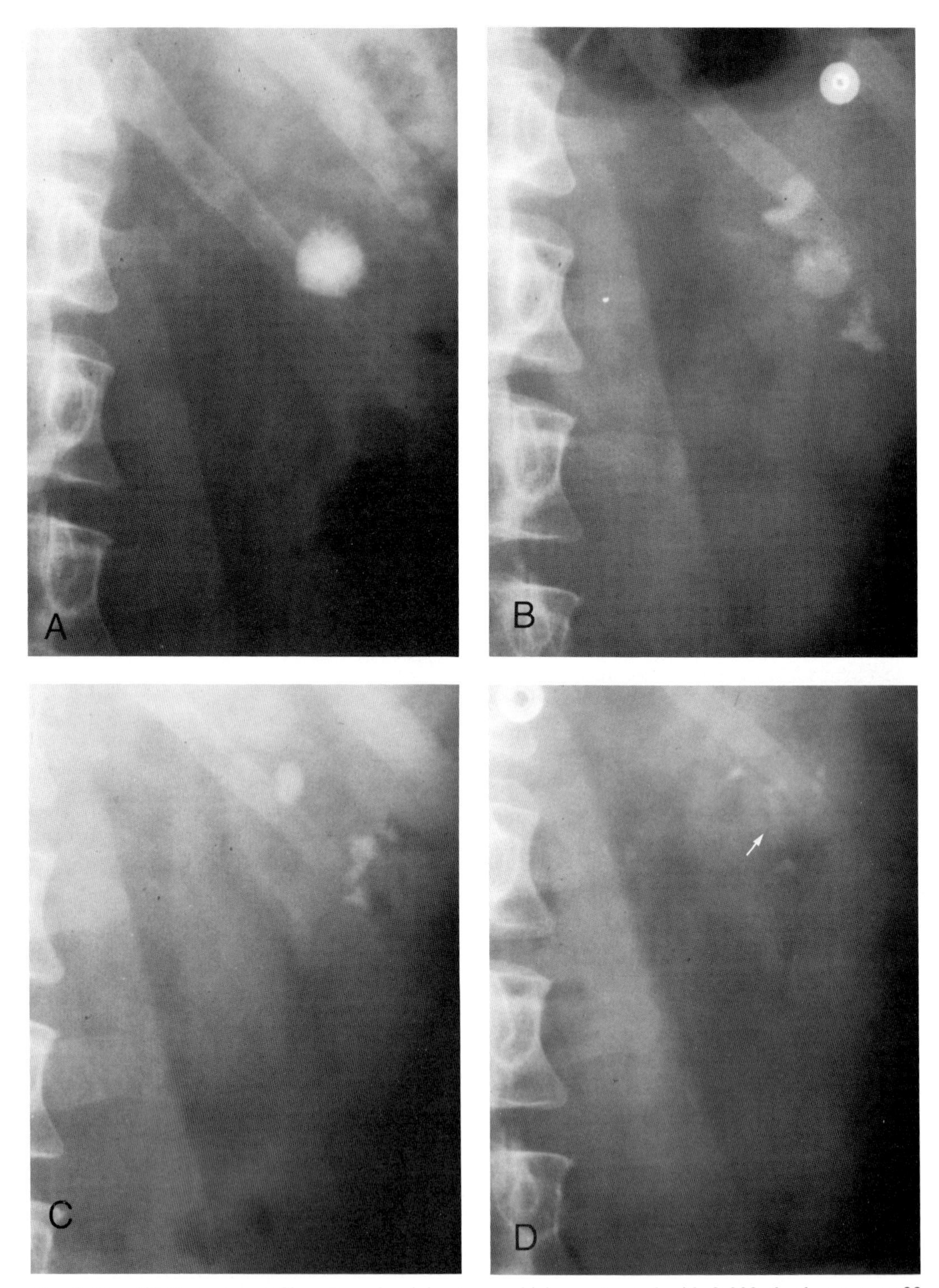


Fig. 8-17. **(A)** Pretreatment film of renal pelvic stone, which was treated with 2,000 shock waves at 22 kV. **(B)** Initial post-treatment film shows good initial disintegration. **(C)** KUB film at 3 months shows residual fragments. **(D)** After second ESWL treatment, good disintegration was demonstrated.

from 1 to 17 years of age (average, 11 years). Rate of complication and colic was no different than that reported in adults. With an 82 percent three month follow-up, 18 kidneys were found stone free (72%).[10]

From Katharinen Hospital in Stuttgart, West Germany, Bub and associates reported 15 children (2 to 17 years old, average 11 years) treated as follows: Eight children, ESWL monotherapy; 2 children, ESWL and percutaneous combined therapy; and 5 children, percutaneous monotherapy. There were no major complications, and all but one child, who underwent percutaneous therapy alone, were rendered stone free.[18]

Other centers have confirmed that, with a few modifications, ESWL of symptomatic stones in children can be performed safely and effectively.[19]

## Retreatment of Residual Fragments

Retreatment is rarely indicated within the first weeks. Demonstrated renal edema and ecchymosis after ESWL seems to resolve after 3 to 4 weeks and spacing of treatments therefore seems prudent. At some centers, though, the decision to retreat often is made during the initial hospitalization. However, the inability to distinguish radiologically the difference between sand in a calyx and residual fragments in the same location would seem to support a more conservative "watch and wait" philosophy (Table 8-3). At the New York Hospital, a retreatment rate of 5 percent (26/518) was reported in a series with an overall stone-free rate of 75 percent.[5] Only symptomatic patients with fragments larger than 5 mm at 3-month follow-up were offered retreatment. Certainly a more aggressive retreatment or endoscopic secondary procedure rate may have increased the 3-month stone free rate, but the fate of patients with residual renal fragments at 3 months (23% in the New York Hospital series) is yet to be determined. Residual disintegrated particles in a dilated lower calyx are rarely symptomatic, unless positive urine cultures accompanying infection stones persist. Fragments larger than 5 mm diameter could become mobile and symptomatic and, therefore, should be retreated or extracted (Fig. 8-17).

**Table 8-3. Residual Fragments (70 Treatments)**

| | |
|---|---|
| Single Location | 51/70 |
| Inferior calyx | (24/51) |
| Renal pelvis | (13/51) |
| Ureter | (7/51) |
| Superior calyx | (4/51) |
| Middle calyx | (2/51) |
| Bladder | (1/51) |
| Multiple Locations | (19/70) |
| Middle | |
| Inferior calyx | |

It has been noted that particles theoretically can become epitaxy centers for further stone formation and growth. Long-term follow up of stone activity in ESWL treated patients will be necessary to assess the effect of this variable on clinical stone events.[5]

## REFERENCES

1. Wilbert DM, Muller SC, Thuroff JW et al: Our experience with extracorporeal shock wave lithotripsy and special aspects in more than 1000 cases (abstract). J Urol 133:310A, 1985.
2. Study of safety and clinical efficiency of current technology of percutaneous lithotripsy and noninvasive lithotripsy, AUA ad hoc committee, DL McCullough, Chairman, May 16, 1985, Baltimore, MD
3. Chaussy C, Schmiedt E: Shock wave treatment for stones in the upper urinary tract. Urol Clin N Am 10:743, 1983
4. Chaussy C: Extracorporeal Shock Wave Lithotripsy. New Aspects in the Treatment of Kidney Stone Disease. Karger, Basel, 1982
5. Riehle R, Fair W, Vaughan E: Extracorporeal shock wave lithotripsy for upper urinary tract calculi. JAMA 255:2043, 1986
6. London R, Kudlak T, Riehle RA: Immersion anesthesia for extracorporeal shock wave lithotripsy. Urology 28:86, 1986
7. Drach G, Dretler S, Fair W et al: Report of the United States Cooperative Study of extracorporeal shock wave lithotripsy. J Urol 135:1127, 1986
8. Breaux E, Burns J: Outpatient extracorporeal shock wave lithotripsy. J Urol (in press)

9. Prevention of bacterial endocarditis, Med Lett 26:3, 1984
10. Newman D: ESWL experience in the pediatric age group. Presented at the second Extracorporeal Shock Wave Lithotripsy Symposium, Indianapolis IN, March 1986
11. Drach WG, Penn R, Jacobs S: Outpatient evaluation of patients with calcium urolithiasis. J Urol 121:564, 1979
12. Pak CYC, Buttone F, Peterson R et al: Ambulatory evaluation of nephrolithiasis: Classification, clinical presentation and diagnostic criteria. Am J Med, 69:19, 1980
13. Menon M, Krishman CS: Evaluation and medical management of the patient with calcium stone disease. Urol Clin N Am 10:612, 1983
14. Eisenberger F: ESWL of staghorn calculi. Read before the Third World Congress on Endourology, New York, 1985
15. Fuchs G, Miller K, Rassweiler J et al: Extracorporeal shock wave lithotripsy: One year experience with Dornier lithotripter. Eur Urol 11:145, 1985
16. Chaussy C, Schmiedt E, Jocham D: Extracorporeal shock wave lithotripsy (ESWL): Change in the management of stone patients. Read before the 20th Congress of International Society of Urology, Vienna, 1985
17. Riehle R, Näslund E: One hundred ureteral stones treated with extracorporeal shock wave lithotripsy. J Urol (in press)
18. Bub P, Clayman R, Fuchs G, and Eisenberger F: Change of indications in the treatment of urinary stones in children. Presented at the second Extracorporeal Shock Wave Lithotripsy symposium, Indianapolis, March 1986
19. Sigman M, Laudone V, Jenkins A et al: Initial experience with extracorporeal shock wave lithotripsy in pediatric patients. J Urol (in press)
20. Riehle R, Carter B, Vaughan E: Quantitative and crystallographic analysis of voided stone fragments after ESWL. J Urol (in press)

9

# Secondary Procedures after ESWL

Keith N. Van Arsdalen

Secondary procedures performed after extracorporeal shock wave lithotripsy (ESWL) include percutaneous, endoscopic, and surgical techniques. These procedures are usually therapeutic and are sometimes anticipated as part of the planned management of a given calculus. Alternatively, they may be instituted in the management of complications following treatment.

This chapter will discuss (1) the planned diagnostic and therapeutic procedures that follow shock wave lithotripsy of different types of stones, and (2) the antegrade and retrograde techniques used to manage complications of ESWL, particularly ureteral obstruction resulting from passage of fragments.

However, before considering the different adjunctive management techniques in detail, secondary retreatment of patients with the lithotripter is discussed.

## RETREATMENT

Although not truly a secondary procedure as defined above, retreatment remains one of the more common therapeutic techniques performed after intitial ESWL treatment. A compilation of data for the American Urological Association (AUA) Ad Hoc Committee report pertaining to ESWL indicated that approximately 88 percent of patients required one treatment, 11 percent (range 7 to 20 percent) required two treatments, and 1 percent (range 0 to 5 percent) required three treatments of more.[1]

At the Hospital of the University of Pennsylvania, from June through December 1985, 52 of 588 patients (8.8 percent) were treated two or more times. The indications for retreatment in this series were based primarily on radiographic evidence of incomplete fragmentation of the calculus into particles felt to be too large to pass spontaneously. Retreatment was considered when plain radiographs, tomograms, or retrograde pyelograms demonstrated particles larger than 5 to 7 mm. Other considerations were the number of large particles, their strategic location, and failure of fragments to progress into or down the ureter. Serial radiographs were usually obtained over 2 or more days as delayed dispersal of stone fragments was noted. If no significant change in the configuration or apparent density of the particles in question was noted over that period of time, retreatment was undertaken.

In most cases, the decision to re-treat was made during the patients' initial hospitalization and, in contrast to the indications discussed below for the secondary diagnostic and therapeutic measures, the decision to re-treat was often made in the absence of clinical symptoms or evidence of obstruction. In this regard, retreatment was performed in an attempt to prevent these problems. Occasionally, with borderline cases, patients were sent home in the hope they would pass their stones spontaneously, though it was expected they might need readmission and delayed retreatment. Clearly, in most cases, retreatment was more immediate than delayed. At the Hospital of the University of Pennsylvania, our overall retreatment rate in this group of patients is actually at the low end of the range in the AUA report.[1] The low rate is probably attributable to two factors—initial treatment of less complex stones and also treatment of many of these stones to the allowable maximum number of shock waves. Our average number of shocks per treatment (1,600 to 1,700) also exceeds the average number of shocks per treatment reported in the AUA report (1,300). However, at other centers, the decision to re-treat is rarely made during the initial hospitalization. The inability to distinguish radiologically the difference between sand and residual fragments often dictates a more conservative "watch and wait" philosophy. At New York Hospital–Cornell Medical Center, a retreatment rate of 5 percent (26/518) was reported in a series with an overall stone-free rate of 75 percent.[2] Only patients with fragment larger than 5 mm at 3-month follow-up were offered retreatment if symptomatic. Certainly a more aggressive retreatment or endoscopic secondary procedure rate may have increased the 3-month stone-free rate, but the fate of patients with residual renal fragments at 3 months (23 percent in the Cornell series) is yet to be determined. Residual disintegrated particles in a dilated lower renal calyx are rarely symptomatic, unless positive urine cultures accompanying infection stones persist. Fragments larger than 5 mm diameter could become mobile and symptomatic and, therefore, should be re-treated or extracted.[2]

## Planned Retreatments

The initial size, location, composition, and configuration of the stone or stones can be used as guides in planning treatment, and in informing patients in advance that they may need more than one treatment. Clinically it is obvious that it takes more shock wave energy to fragment large stones than small stones of the same composition.[3] The composition of the stone also determines how much energy will be necessary to fragment it. Depending on the initial stone burden, the operator may actually plan two or more treatments as a staged approach to the stone or stones in the collecting system. For stones situated entirely within the kidney, the renal pelvic component is generally treated first to clear this area, especially near the ureteropelvic junction outlet. Residual pelvic fragments and other stones in the collecting system may require an additional session or sessions in order to complete fragmentation.

The management of simultaneous upper ureteral and ipsilateral renal calculi may also be approached in a staged fashion, depending on the number of shock waves delivered initially to the ureteral stone. If most of the allowable maximum number of shocks per treatment are necessary to fragment a ureteral calculus, few shocks are left to treat stones in other areas. When the degree of fragmentation of either the ureteral or renal calculi is incomplete after the first treatment, additional sessions may be scheduled after 24 to 48 hours.

The composition and configuration of the stone(s) are also important. Cystine and brushite stones, particularly larger ones, are more difficult to fragment into particles that will pass spontaneously. Before the first ESWL treatment, patients are advised that repeat treatment and adjunctive techniques are often required. The same must be said for struvite calculi of various configurations and increasing stone burden. The role of staged therapy for staghorn calculi (both pretreatment percutaneous nephrostolithotomy and post-treatment nephrostomy and/or shock-wave retreatment) must be outlined for the patient in detail before the treatment plan begins. Adjunctive techniques useful in the management of metabolic and struvite calculi are considered below.

# DIAGNOSTIC AIDS AFTER ESWL

Abdominal radiographs, intravenous urograms, sonograms, and renal scans are common diagnostic procedures used at various times in the postoperative period to monitor the stone status and function of the urinary tract. All patients receive one or more of these studies for routine follow-up.

However, for nonopaque or faintly opaque calculi, kidney–ureter–bladder (KUB) films alone are not sufficient. At the University of Pennsylvania, retrograde pyelography (via a retrograde catheter) either alone or in combination with plain renal tomography has been used as a diagnostic technique in the immediate postoperative period; the intent is to assess the degree of fragmentation and to decide whether further treatment is necessary. Nonopaque and faintly opaque stones may be localized intraoperatively by instilling dilute contrast medium through a previously placed retrograde ureteral catheter. After completing treatment, the retrograde catheter is left in place and retrograde pyelography with fluoroscopy is performed on the first postoperative day (Fig. 9-1). The completeness of fragmentation is noted as is the presence of large particles. For faintly opaque calculi, renal tomography may give the same information without the need for the more time-consuming fluoroscopic procedure.

For nonopaque and faintly opaque calculi, the results of the above studies determine the need for retreatment and/or the need for adjunctive procedures and techniques. If the particles are all small and appear likely to pass spontaneously, the retrograde catheter is removed. If a large number of particles remain, or if the particles are borderline size (>5 mm) with regard to their chance of spontaneous passage, the retrograde catheter is exchanged fluoroscopically or cystoscopically for an internal double-pigtail ureteral stent. To facilitate this exchange, an open-ended retrograde catheter is the type selected for placement before ESWL in these patients with nonopaque calculi. Once placed, the double-pigtail stent prevents obstruction. Small particles may pass around the stent, and oral alkalinizing agents[4,5] or other medications[6] may be administered to hasten dissolution of certain particles (uric acid, cystine), even in cases where alkalinization was not successful prior to ESWL.

## Secondary Procedures

### CHEMOLYSIS

Dissolution is a surface phenomenon and the difference in efficacy before and after treatment is attributed to the dramatic increase in the overall surface area of the particles compared to the original stone. Furthermore, some calculi are lightly coated with calcium, and this outer layer prevents dissolution of the underlying crystalline structure before treatment. After ESWL, the internal structure of the stone is completely exposed for more successful chemolysis.[7]

For large retained particles of uric acid, cystine, or struvite calculi, direct irrigation may be more useful than oral medication. Again, because of the aforementioned factors, the chance of successful dissolution is increased. Irrigation may be performed with retrograde or antegrade techniques. An additional option, therefore, in the patients initially managed with placement of an open-ended retrograde catheter, is to exchange the single catheter for two retrograde catheters so that one may be used for irrigation and the other for drainage. The system employed always uses an overflow tube as a pop-off valve set to spill at 20 to 25 cm $H_2O$ to prevent high pressures and fluid absorption if outflow obstruction occurs.[8] The available solutions appropriate for dissolution of specific calculi have been summarized elsewhere.[8]

If dissolution occurs rapidly, as is generally the case with uric acid calculi, the entire process may be accomplished in a retrograde fashion (Fig. 9-2). Longer irrigation may be required to dissolve cystine or struvite particles, however, and this process may be better tolerated by the patient and more easily managed by the staff with the use of a percutaneous nephrostomy (PCN) catheter or catheters. The patient is certainly more ambulatory, has less risk of catheter dislodgement or slippage, and is generally more comfortable and compliant without several catheters exiting through the urethra. Furthermore, a larger-diameter catheter can be placed percutaneously allowing better drainage with less risk of obstruction. Insertion of these tubes may be planned as part of the secondary management

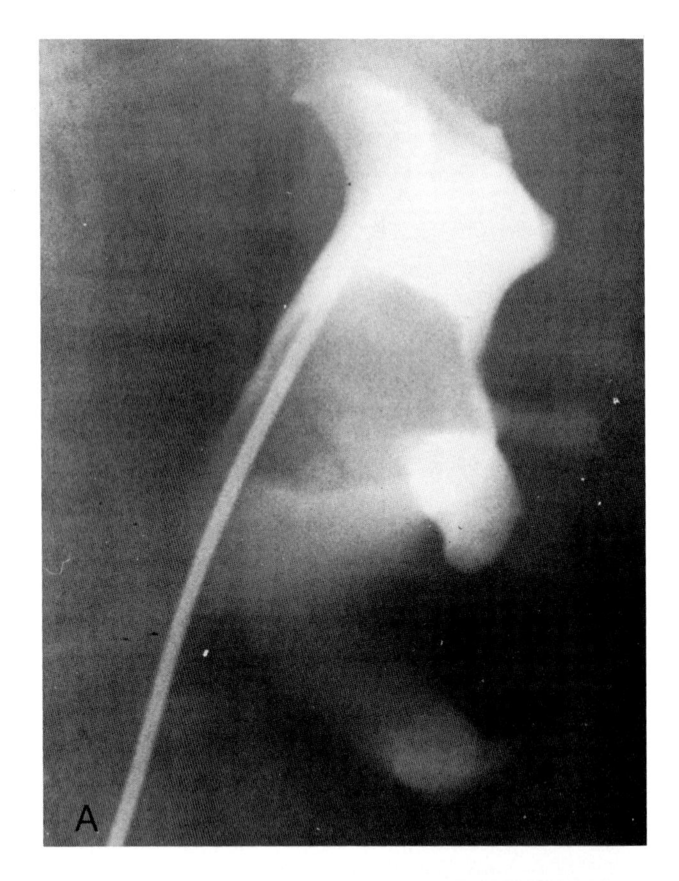


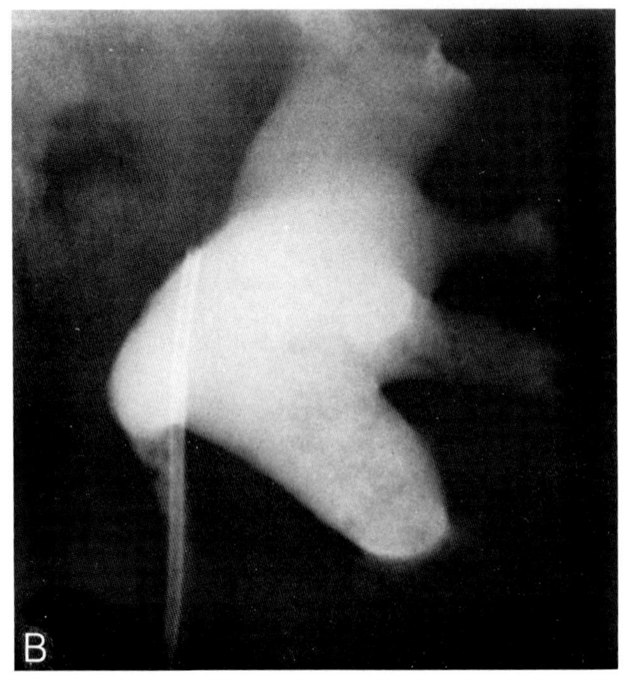


Fig. 9-1. **(A)** Demonstration of nonopaque calculus before ESWL. **(B)** Retrograde pyelography after ESWL confirmed adequate fragmentation, and the catheter was removed.

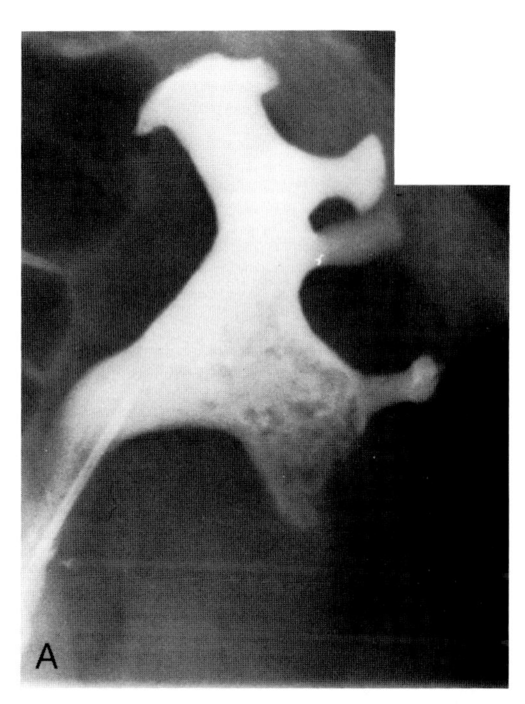


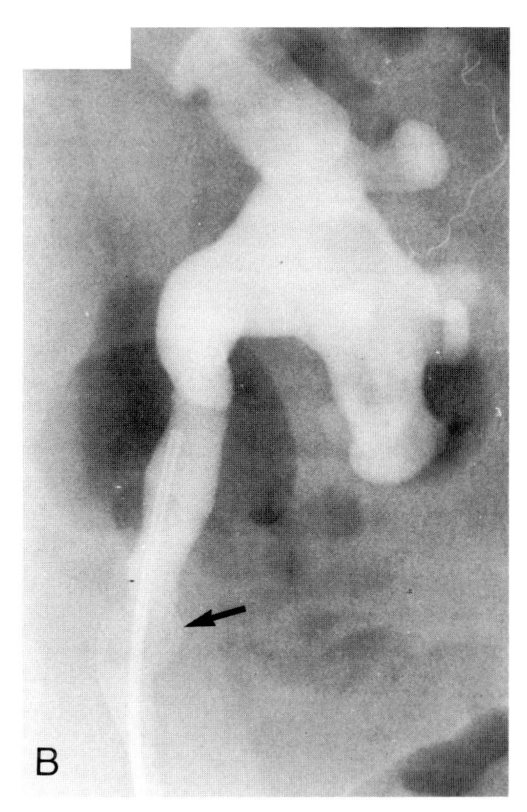


Fig. 9-2. **(A)** Large amount of nonopaque stone debris after ESWL. **(B)** Complete dissolution after 72 hours of retrograde irrigation with sodium bicarbonate solution. Note slippage of retrograde catheters (arrow).

of a calculus after initial fragmentation with ESWL. Again, a hydrostatic overflow valve is inserted in this system to prevent high pressures and associated complications. The same solutions described above may be used for percutaneous irrigation. Periodic nephrostograms confirm the degree of dissolution and the patency of the urinary tract.

## Nephrostomy: Planned Combined Techniques

Large-bore catheters inserted percutaneously may be particularly important in managing struvite calculi. Most ESWL operators now recommend percutaneous nephrostomy placement and stone debulking before ESWL for large staghorn calculi.[14–17] At times, however, antegrade nephrostomy insertion and placement of a guide wire into the ureter for subsequent tract dilatation is not possible because the guide wire or access catheter cannot be maneuvered around a calculus that fills the collecting system. In these cases, a retrograde ureteral catheter is inserted to prevent ureteral obstruction by fragments and ESWL is then directed primarily to the lower pole extension and pelvic portion of the calculus. If necessary, this procedure may be followed by transfer of the patient to the uroradiology suite for placement of a percutaneous nephrostomy. When possible, epidural anesthesia is used for both procedures and the nephrostomy tube is therefore inserted with the benefit of the anesthetic. This also makes it possible to dilate the tract in the uroradiology suite with a high-pressure balloon and insertion of a large-bore (24 F) Malecot catheter, allowing drainage of the stone and mucoid debris that is often encountered. In these cases, percutaneous nephrostolithotomy may subsequently be used as a secondary procedure for debulking, or the patient may be re-treated with the lithotripter.

A word of caution is indicated here. At the University of Pennsylvania, a significant risk of sepsis has been associated with PCN placement and dilation immediately after ESWL. In trying to simplify the management of staghorn calculi, and with the hope of decreasing the number of procedures per patient, we followed the above steps as a plan of management in several patients, without attempting to initially debulk the stone. All patients developed a high fever, and one became hypotensive after the nephrostomy tube was inserted. Almost certainly, the tissue penetration and blood vessel disruption during the nephrostomy insertion allowed the bacteria released by ESWL to easily enter the vascular system. The technique described may therefore be useful in limited circumstances but is not recommended for primary management when initial PCN placement and debulking by endoscopic techniques is possible.

## Approach to Ureteral Obstruction

Although other complications have been reported,[1] the problem of ureteral obstruction is by far the most common ESWL sequela requiring secondary intervention. Several points warrant emphasis. Just as retreatment is more often required for larger stones, so obstruction is more common after larger stones are fragmented, even when the initial degree of fragmentation is satisfactory.

Partial intermittent ureteral obstruction is common after ESWL; it has been reported in 31 percent of patients treated.[1] It is less common for smaller stones, occurring in 24 percent of patients with stones $\leq$ 1 cm in diameter and in 50 percent of patients with stones 2 to 3 cm in diameter. Stones in the 1 to 2 cm size range were reported to produce partial obstruction in 34 percent of cases.[1] During the FDA-monitored trials, total obstruction was reported for 1 percent of all stones treated.[1]

The etiologic mechanism responsible for obstruction may involve a single stone particle, but more commonly it involves the development of a ureteral steinstrasse (from German *stone street,* the term used to describe an accumulation of stone fragments producing a characteristic radiographic appearance). This may consist entirely of small fragments or may be associated with a lead fragment that is too large to pass. At the Hospital of the University of Pennsylvania, the development of a ureteral steinstrasse has been noted in one quarter to one third of our patients. Patients with clinical indications who were studied immediately with radionuclide scanning, ultrasonography, or intravenous urography, partial obstruction was noted in most cases, although in several there was complete obstruction. Sepsis as a postoperative complication is usually associated with obstruction; completely obstructed systems associated with infection require prompt intervention.

Fortunately, continued observation may be adequate in most cases of partial obstruction because most of these stones will pass relatively soon after treatment. Some may gradually disperse and be intermittently symptomatic over a longer period of time with eventual clearance of all fragments and no adverse sequelae noted thereafter (Fig. 9-3). Close radiologic follow-up, however, is mandatory as complete obstruction may occur silently and insidiously. This has been noted in completely asymptomatic patients who have returned 4 weeks after treatment for a routine follow-up intravenous urogram or renal scan.

Observation with periodic radiographs may be continued until all fragments pass, until it is evident that additional passage of the calculous debris has become stymied, or until patients become symptomatic. Intervention is indicated in situations that would classically prompt surgical intervention for an ordinary obstructing calculus—complete obstruction, associated fever or infection, unrelenting colic, or significant gastrointestinal distress with persistent nausea and vomiting.

The rate of secondary intervention necessary to manage the clinical problems described above ranges from approximately 6 to 12 percent (Table 9-1). Further breaking down these figures, information from Chaussy and Fuchs indicates

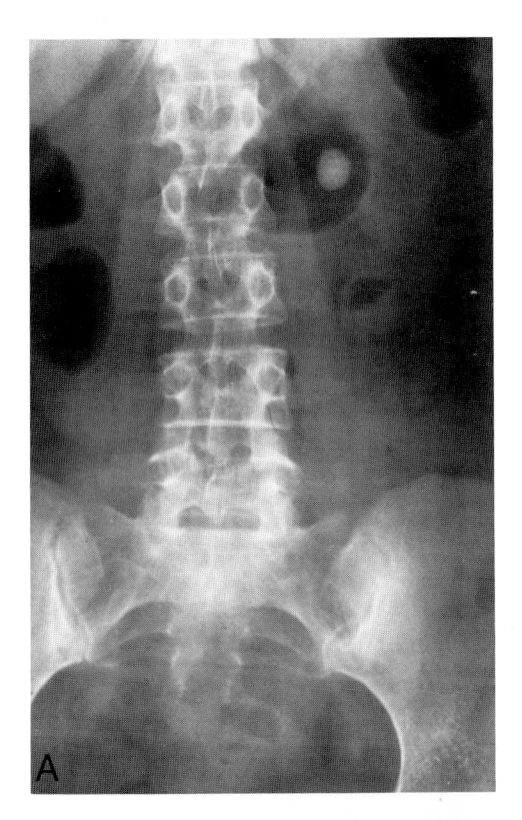


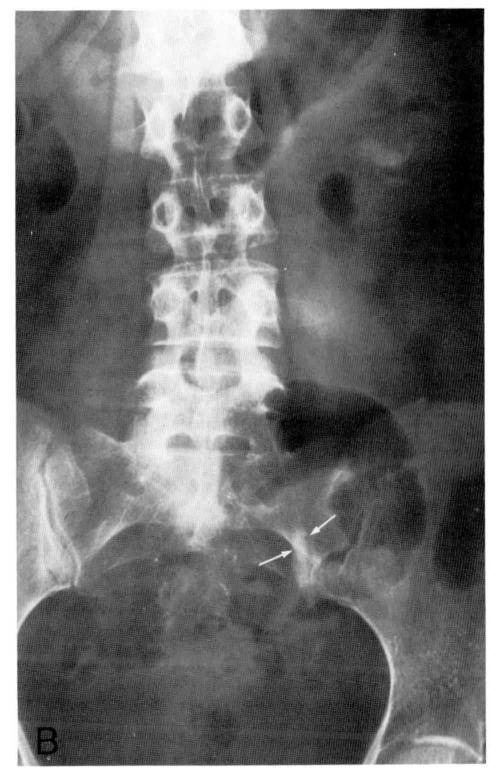


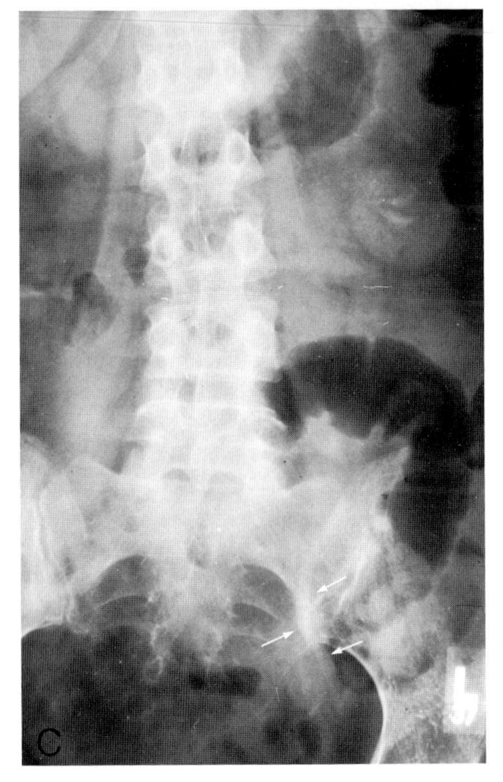


Fig. 9-3. The development and gradual passage of a ureteral steinstrasse is noted in this series of radiographs. The number of days after treatment are as marked. **(A)** Pretreatment KUB. **(B)** Postoperative day 1: steinstrasse develops overlying lower edge of sacrum (arrow). Stone is completely fragmented. **(C)** Postoperative day 2: increase in length of steinstrasse. (*Figure continues.*)

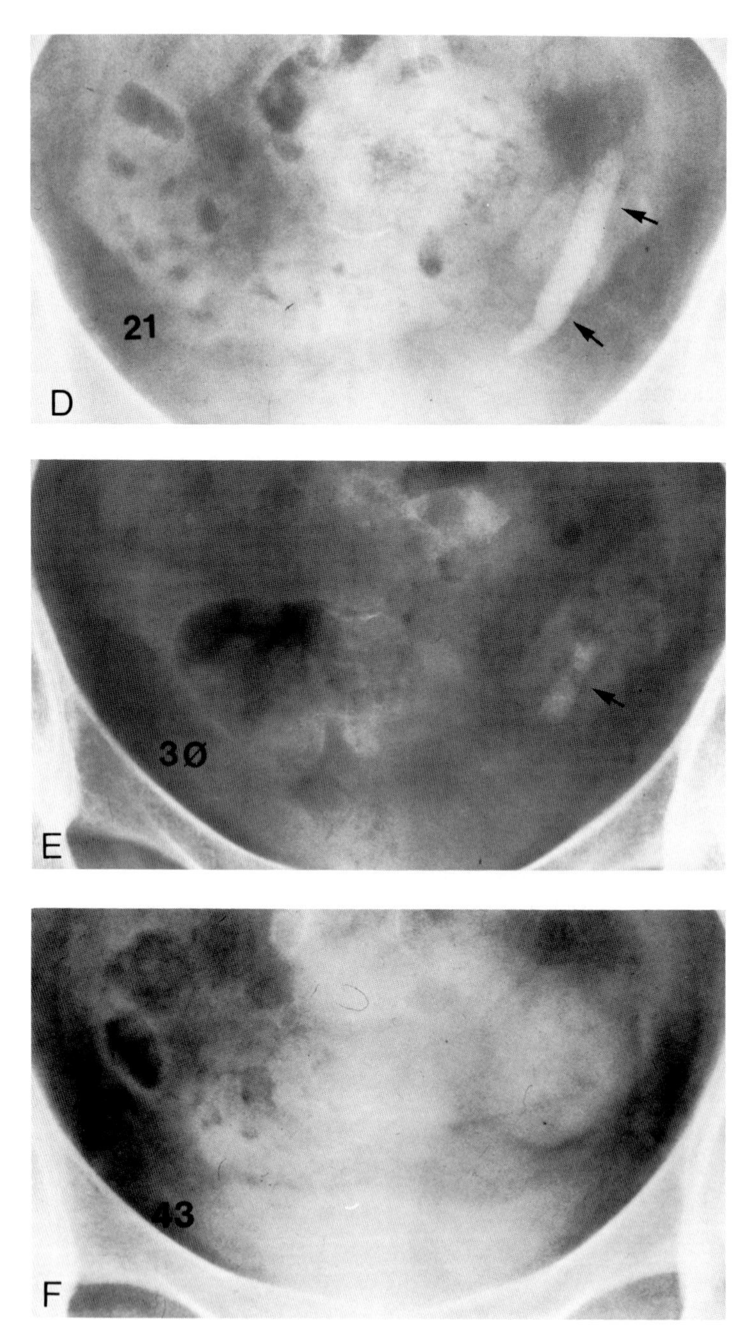


Fig. 9-3 (*continued*). **(D)** Postoperative day 21: passage of steinstrasse to juxtavesical ureter. **(E)** Postoperative day 30: decrease in number of fragments in distal ureter. **(F)** Postoperative day 43: complete clearance of distal ureter. No manipulation was required. Follow-up intravenous urogram showed no residual renal fragments with satisfactory function bilaterally.

**Table 9-1. Secondary Procedures after ESWL**

| Institutions | No. of Patients | Percent Secondary Procedures |
|---|---|---|
| Munich[a] | 2200 | 11.3 |
| Stuttgart[a] | 1800 | 6.3 |
| Sapporo[a] | 500 | 11.4 |
| UCLA[a] | 500 | 4.8 |
| Cornell[2] | 518 | 7.7 |
| Hospital of the University of Pennsylvania | 588 | 9.3 |

[a] Data from Chaussy CG, Fuchs GJ: World experience with extracorporeal shock-wave lithotripsy for the treatment of urinary stones: An assessment of its role after 5 years of clinical use. Endourology 1:7, 1986.

that auxiliary measures may be required in as many as 35 percent of patients with "problem stones." These are stones larger than 2.5 cm, staghorn stones, and some ureteral stones. The secondary procedures which may be used include percutaneous techniques, retrograde manipulation, and rarely open surgery. The type of intervention depends in part upon the location of the obstructing fragments, the philosophy of the responsible physician with regard to the best method of management for a particular patient, and the willingness of the patient to undergo additional "nonsurgical" or "less invasive" procedures.

## Techniques for Intervention

Placement of a percutaneous nephrostomy tube is required in 1 to 5 percent of patients, to relieve symptoms and decompress the system. It also presumably allows for resolution of ureteral edema in the area of obstruction and possibly improves ureteral peristalsis as well, as it has been observed that the ureteral particles were often passed spontaneously after the nephrostomy tube was placed (Fig. 9-4). In these situations, the particles passed even though the urine had been diverted from flowing down the ureter. If spontaneous passage does not occur, the percutaneous access route may be used for various of the endourologic procedures that were developed, refined, and popularized over the past several years, before the introduction of ESWL.[9,10]

Percutaneous nephrostolithotomy with ultrasonic lithotripsy may be of value in managing retained renal fragments that are large or relatively refractory to extracorporeal techniques because of their composition, location, or the patient's body habitus. Direct-vision extraction through the nephroscope, with or without ultrasonic fragmentation, is also useful to remove large fragments from the ureter or to remove upper ureteral calculi that have not disintegrated after one or more attempts with ESWL. Baskets may be passed with fluoroscopic control to extract fragments that have passed further down the ureter,[11] and distal ureteral calculi may be dislodged or flushed into the bladder with antegrade passage of appropriate catheters.[12] Secondary percutaneous decompression or manipulation has been required in 16 of the 588 patients (2.7 percent) treated at the Hospital of the University of Pennsylvania.

Ureteral obstruction from steinstrasse usually develops at the level of the distal ureter. This location lends itself to transurethral manipulation. It is usually impossible to pass retrograde catheters through and beyond a distal ureteral steinstrasse to relieve obstruction; rarely, this may be possible if there are only a few fragments or if the lead fragment can be retrieved. This may be accomplished by milking the lead fragment into the bladder with a cold resectoscope loop, if it is located in the intramural portion of the ureter, or by grasping it directly with alligator forceps, if it is protruding through the ureteral orifice. Redistribution and realignment of the fragments may allow subsequent passage. A Water Pik device attached to a ureteral catheter has been used at the University of Virginia to help disperse the fragments by bombarding them with pulsed jets of fluids (personal communication, Alan Jenkins, M.D.).

A ureteral meatotomy may be performed to retrieve the lead fragment or to gain access to the distal ureter for endoscopic manipulation. The fixed blade of the endoscopic scissors is introduced into the ureteral orifice, taking care to follow the course of the intramural ureter. The movable blade is then repeatedly opened and closed under direct vision, cutting the intravesical portion of the ureter for a total length of 5 to 10 mm, depending upon the length of

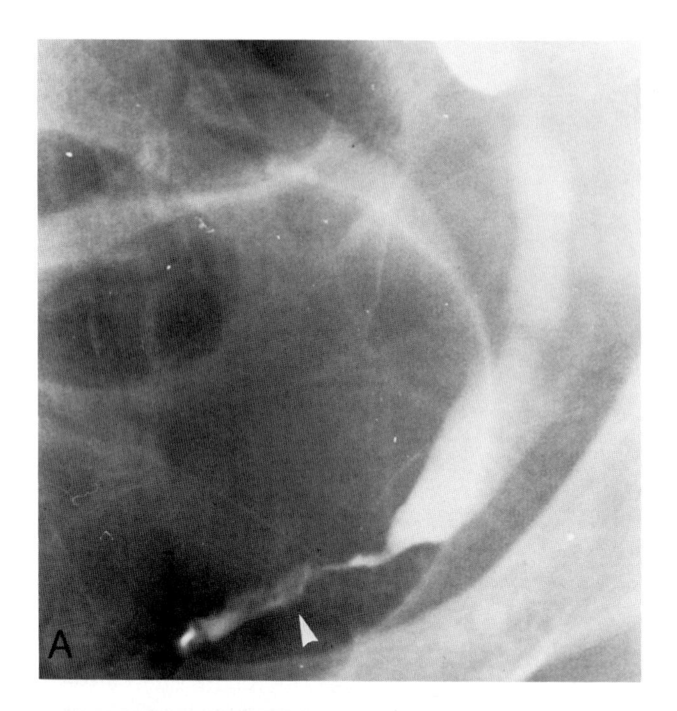


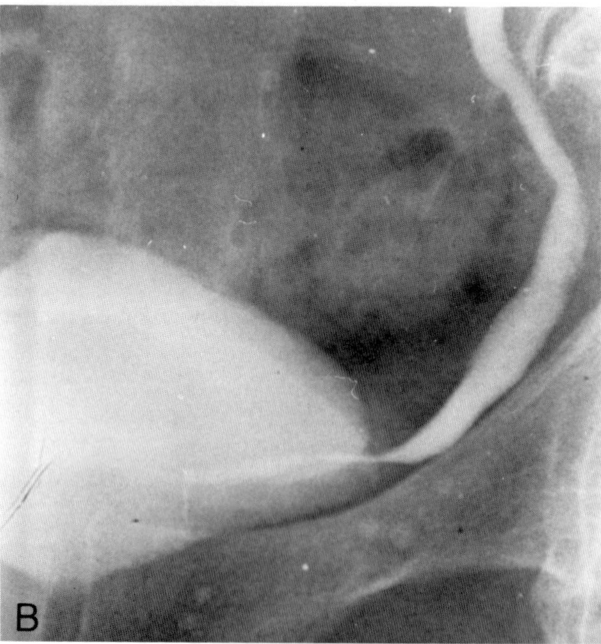


Fig. 9-4. **(A)** A faintly opaque steinstrasse (arrowhead) was noted following treatment of a partial staghorn calculus. This nephrostogram performed after placement of a PCN for fever and pain shows obstruction from fragments. **(B)** Five days later, with no additional manipulation, the fragments have passed. PCN placement may be combined with antegrade manipulation.

narrowing in this area and now much the orifice springs open with each cutting motion. Fragments will often spew from the orifice after the obstructing calculous debris is released.

Care must be exercised with regard to the angle of the meatotomy and the length of the incision, and this technique is only useful when it appears that the extreme distal portion of the ureteral tunnel is in fact the site of obstruction. In this regard, the appearance of a pseudoureterocele configuration of the stone fragments on radiographs (Fig. 9-5) or a bulging intramural tunnel with a small meatus on cystoscopic examination suggests that a meatotomy will be successful. Obviously, the role of a meatotomy is much more limited if the site of the obstructing fragments is above this level.

Direct-vision management with the ureterorenoscope has been the most commonly used technique to eliminate obstructing calculi or fragments. This technique applies to all levels of obstruction; however, an obstructing upper ureteral fragment or steinstrasse (above the pelvic brim) is usually retreated with ESWL, with or without a retrograde catheter in place. If the fragments fail to pass from the upper ureter after a second treatment, however, antegrade techniques or, more commonly, ureteroscopy are used.

When possible, access to the ureter is gained in the same way as for ureteroscopy for any other indication. A guide wire is passed cystoscopically to the highest level possible in the collecting system, and the ureter is dilated under fluoroscopic control with either a semirigid coaxial dilator or a balloon dilator. It may be particularly useful in males to use a 21 F or larger cystoscope, leaving the sheath in the urethra after the wire has been passed. The dilators are then passed over the guide wire which exits through the sheath. This gives a straight path to the ureteral orifice and produces less urethral trauma. After dilation is complete, the ureterorenoscope is then also passed through the cystoscope sheath adjacent to the guide wire (Fig. 9-6). This allows repeated rapid access to the ureteral orifice (which is necessary when removing a large amount of calculous debris) with minimal urethral damage. The cystoscope sheath is only useful in accessing the lower half of the ureter and cannot be used to reach the upper ureter because the ureterorenoscope hub abuts the cystoscope sheath, the tip of which is at the ureteral orifice. At this point, the distance between the ureteral orifice and the external urethral meatus is fixed by the length of the cystoscope sheath, and no additional compression can be applied to allow further passage of the ureterorenoscope.

In patients with stone fragments in the intramural ureter or the juxtavesical ureter, it may not be possible to dilate the ureter in a coaxial fashion as described above, usually because a guide wire cannot be passed high enough to keep it from being dislodged repeatedly during manipulation. In these instances, the ureteral orifice and tunnel may be directly dilated cystoscopically with unguided conical metal dilators, although the risk of injury to the ureter is somewhat increased with this method;[13] a dilation balloon can be used or, occasionally, the ureterorenoscope can be passed directly. If the obstructing fragments are in the intramural ureter, a ureteral meatotomy as outlined above may actually facilitate direct passage of the ureterorenoscope. In these instances, the intramural and distal ureter have already been dilated by the advancing stone fragments and the elevated pressures associated with the obstruction.

Once access has been gained, repeated passes with the ureteroscope may be required to remove the fragments. This may be accomplished with direct-vision extraction using wire baskets or, when necessary, in combination with direct-vision ultrasonic lithotripsy. An attempt should be made to clean the ureter of as much debris as possible to prevent reaccumulation of fragments and recurrent obstruction. In one case at the Hospital of the University of Pennsylvania, the ureter was rendered stone free, but subsequent passage of residual renal fragments produced an almost identical ureteral steinstrasse. Of interest, a significant lead fragment was not found at the time of ureteroscopy, and there was no ureteral narrowing or intrinsic pathology evident at the site of obstruction to account for the holdup of these small fragments. If large fragments are encountered endoscopically in the

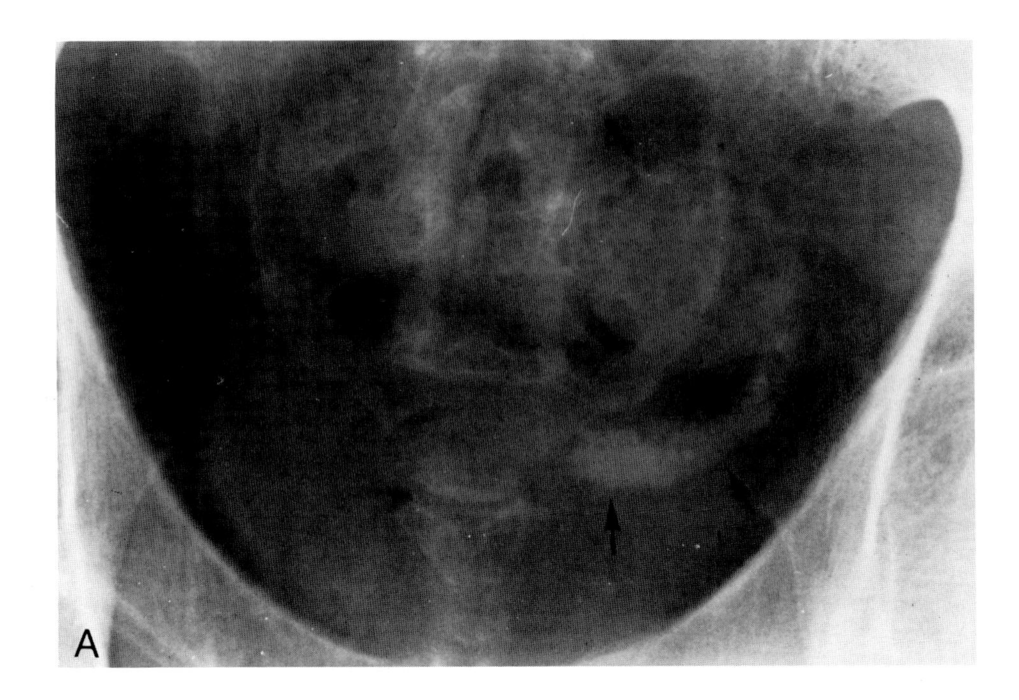


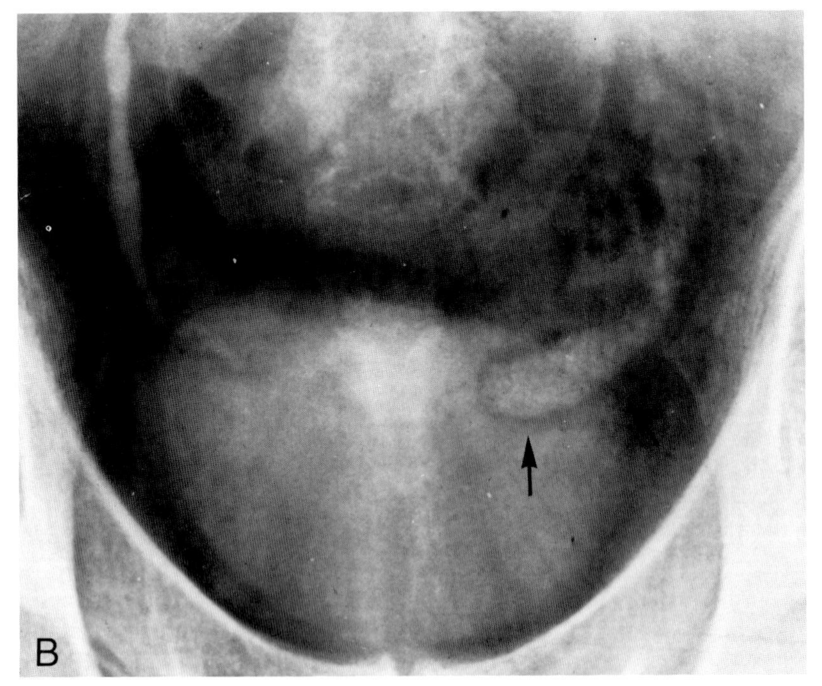


Fig. 9-5. Plain radiograph **(A)** and contrast-filled bladder **(B)** demonstrate pseudoureterocele configuration of steinstrasse (arrows). Ureteral meatotomy and ureteroscopy relieved obstruction.

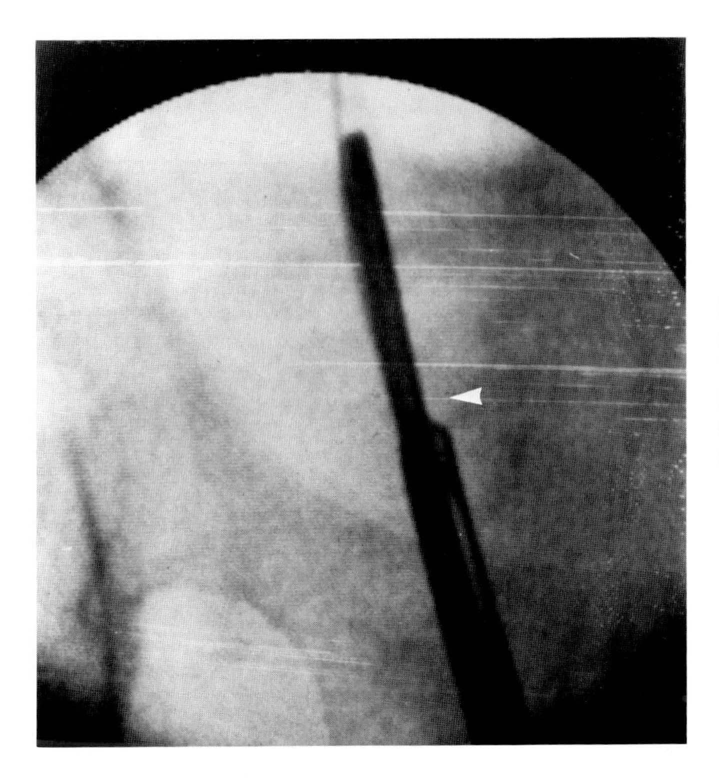

Fig. 9-6. Image taken off fluoroscopic monitor during ureteroscopy. The ureteroscope (arrowhead) is passed through a cystoscope sheath adjacent to the safety wire to traverse the urethra.

proximal ureter, consideration may be given to retrograde displacement into the renal collecting system for subsequent retreatment with ESWL.

Following ureteroscopy, one or two retrograde catheters are left in place. A single 5 F or 6 F catheter is used when the dilation process and extraction have been rapid with minimal ureteral trauma. Two 5 F catheters are left in place if a ureteral meatotomy has been performed, or if multiple passes of the scope have produced significant mucosal edema or injury. When used in this fashion, the first catheter is inserted through the ureteroscope under direct vision. The second is placed over the safety guide wire under fluoroscopic control. The retrograde catheters remain for 36 to 48 hours after the procedure.

## Surgery after ESWL

Finally, consideration must be given to open surgery as a secondary procedure following ESWL. The success of ESWL as a primary procedure and of the secondary adjunctive endourologic procedures discussed above have made surgery a rarity; however, it is still sometimes required as a tertiary approach.

Certainly the frequency of surgery after ESWL depends on two factors—the treating urologist's familiarity with complex endourologic techniques and the tolerance of the patient for yet another closed procedure. At New York Hospital–Cornell Medical Center, in a review of 44 secondary procedures after 518 treatments, three ureterolithotomies were performed for ESWL "failure" by urologists not involved with a shock wave unit. At surgery, two of the three stones were found to be disintegrated, but fragments had not moved. It is possible that these cases could have been handled endoscopically.[2] Patients who have had ureteral stents, two or three ESWL treatments, and prolonged hospitalization may prefer definitive surgery rather than another series of options—ureteroscopy, and then possibly repeat ESWL or percutaneous extraction, and then perhaps irrigation for chemolysis and/or flexible nephroscopy for retained fragments.[2]

## CONCLUSION

Worldwide experience with ESWL is broadening at an astounding rate as most calculi which would be treated surgically are being managed with this new technology. The sound of the shock waves is not nearly as rewarding to the operator as the crisp ping of a surgically removed calculus as it drops into the scrub nurse's stainless steel basin. The major redeeming feature of ESWL, however, is that it works and, apparently, with minimal complications.

However, ESWL is not a panacea. Much of the available information regarding complications and secondary management has been passed by word of mouth, with few formal publications found to date. Hopefully, as more information becomes available it will become clear which patients are better served primarily by percutaneous, ureteroscopic, or surgical techniques, alone or in combination with ESWL, to decrease the overall number of procedures per patient.

It should be emphasized that before the introduction of ESWL, oral chemolysis and direct irrigation of stones had been developed and used with varying degrees of success. ESWL often makes these methods more successful by increasing the surface area of the stone and breaking up resistant coatings. The secondary measures employed in the management of obstruction are based on the endourologic techniques developed or refined in the recent past. These antegrade and retrograde techniques are necessary after ESWL often enough that operators using an extracorporeal shock wave lithotripter must have access to all of them and be proficient in their use. As larger, complex stones are treated noninvasively, more secondary procedures will be required. New equipment and techniques will almost certainly evolve to decrease the number of complications or to aid in their management.

## REFERENCES

1. Report of American Urological Association ad hoc committee to study the safety and clinical efficacy of current technology of percutaneous lithotripsy and noninvasive lithotripsy. May 16, 1985
2. Riehle RA, Fair WR, Vaughan ED: Extracorporeal shock-wave lithotripsy for upper urinary tract calculi: One year's experience at a single center. JAMA 255:2043, 1986
3. Chaussy C (ed): Extracorporeal Shock Wave Lithotripsy. Karger, Basel, 1982
4. Freed SZ: The alternating use of an alkalizing salt and acetazolamide in the management of cystine and uric acid stones. J Urol 113:96, 1975
5. Uhlir K: The peroral dissolution of renal calculi. J Urol 104:239, 1970
6. Crawhall JC, Scowen EF, Watts RW: Effect of penicillamine in the treatment of cystinosis. Br Med J 1:588, 1963
7. Schmeller NT, Kersting H, Schuller J, Chaussy C, Schmiedt E: Combination of chemolysis and shock wave lithotripsy in the treatment of cystine renal calculi. J Urol 131:434, 1984
8. Sheldon CA, Smith AD: Chemolysis of calculi. Urol Clin N Am 9:121, 1982
9. Fernstrom I, Johansson B: Percutaneous pyelolithotomy—a new extraction technique. Scand J Urol Nephrol 10:257, 1976
10. Segura JW: Endourology. J Urol 132:1079, 1984
11. Banner MB, Pollack HM: Percutaneous extraction of renal and ureteral calculi. Radiology 144:753, 1982
12. Spataro RF, McLachlan MSF, Davis RF et al: Percutaneous antegrade extrusion of ureteral stones. Radiology 139:725, 1981
13. Huffman JL, Clayman RV: Endoscopic visualization of the supravesical urinary tract: Transurethral ureteropyeloscopy and percutaneous nephroscopy. Semin Urol 3:60, 1985
14. Eisenberger F, Fuchs G, Miller K et al: extracorporeal shock-wave lithotripsy (ESWL) and endourology: an ideal combination for the treatment of kidney stones. World J Urol 3:41, 1985
15. Kahnoski RJ, Lingeman JE, Coury TA et al: Combined percutaneous and extracorporeal shock-wave lithotripsy for staghorn calculi: an alternative to anatrophic nephrolithotomy. J Urol 135:679, 1986
16. Schulze H, Hertle L, Graff J et al: Combined treatment of branched calculi by percutaneous nephrolithotomy and extracorporeal shock-wave lithotripsy. J Urol 135:1138, 1986
17. Chaussy CG, Fuchs GJ: World experience with extracorporeal shock-wave lithotripsy for treatment of urinary stones: An assessment of its role after 5 years of clinical use. Endourology 1:7, 1986

# 10

# Treatment of Ureteral Stones

Gerhard J. Fuchs
Christian G. Chaussy
Robert A. Riehle, Jr.

Until recently, symptomatic upper ureteral stones were treated surgically or endoscopically with percutaneous retrograde manipulation or ureteroscopic extraction. Lithotripsy via these endoscopes required direct contact of probe with stone. At this time, extracorporeal shock wave lithotripsy (ESWL) has been shown effective as a noncontact, noninvasive means to disintegrate upper urinary tract calculi of all chemical compositions. Because endoscopic manipulation of these calculi before ESWL improves both the rate of effective disintegration and the stone-free rate, currently over 90 percent of patients with symptomatic ureteral stones above the pelvic brim can be treated using ESWL.

## HISTORY OF SHOCK WAVE LITHOTRIPSY OF URETERAL STONES

After shock wave lithotripsy was shown to be effective for solitary renal stones, attempts were made to use it for ureteral stones.[1–3] Because of technical limitations, ESWL was used only for stones situated above the iliac crest (Table 10-1). Stones below the iliac crest were not primarily amenable to ESWL, as most of the shock wave energy is absorbed by the bony pelvis. Furthermore, both x-ray localization and assessment of the degree of disintegration are difficult for stones overlying bony structures. Also, the current design of the patient support does not allow the patient to be positioned appropriately to localize stones in the prevesical ureter. However, if ESWL is used in conjunction with minimally invasive endourologic techniques, some of these stones prove indeed to be amenable to shock wave treatment.

Of the ureteral stones initially accepted for ESWL, many did not respond well to the shock wave energy. The initial Munich results of 1980 were not encouraging.

The first two stones treated this way did not disintegrate sufficiently, and surgical removal was performed.[1] The cause of failure was found to be that only the outer shell of the stone was completely disintegrated, whereas the core of the stone was incompletely broken. Figure 10-1A shows such a stone which was removed surgically after a failed ESWL treatment.

At this point, it was theorized that the degree of stone impaction and a missing fluid interface between the stone and the ureteral wall were the reasons for insufficient disintegration. These concepts are illustrated by the following case study.

**Table 10-1. History of ESWL of Ureteral Stones**

| | |
|---|---|
| Feb 1980 | First clinical application of ESWL for kidney stones (Munich, West Germany) |
| Aug 1980 | First attempts at using ESWL for treatment of ureteral stones |
| 1981–1985 | In situ ESWL treatment of ureteral stones. Selection criteria:<br>Stone is above iliac crest<br>Stone does not completely block ureter<br>Stone history <6 weeks in ureter<br>Primary success rate, of 60% (Chaussy, 1982) |
| Mar 1985 | Combined treatment for all ureteral stones (UCLA, Cornell); stone manipulation + ESWL with success rate >95% |
| Sept 1985 | Differeniated approach depending on radiographic features of IVP films (UCLA); in situ ESWL/combined approach >95% |

## CASE 1

In this patient, a large stone was situated at the same site in the upper ureter for more than 2 months, and in-situ ESWL treatment failed (Fig. 10-2). After a first ESWL session with 1,800 shock waves at an average of 22 kV, the stone only partly disintegrated, and the patient passed a small amount of debris (Fig. 10-2C). In a second session, 2 days after the first, the remaining stone was dislodged with a ureteral catheter and pushed back into the renal collecting system (Fig. 10-2D). ESWL was performed with 1,200 shock waves at an average of 20 kV, and resulted in complete disintegration of all residual stone parts in the kidney and ureter (Fig. 10-2E). The patient became stone free within 2 weeks (Fig. 10-2F).

Based on this initial experience, it was postulated that prerequisites for successful ESWL treatment were the following: (1) absence of complete blockage of the ureter, and (2) no ureteral narrowing at the stone site. These criteria usually correlated with a proven stone history of less than 6 weeks at the same location so that the selection for in-situ ESWL treatment of ureteral stones was made on the basis of stone history (Table 10-1).[1–3]

Owing to the tremendous demand by patients for ESWL and the resulting long backlog, the initial German centers, Munich and later Stuttgart, had waiting times of 2 and 1.5 years respectively. Therefore, only a few patients (selected according to the above criteria) could be accepted for ESWL of ureteral stones. Most of these patients had initially been accepted for ESWL of renal stones, and the stones dropped into the ureter during their wait.[1–7]

Ureteral stones treated with ESWL under these conditions revealed a primary success rate of approximately 60 percent. In conjunction with post-ESWL auxiliary procedures, the success rate was finally in the range of 80 to 85 percent (Table 10-2).[2–7]

Auxiliary procedures used were mainly retrograde repositioning of the primarily unsuccessfully treated stone into the renal collecting system with consecutive repeat ESWL, ureteroscopic stone extraction or disintegration, and if all other methods failed, open surgery (Table 10-2).[2,3,6,7]

In a recent study the Mainz group experimentally proved the hypothesis of the necessity of a fluid interface between the stone and the ureteral wall for successful in-situ ESWL treatment of ureteral stones.[8]

## The Treatment of Ureteral Stones

At present, there are basically two techniques for treating ureteral stones: in-situ disintegration and the "push-and-bang" method. Both approaches have advantages and shortcomings. For the time being, the final criteria for a staged approach are not yet defined. The combined approach of ureteral manipulation and ESWL

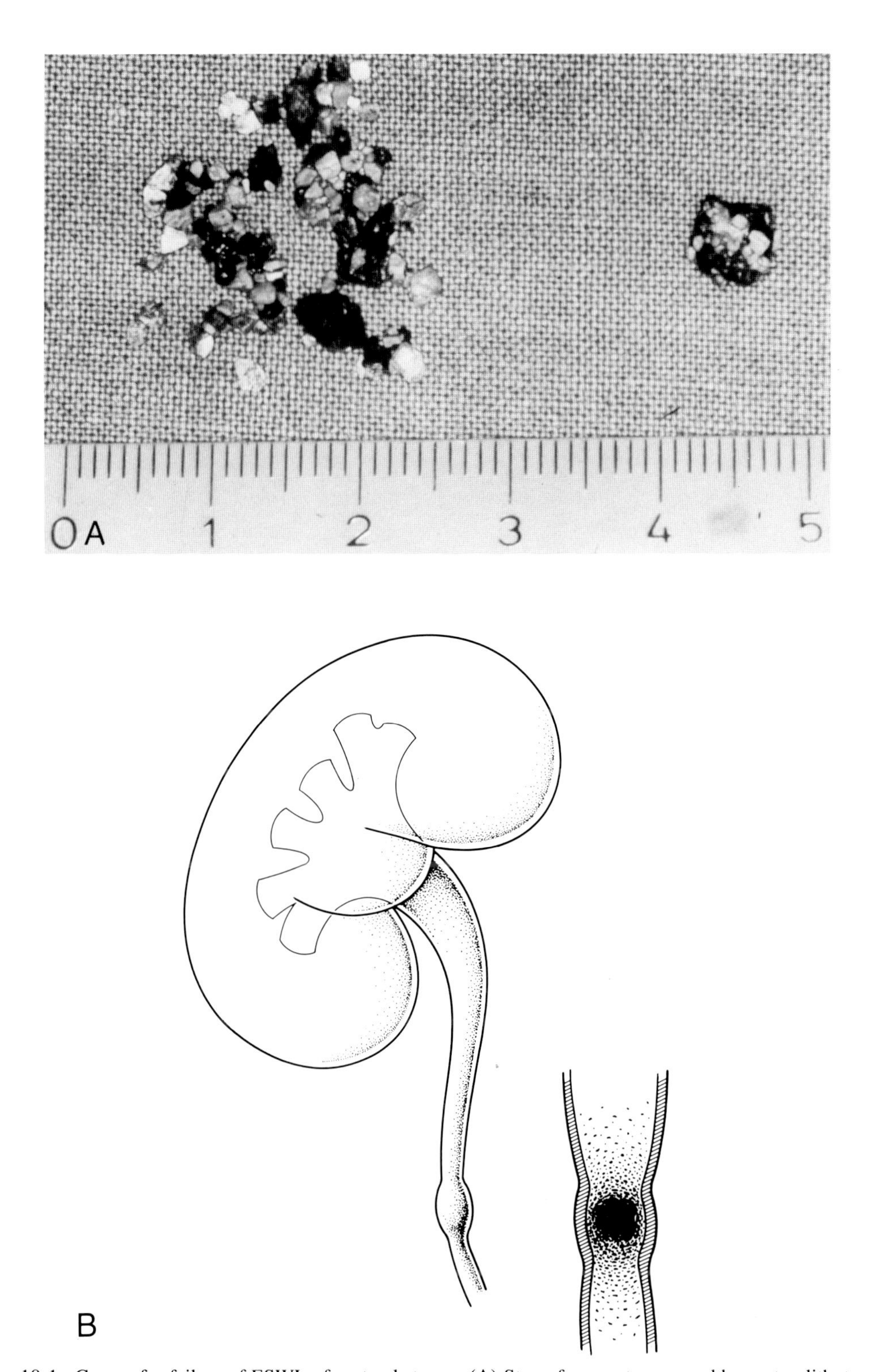


Fig. 10-1. Causes for failure of ESWL of ureteral stones. **(A)** Stone fragments removed by ureterolithotomy after failed ESWL. To the left are the stone fragments from the outer shell, which show various degrees of disintegration. To the right is the intact stone core. **(B)** Schematic drawing. At surgery the entire stone was found to be held in an edematous stone bed which prevented dispersal and passage down the ureter.

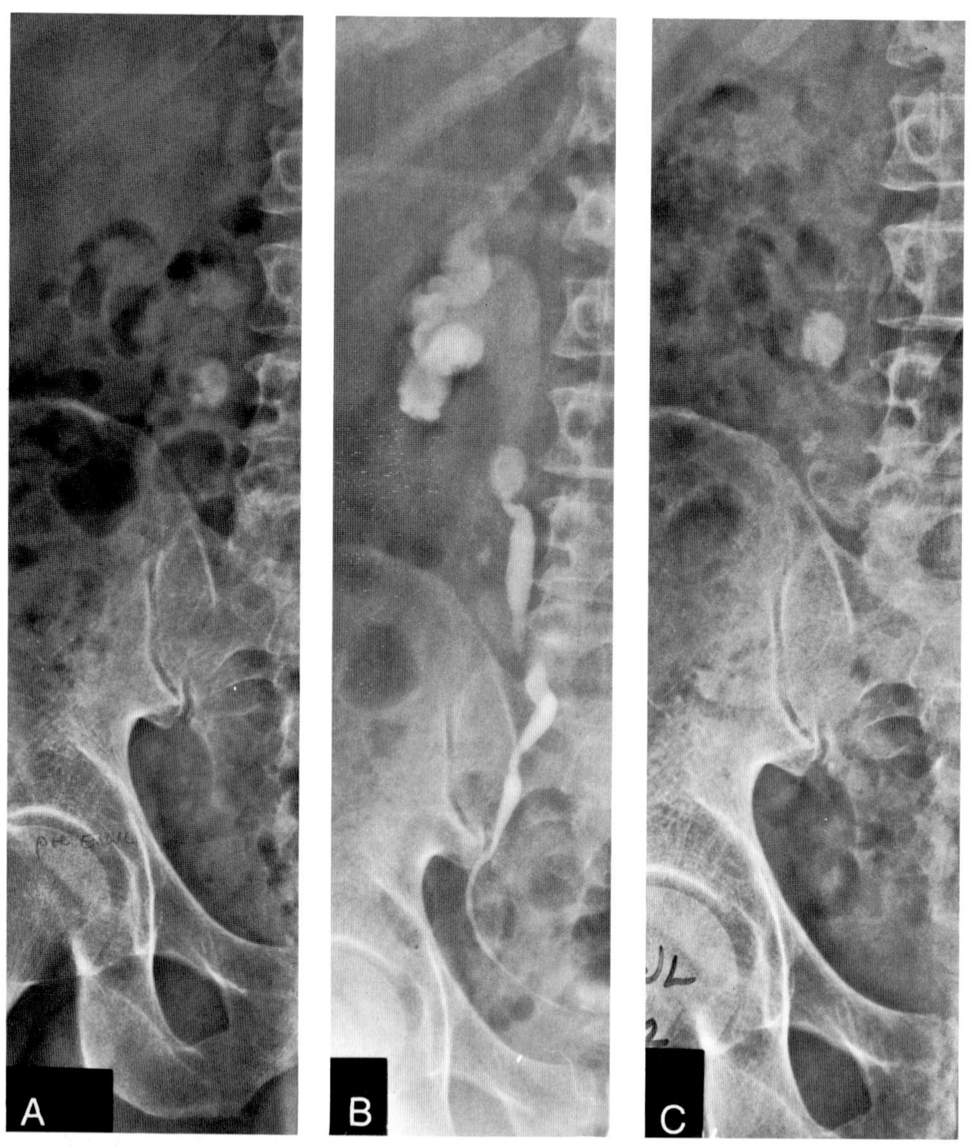


Fig. 10-2. Management of ureteral stone after failed ESWL treatment in-situ. (**A**) Pre-ESWL KUB film shows a large right ureteral stone at the level of L4. (**B**) The IVP shows considerable narrowing at the stone site and gross hydronephrosis proximal to the stone. (**C**) Post-ESWL KUB film shows the stone almost unchanged (*Figure continues*).

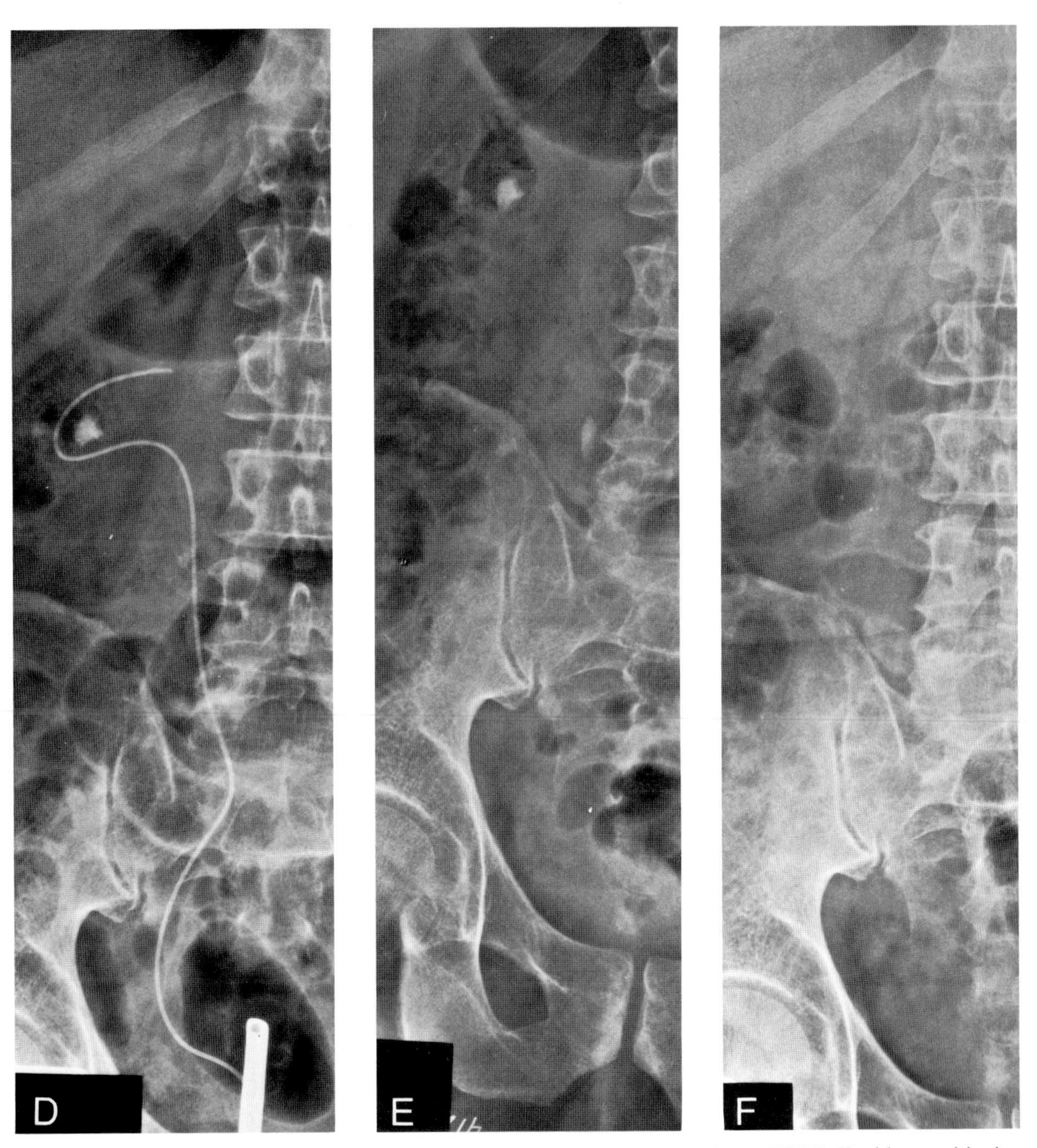


Fig. 10-2 (*continued*). **(D)** Stone manipulation using a 4 F ureteral stent prior to ESWL II with repositioning of the stone into the renal collecting system. **(E)** Situation after second ESWL showing disintegrated stone particles in the lower calices and in the ureter at the level of L4. **(F)** At the 2 week follow-up, the patient was found to be completely stone free.

**Table 10-2. Modalities Used for Successful Treatment of Ureteral Stones**

| Treatment Modality | Percent of Patients Treated by | |
|---|---|---|
| | In Situ ESWL (Stones above Iliac Crest) | Manipulation + ESWL (All Ureteral Stones) |
| Noninvasive | | |
| ESWL alone | 60% | 0 |
| Minimally invasive | | |
| Stent manipulation (+ repeat ESWL) | 25% | 91.4% |
| Ureteroscopy + ESWL | 2% | 4.0% |
| Ureteroscopic lithotripsy | 7% | 4.6% |
| Invasive | | |
| Surgery | 6% | 0 |
| Average number of ESWL sessions | | |
| One | 60% | 95% |
| Two | 33% | 4% |
| Three | 7% | 1% |
| Average number of shock waves | 2480 | 1720 |
| Average hospital stay | 2.6 days | 1.4 days |

seems to be most reliable for most ureteral stones. This approach applies to all ureteral stones, irrespective of their location. It also gives reliable results with minimal invasiveness.[9–11,12]

Based on our experience with more than 800 ureteral stones (Munich, C.G.C., 1980 to 1984; Stuttgart, G.J.F., 1983 to 1985; New York, R.A.R., 1984 to present, UCLA 1985 to present), the following guidelines for the treatment of ureteral stones can be drawn.

## In-Situ Treatment

Review of radiographic films of patients who had ureteral stones treated in situ with ESWL between 1980 and 1985 disclosed certain characteristic findings distinguishing successfully from unsuccessfully treated patients. From this review, it appears that neither stone size nor stone position (mid- or upper ureter) is a major determinant for successful disintegration. The single most important variable appears to be the presence of an expansion space around the stone for particle dispersion (Table 10-3).

**Table 10-3. Criteria for Successful in-Situ ESWL Treatment of Ureteral Stones**

| Criteria |
|---|
| Stone is above iliac crest |
| No urinary tract infection |
| Existing expansion chamber |
| IVP does not show narrow stone bed |
| Contrast passes around the stone |

Approximately 10 to 15 percent of patients with symptomatic ureteral stones qualify for ESWL treatment in situ. General selection criteria are location of the stone above the iliac crest and the absence of urinary tract infection. Further criteria are the presence of a natural expansion chamber, detected by the absence of ureteral narrowing on the intravenous pyelogram (IVP) film and the presence of a fluid interface between the stone and the ureteral wall. The latter is demonstrated on the IVP film by the contrast dye passing alongside the stone (Fig. 10-3).

### CASE 2

A 68-year-old woman presented with intermittent flank pain. Because she had passed multiple small stones in the past year, an exact stone history could not be established. Therefore, new kidney–ureter–bladder (KUB) and IVP films were obtained (Fig. 10-4A,B). Because of the absence of ureteral narrowing and a normal contrast delineation of the ureter, the patient qualified for ESWL treatment in situ.

The ureteral and renal stones were all successfully treated in one session of ESWL. The three ureteral stones received 1,800 shock waves at an average of 21 kV (range 24 to 19 kV), and the two upper and mid-caliceal stones in the kidney received 1,000 shock waves at an average of 19 kV.

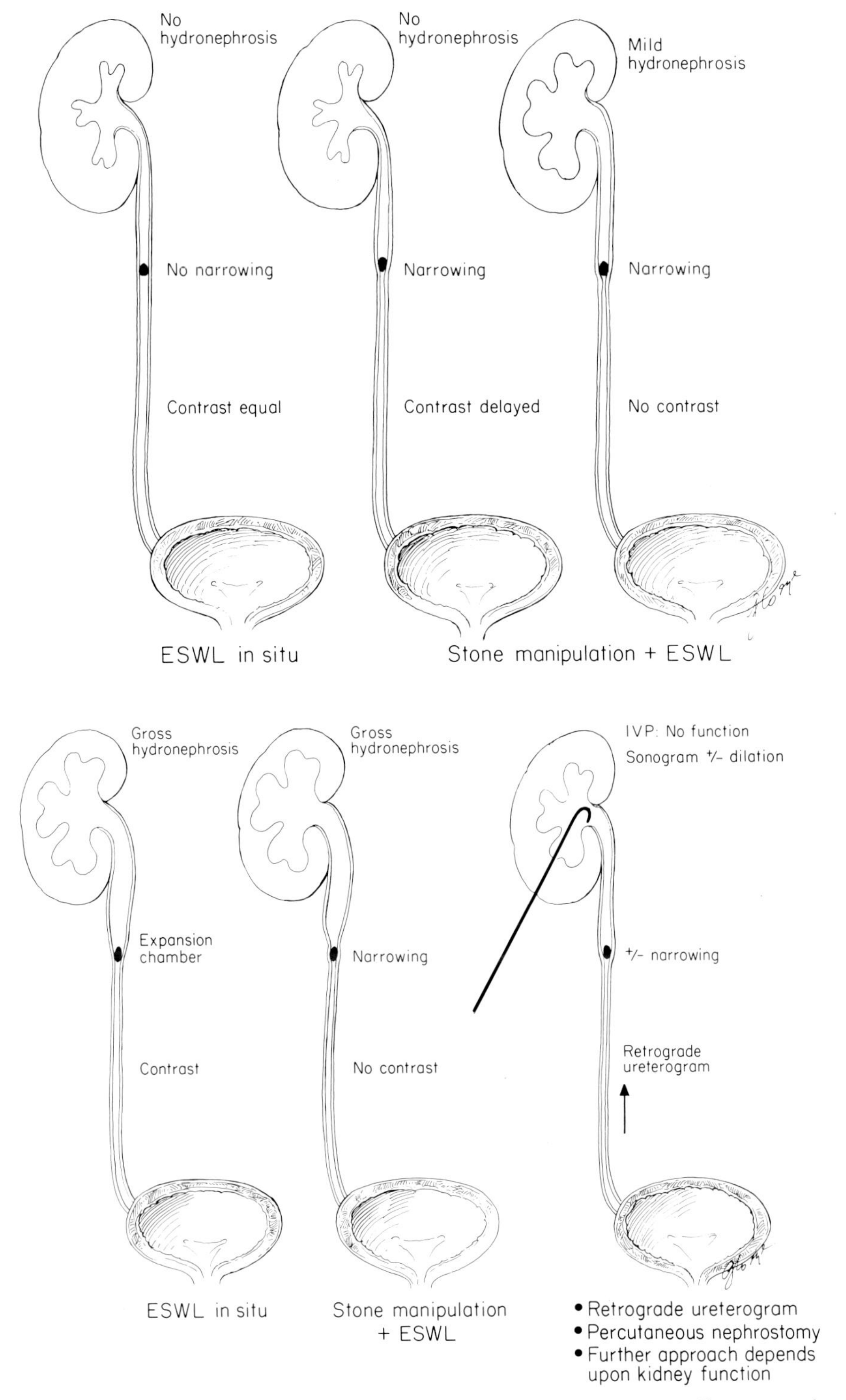


Fig. 10-3. Criteria for the differential indications in ureteral stone management (for stones above iliac crest).

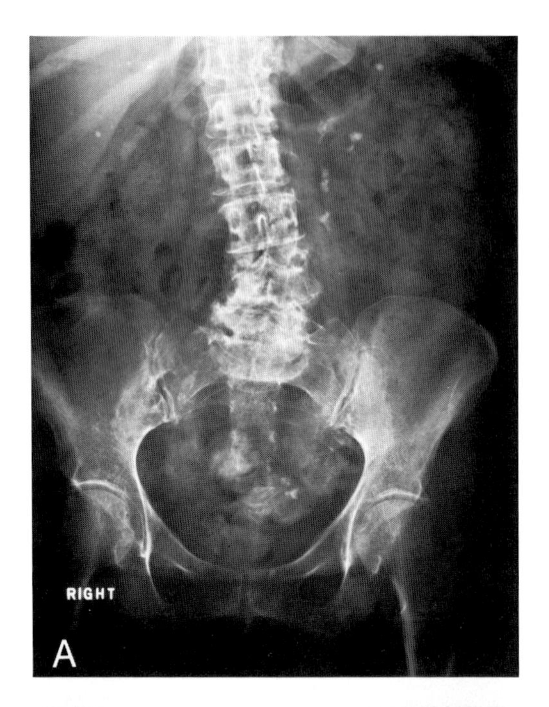


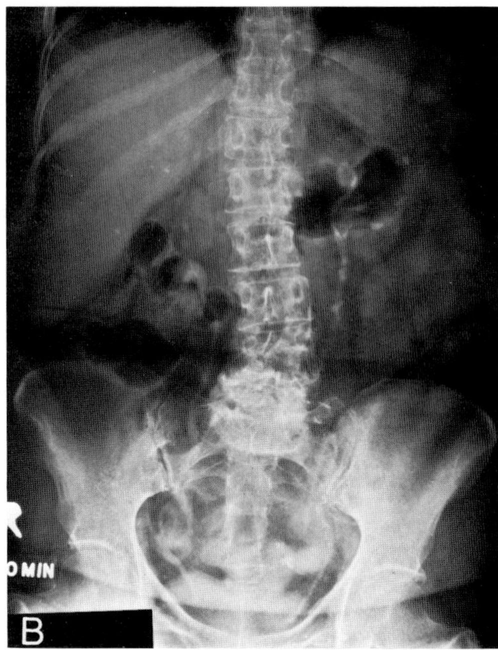


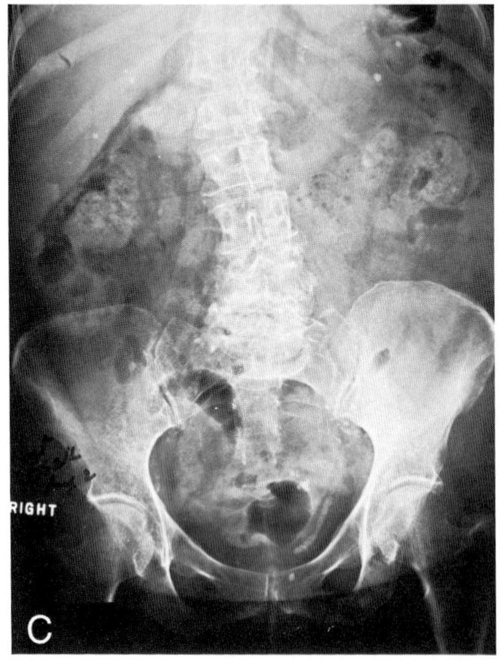


Fig. 10-4. Current indication for in-situ ESWL treatment of ureteral stones. **(A)** KUB film showing multiple ureteral stones in the upper third of the left ureter. **(B)** The IVP film demonstrates a nondilated renal collecting system, and contrast dye passes easily alongside the ureteral stones. In addition to the ureteral stones, several renal stones are also visible. **(C)** KUB film taken on the first day after treatment demonstrates absence of any stone material in the kidney and a long steinstrasse in the left terminal ureter.

### CASE 3

Case 3 (Fig. 10-5) illustrates another indication for in-situ treatment of a ureteral stone. A 38-year-old man presented with intermittent flank pain and microhematuria which had lasted for 5 weeks. An IVP revealed an upper ureteral stone and mild hydronephrosis proximal to the stone. There was no indication of ureteral narrowing at the stone site and the excretion of contrast was not delayed (Fig. 10-5B). These findings suggested the presence of a sufficient expansion chamber, so in-situ ESWL was performed using 1,600 shock waves at an average of 22.5 kV (range 24 to 21).

## Technical Considerations with in-Situ ESWL

When treating ureteral stones above the iliac crest with ESWL in situ, with or without placement of a ureteral stent(s) alongside the stone, various technical factors have to be considered in regard to patient positioning and shock wave application.

Both visualizing the stone and positioning it in the F2 focus are more difficult with ureteral than with renal stones. In obese patients, they may be impossible. Ureteral stones are hard to visualize for ESWL because the oblique angle of the x-ray tubes superposes the medial course of the upper ureter onto the spine or transverse processes on one or both image intensifiers. The stone is hard to see against a bony background. To overcome this bothersome situation, the patient must be rotated in the support system until the stone image lies free of bone (Fig. 10-6B). Furthermore, to localize stones at the level of L4-5, the patient must be moved cranially on the support so that the posterior flank does not contact the ellipsoid. To achieve the proper degree of patient rotation and to maintain this position throughout the procedure, Styrofoam pads are positioned under the contralateral shoulder and thorax, and also under the contralateral thigh. The thigh support on the treated side is always lowered and the leg support accordingly is somewhat raised. On the contralateral side, the thigh support is somewhat raised. These measures are taken to better support the patient's rotation, and thus facilitate stone localization.

Even when the criteria for in-situ treatment with ESWL are met, localization of relatively small and/or less radiodense stones may be particularly difficult. In these situations a "dry" simulation to verify visualization and localization of the stone should be performed 1 day before treatment. If the stone cannot be visualized, placement of a ureteral stent is helpful as it allows for detection of the course of the ureter, and if required, for the administration of contrast.

Occasionally, (particularly in the obese patient) it is not possible to localize the in-situ ureteral stone in the focal area. The stone must then be repositioned into the renal collecting system if it is to be treated by ESWL.

ESWL of in-situ ureteral stones requires higher energy and more shock waves to disintegrate the stone satisfactorily. For in-situ ESWL, treatment should begin at 22 to 24 kV and the energy should be decreased when disintegration of the stone is seen on the monitors. Frequently, the typical pattern of disintegration is not detected during the procedure. Thus, the duration of the treatment and the energy levels employed have to be arbitrarily chosen, based on previous experience.

## Combined Treatment by Stone Manipulation and ESWL

Ureteral stones located below the iliac crest are not always eligible for in-situ ESWL treatment because much of the energy is absorbed by bone and because the stone cannot always be positioned at F2 (Table 10-4). Unless they can be repositioned into an area more amenable to ESWL—the upper ureter or the renal collecting system—success rates are lower, and these stones are usually approached via ureteroscopy.

One of the causes for failure of in-situ ESWL is the lack of a sufficient expansion chamber around stones wedged against the ureteral wall. Experience has shown that in-situ disintegration of ureteral stones is dependent upon the degree

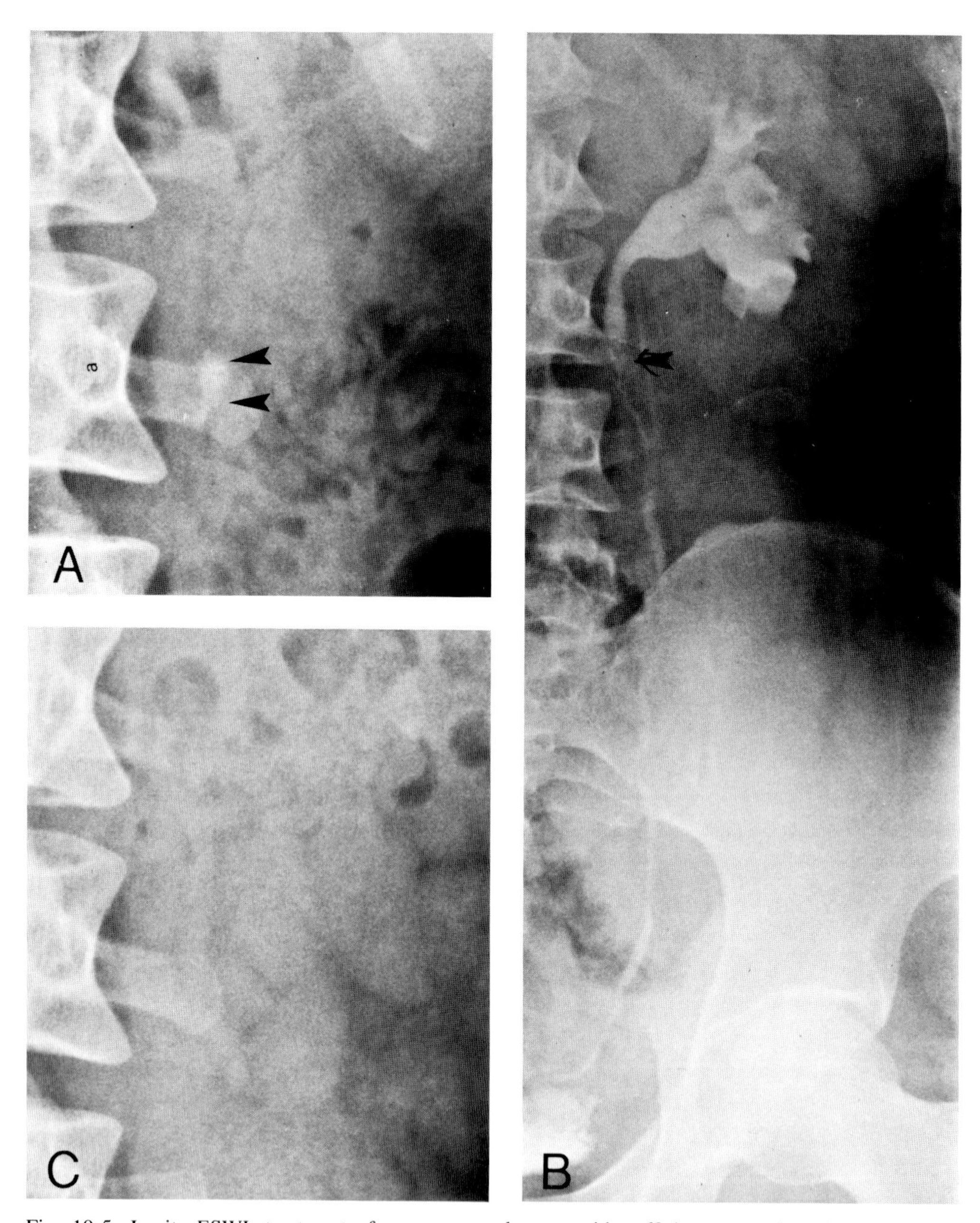


Fig. 10-5. In-situ ESWL treatment of upper ureteral stone with sufficient expansion chamber. **(A)** KUB film showing left ureteral stone at the level of the lateral process of L3. **(B)** The IVP film reveals that contrast flows around the stone. No narrowing is seen at the stone site, and proximal to the stone mild hydronephrosis is demonstrated. **(C)** KUB film immediately after ESWL confirms complete stone disintegration.

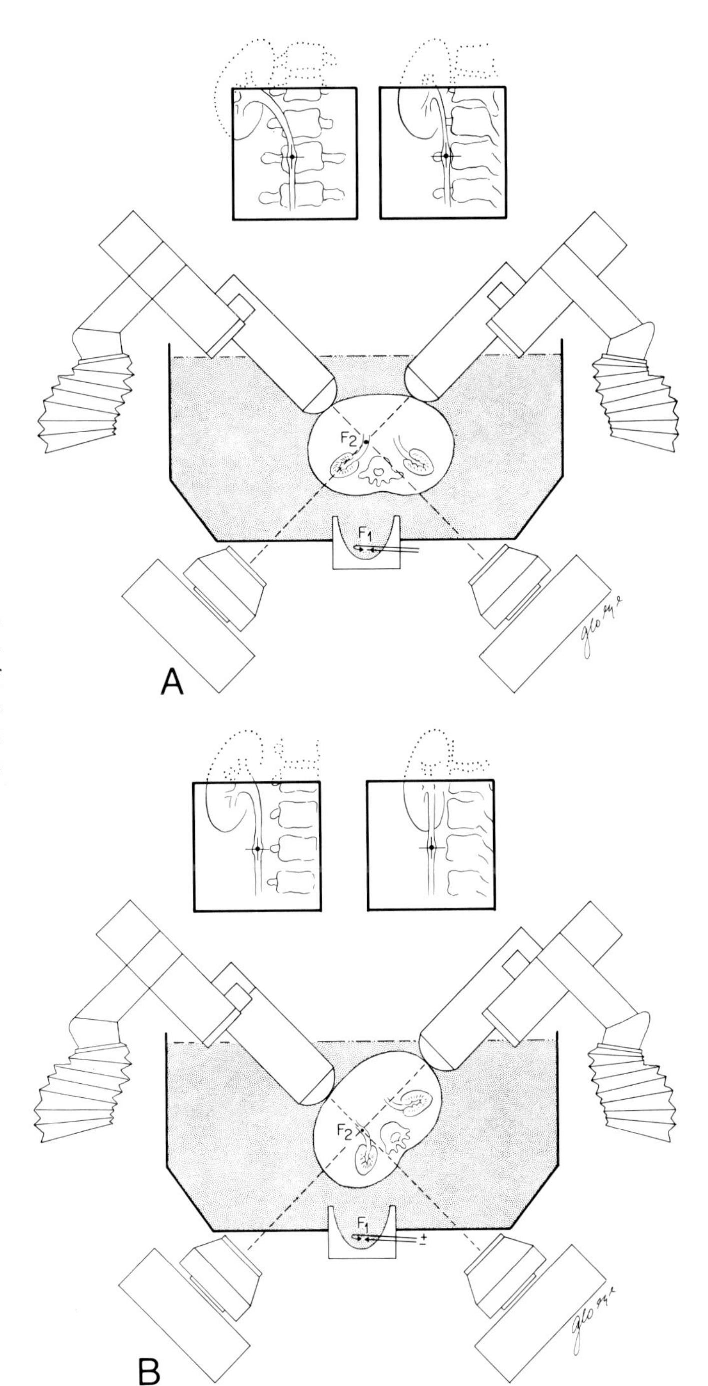


Fig. 10-6. Influence of patient positioning on x-ray localization and imaging. **(A)** Schematic drawing of monitor images of a ureteral stone when patient is supine. **(B)** Rotation of the patient towards the treated side facilitates stone visualization.

**Table 10-4. Tactical and Technical Considerations in Ureteral Stone Management**

| Stone Manipulation and Immediate Treatment | Two-Stage Procedure, If Acute Infection |
|---|---|
| 1. Stones above iliac crest without expansion chamber | 1. Percutaneous nephrostomy and antibiotics |
| 2. In situ ESWL possible when stone in position <4–6 weeks | 2. ESWL |
| 3. Stones below iliac crest—stone manipulation required | |

of physical contact between the calculus and the ureteral wall. When the ureteral wall is in close proximity to the calculus, it usually means that the stone is impacted in an edematous, spastic ureter, and insufficient disintegration will occur because no expansion space is available for stone fragmentation.[7,8] This finding is regardless of the amount of energy delivered.[8] Therefore, it was believed that successful disintegration of ureteral stones by ESWL depends on the presence of an expansion chamber.[10,11]

For all ureteral stones these requirements can be met by two means: (1) relocating the stone into the renal pelvis (natural expansion chamber)—the ideal method (Fig. 10-7)—or (2) relocating the stone into a ureteral segment amenable to ESWL treatment, i.e., the dilated mid or upper ureter, or creation of an artificial expansion chamber greater than 6 F in size (Fig. 10-8).

## Technique for Ureteral Stone Manipulation

Stones that are impacted or associated with high-grade obstruction usually cannot be relocated by simple advancement of a ureteral catheter.[6,7,10,11,13,14] The catheter abuts against the stone trapped in an edematous ureteral bed, and cannot bypass it into the upper ureter. If a guide wire cannot be negotiated past the stone, more forceful measures should not be undertaken, as they may easily perforate the ureter.

However, maneuvers using ureteral lubrication have been found to facilitate stone repositioning.[10,11,15] Ureteral lubrication is achieved using a 2 percent water-soluble xylocaine jelly mixed with normal saline or distilled water. The proportion of jelly to solvent is such that the solution flows easily when injected via a 6 F ureteral catheter. In general, the mixture is 20 to 30 percent jelly to 70 to 80 percent solvent. It takes a few minutes of thorough stirring to obtain good dispersion of solute into solvent. A 6 F or 8 F Braasch catheter is advanced to the level of the stone. Five to ten ml of solution are forcefully injected, and the stone is "probed" by advancing and retracting the scope and the catheter as a unit for a few millimeters. In most instances, this maneuver is sufficient to displace the stone and relocate it into the renal collecting system.[10,11,15]

### CASE 4

Case 4 (Fig. 10-7A–D) involved a 72-year-old man with a relatively large stone in the right terminal ureter and two stones in the right lower calyx (Fig. 10-7A). Passage of an 8 F Braasch and a 6 F spiral-tip catheter failed to reposition the stone above the iliac crest (Fig. 10-7B). Ureteroscopic stone extraction was then attempted, but the stone was too large to be engaged in the basket. Subsequently, the stone was successfully relocated into the renal pelvis from where it fell into the lower calyx (Fig. 10-7C,D). ESWL treatment with 1,500 shock waves was performed and the patient became stone free after 3 weeks.

In cases where the stone cannot be repositioned into the renal pelvis, a different approach is followed. When the stone is located or repositioned in the mid or upper ureter and cannot be advanced into the renal pelvis, the aim is to create an artificial expansion chamber of at least 6 F in size for successful ESWL disintegration.

To create this artificial expansion chamber, an array of catheter combinations is employed—a 6 F whistle-tip and 6 F spiral-tip or two 4 F spiral-tip catheters can be employed. When pos-

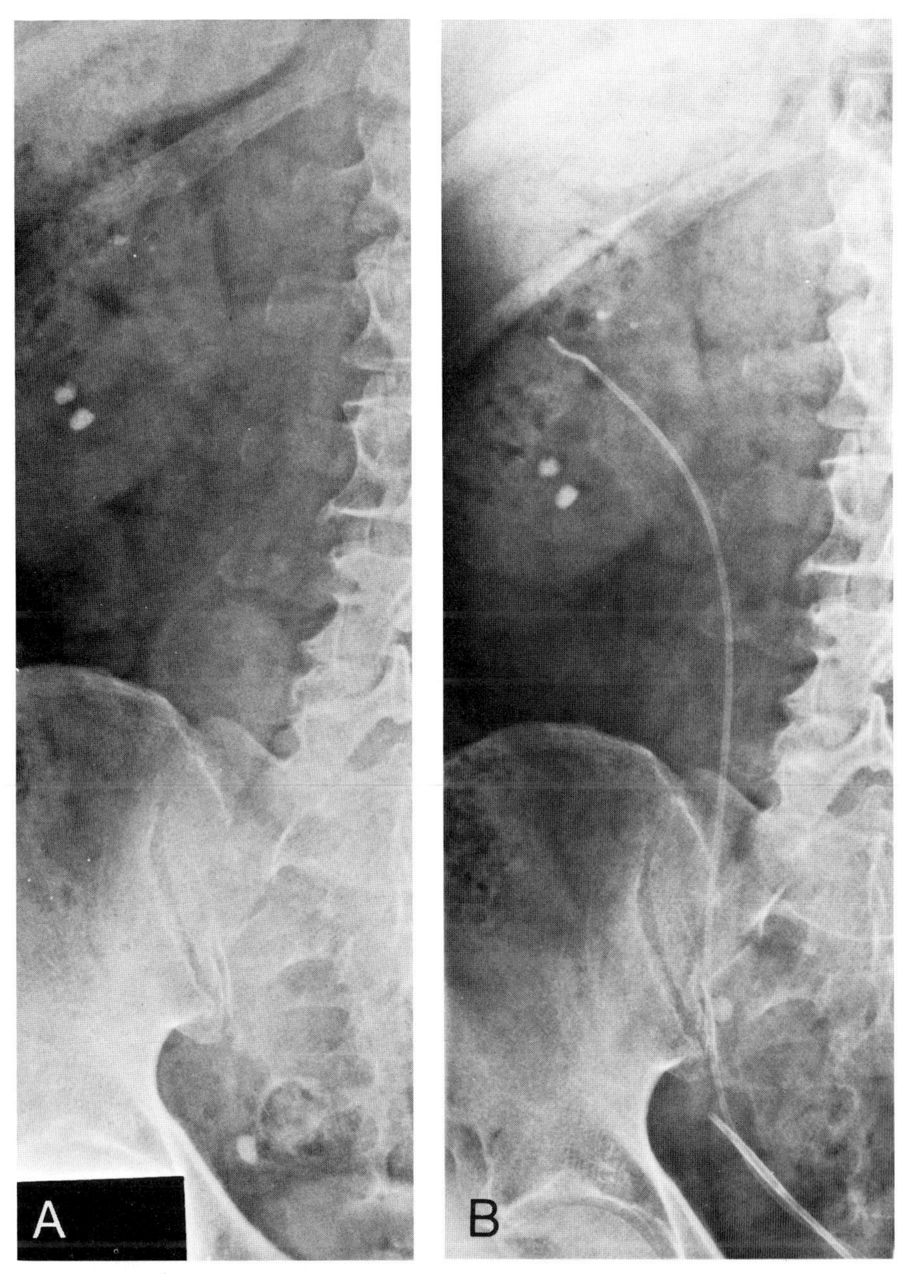


Fig. 10-7. Management of ureteral stone in the terminal ureter by repositioning into renal collecting system. **(A)** KUB film showing two lower caliceal stones in the right kidney and a relatively large stone located in the right terminal ureter. **(B)** A 5 F spiral-tip catheter is advanced into the kidney and the lower ureteral stone is moved upwards to the level of the pelvic brim. (*Figure continues.*)

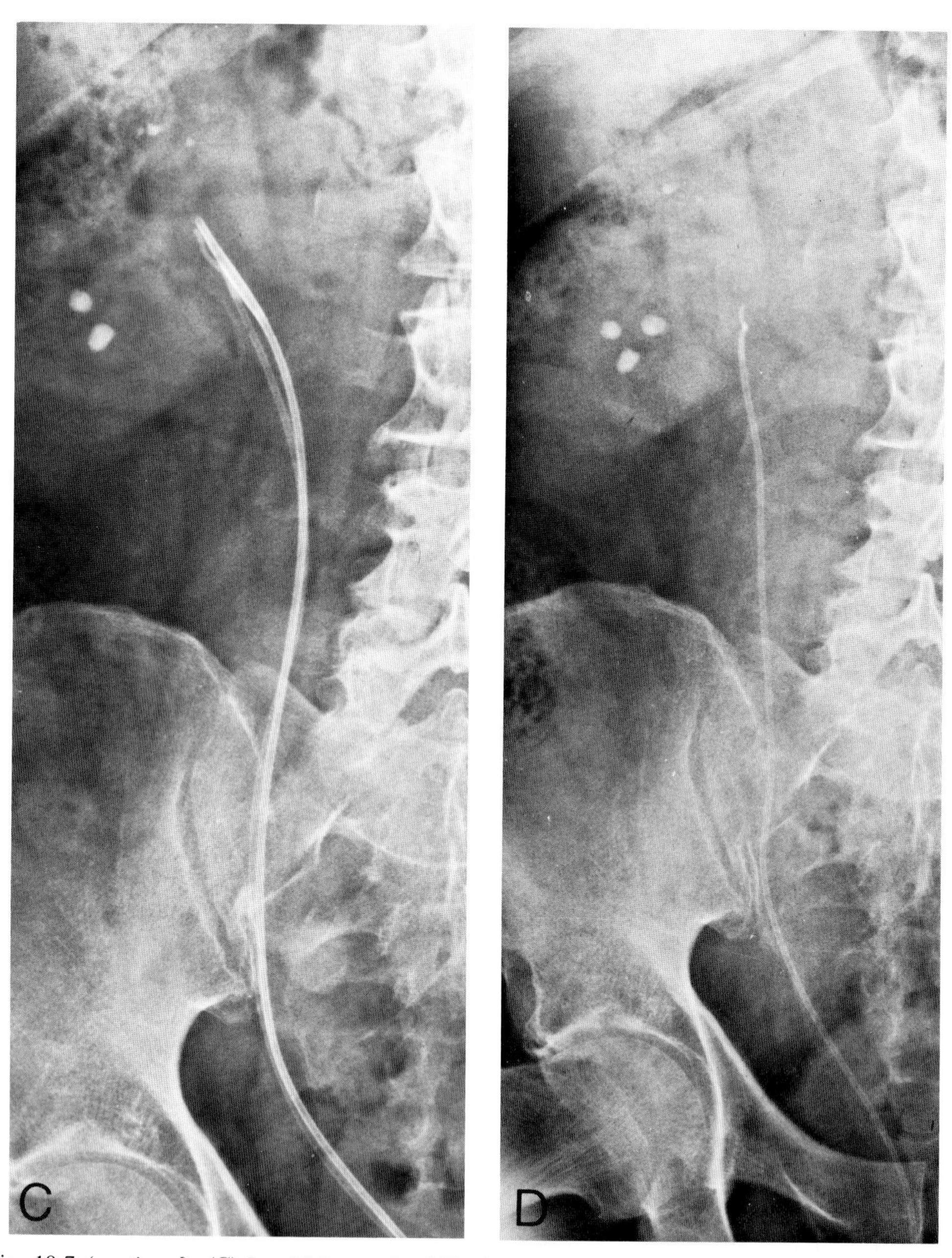


Fig. 10-7 (*continued*). **(C)** In addition to the 5 F spiral-tip a 6 F straight ureteral stent is advanced into the renal pelvis. **(D)** Using a ureteroscope, the stone is repositioned up into the renal collecting system and is now located in the lower calyx. (*Figure continues.*)

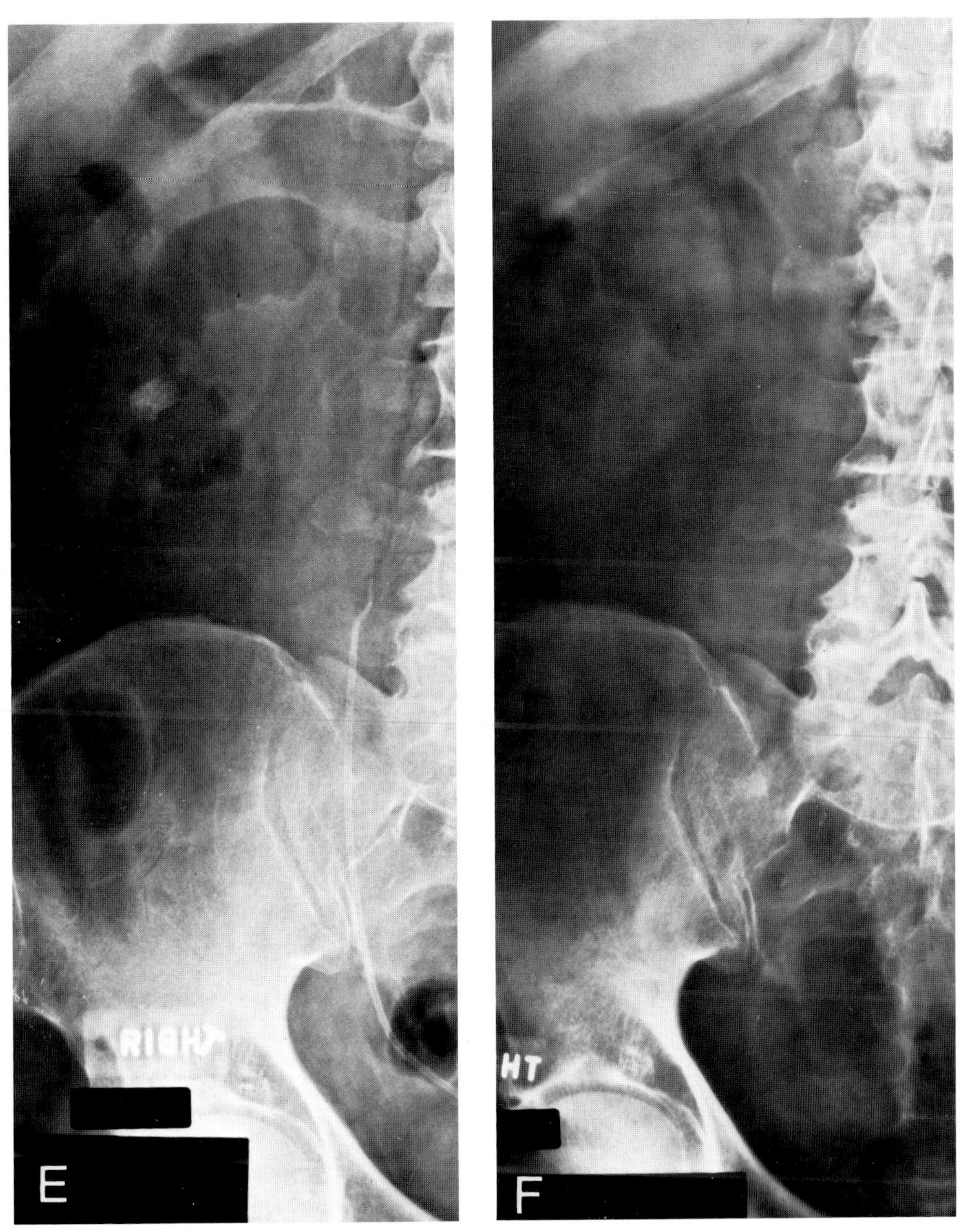


Fig. 10-7 (*continued*). **(E)** Immediately after ESWL the KUB confirms complete disintegration of all three stones. **(F)** Two weeks after the initial treatment no more stone material is detected on the follow-up KUB.

sible, a 6 F spiral catheter is advanced alongside the Braasch catheter to the level of the stone. Lubrication and stone probing is repeated several times. When the spiral-tip catheter passes the stone, it is advanced into the renal pelvis, and the Braasch catheter is left below the stone to prevent migration of the stone. The catheter(s) is then securely attached to a 14 F or 16 F Foley catheter left indwelling in the bladder.[10–12,15]

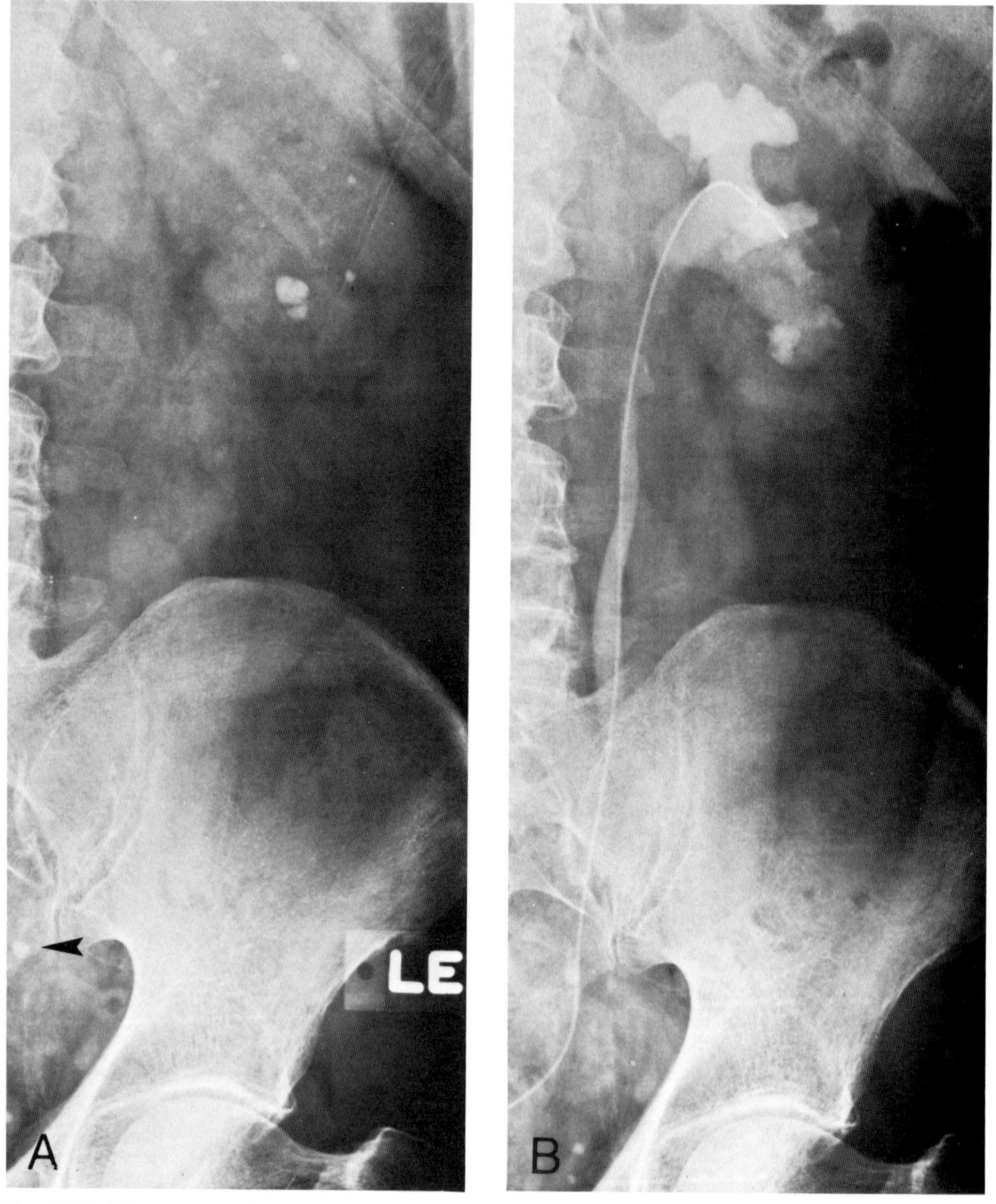


Fig. 10-8. Management of ureteral stone in the terminal ureter by repositioning into upper ureter and creating of an artifical expansion chamber. (*Figure continues*.)

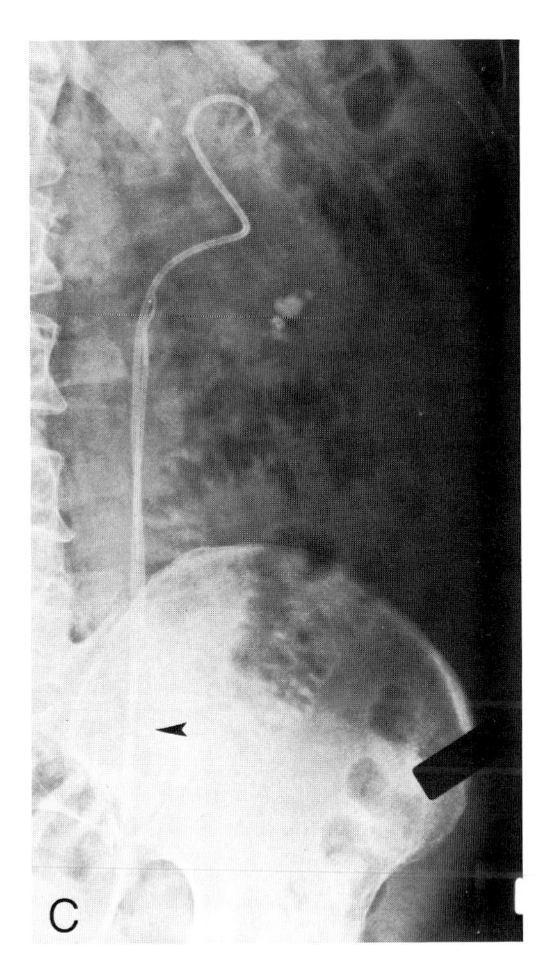


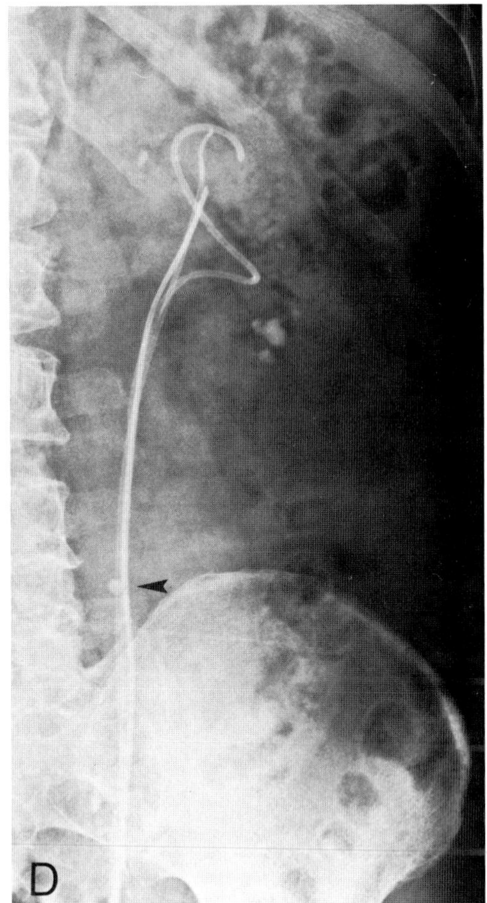


Fig. 10-8 (*continued*).

## CASE 5

This 62-year-old man had a stone located in the left terminal ureter and multiple stones in the upper, mid, and lower calices (Fig. 10-8A). Endoscopic manipulation of the ureteral stone was undertaken. At first it was only possible to negotiate a Benson guide wire alongside the stone and advance it up into the kidney. After thorough lubrication, a 6 F Braasch catheter was passed over the guide wire (Fig. 10-8B). Passage of a 6 F spiral-tip catheter failed to reposition the stone above the iliac crest. However, the stone could be dislodged and moved up to the middle of the sacroiliac joint (Fig. 10-8C). Finally, the stone was repositioned above the iliac crest using another 4 F ureteral stent (Fig. 10-8D). In this position, the stone could be localized with the lithotripter fluoroscopic localization system. The presence of an expansion chamber of 6 F size was sufficient to allow disintegration of the stone, which received 1,200 shock waves at an average of 21 kV.

Even unusually large stones located in the lower ureter can be relocated and successfully treated using this method.

## CASE 6

This patient's 25 × 16 mm stone was obviously too large for endoscopic extraction without ureteroscopic lithotripsy, and its location precluded primary ESWL treatment (Fig. 10-9A). Using a 6 F spiral-

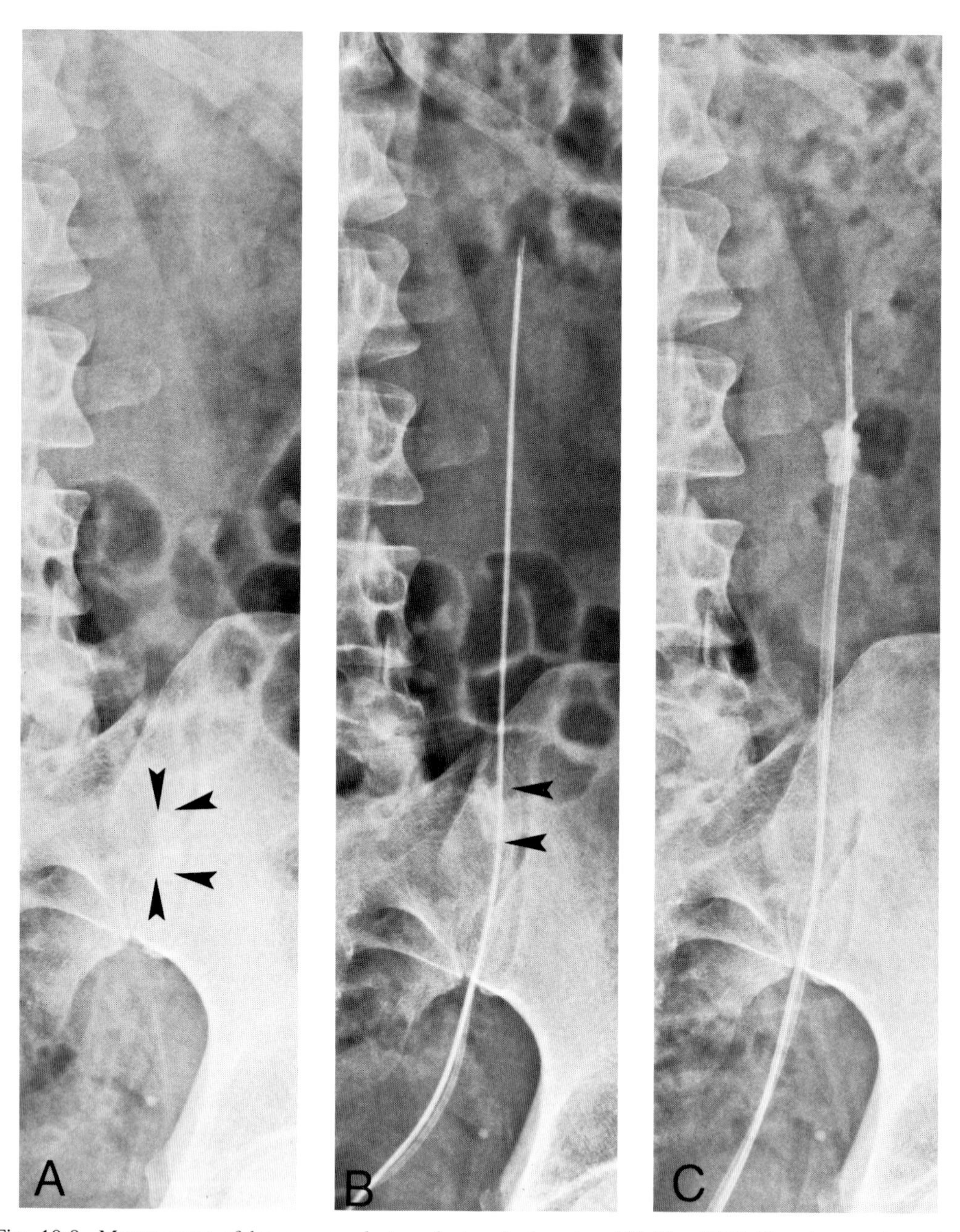


Fig. 10-9. Management of large ureteral stone in terminal ureter. **(A)** The KUB film shows a huge left ureteral stone located in the mid of the sacroiliac joint (arrows). **(B)** A 5 F straight ureteral stent is passed alongside the stone and a second catheter, a 6 F spiral tip, is used to dislodge the stone. **(C)** The stone has been moved up to the level of the lower margin of L3. (*Figure continues.*)

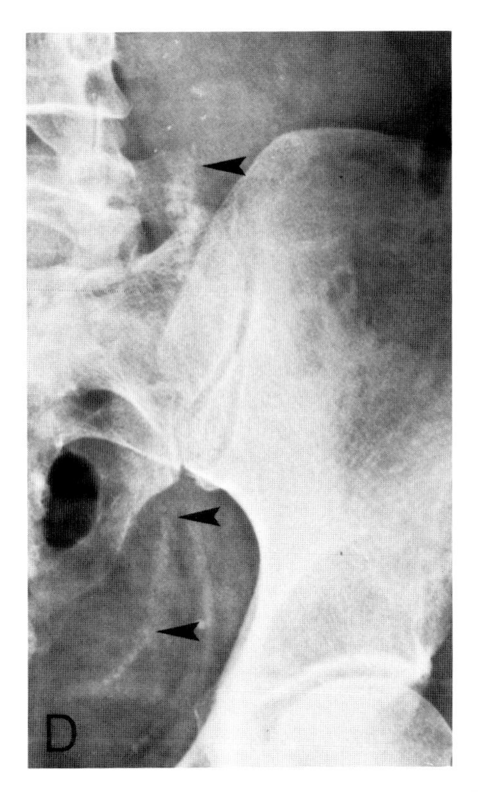


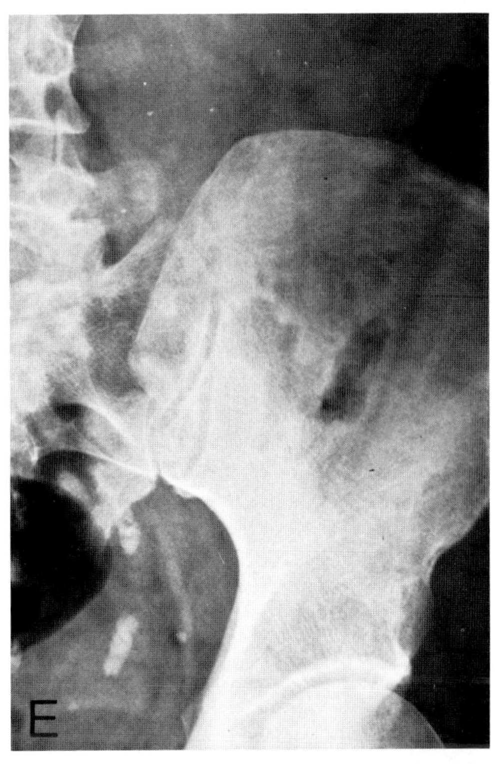


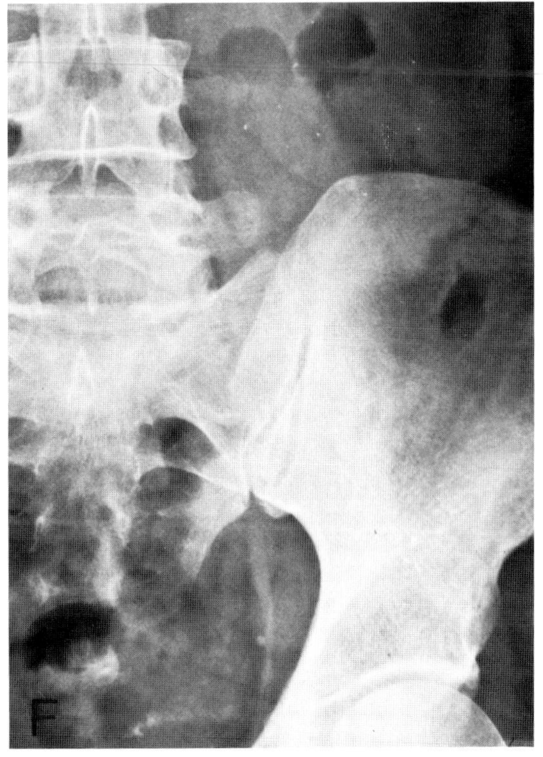


Fig. 10-9 (*continued*). **(D)** Immediately after ESWL and withdrawal of the stents, the KUB film shows complete disintegration of the stone; the particles are already beginning to pass. **(E)** On day 1 after ESWL two lower ureteral steinstrassen are seen on the KUB film. **(F)** At the 2-week follow-up, no more fragments are detected in the ureter.

tip and a 5 F whistle-tip catheter, the stone was repositioned in the mid-ureter. A 5 F whistle-tip catheter was passed alongside the stone, and a 6 F spiral catheter was placed below the stone (Fig. 10-9B). ESWL treatment was successful with 1,800 shock waves at 22 kV (Fig. 10-9C). One day after treatment, a steinstrasse was seen in the terminal ureter (Fig. 10-9D) and by the next day the patient was already stone free (Fig. 10-9E).

When the ureteral stone is completely impassable, placement of a 6 F spiral-tip catheter may prove helpful. The tip of the catheter inserts itself between the stone and the ureteral wall, and the spiral part of the catheter cups itself under the stone, creating an expansion chamber.

### CASE 7

An IVP obtained on this 48-year-old woman showed an obstructing ureteral stone and hydronephrosis (Fig. 10-10A). This stone had a history of more than 8 months and all attempts to reposition the stone or to pass a stent alongside it had failed. It was only possible to advance two spiral tip catheters (4 F and 6 F) beneath the stone (Fig. 10-10B). This created an expansion chamber sufficient for successful stone disintegration, which was achieved with 1,900 shock waves at an average of 22 kV.

When the stone cannot be manipulated at all with ureteral stents, ureteroscopic extraction or repositioning is employed as the next step. If this cannot be achieved, ultrasonic disintegration through the ureteroscope is performed.

## Infected Ureteral Stones

Urinary tract infection and obstructive pyelonephritis are factors complicating ureteral stone management. Therefore, these situations deserve special consideration.

Experience has shown that patients with kidney stones and existing urinary tract infection can safely be treated with ESWL if appropriate antibiotic treatment is initiated at least 24 hours before ESWL, and decompressive stents or nephrostomy tubes are used as indicated. For ureteral stones associated with obstructive pyelonephritis, a percutaneous nephrostomy tube is placed first to relieve the obstruction, and antibiotic treatment is instituted to minimize the risk of septic complications. Once the symptoms have subsided, ESWL is performed according to the aforementioned guidelines.

## Ureteral Stones in Patients with Supravesical Urinary Diversion

Among the more common complications of this type of supravesical urinary diversion are recurrent urinary tract infections and secondary urinary stone formation. Most stones found in this patient group are of the struvite variety (infection stones), which fragment at lower energies than other stones and disintegrate into smaller particles. Because of this response, more struvite ureteral stones qualify for in-situ treatment than do stones of other chemical compositions.

In most instances of urinary diversion (ileal conduit, Kock pouch, Camey procedure, ureterosigmoidostomy) the retrograde route is not easily accessible for ureteral instrumentation. Therefore when required, manipulation of ureteral stones has to be performed via an antegrade approach.

These patients very often present with obstructive pyelonephritis requiring placement of a percutaneous nephrostomy tube for relief of obstruction. Once the acute symptoms have subsided, this route can be used for antegrade manipulation of the stone.

### CASE 8

A 50-year-old man with an ileal conduit for urinary diversion after cystectomy presented with an upper ureteral stone and pyelonephritis. First, a percutaneous nephrostomy was inserted, and antibiotic therapy was started. ESWL was performed 1 week later following successful manipulation of the stone into the renal pelvis in an antegrade fashion using a balloon catheter.

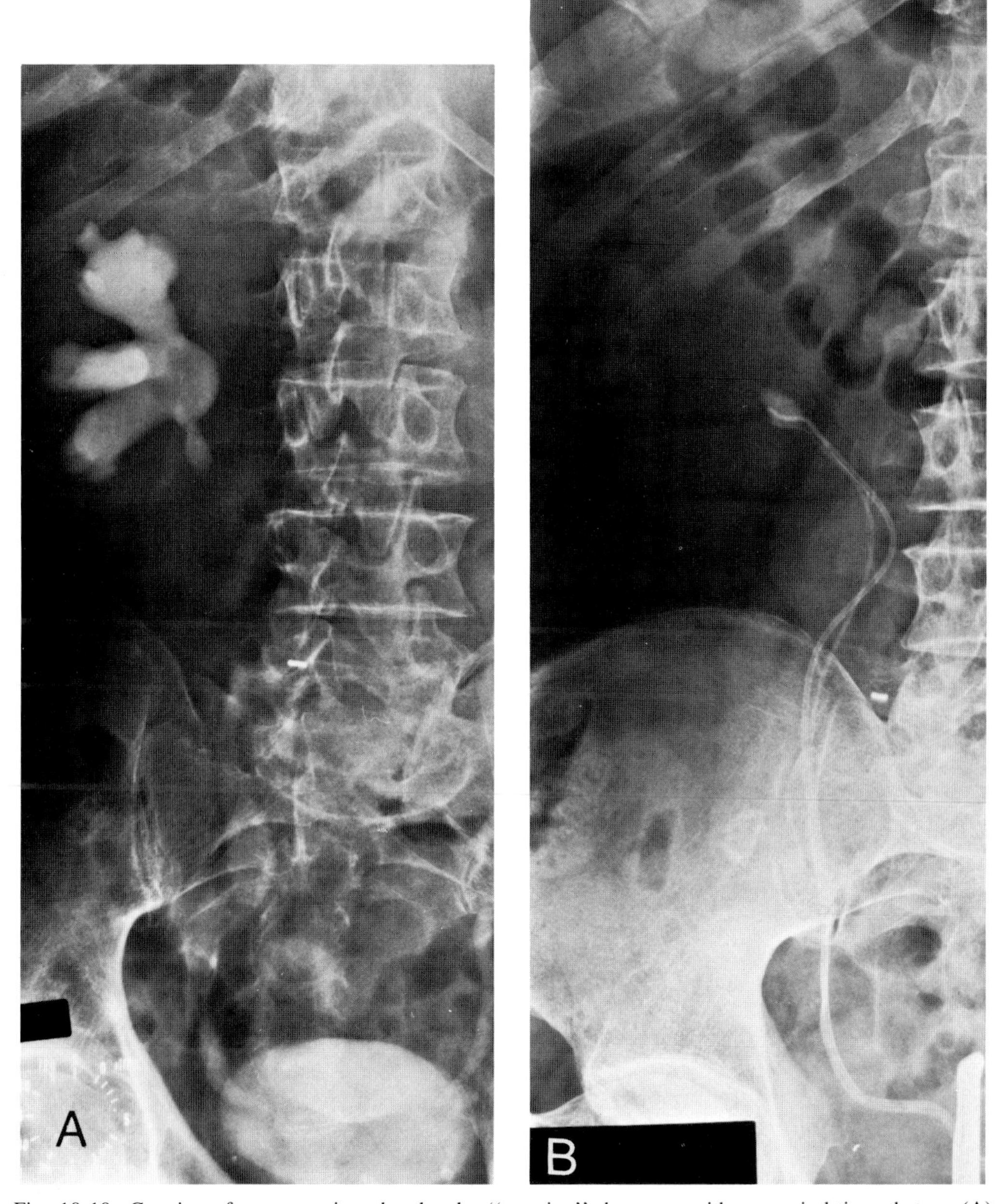


Fig. 10-10. Creation of an expansion chamber by "cupping" the stone with two spiral-tip catheters. **(A)** Upper ureteral stone, completely obstructing. **(B)** A 4 F and a 3.6 F spiral-tip catheter are placed around the stone.

**Table 10-5. Relative Invasiveness of Treatments for Ureteral Stones**

| Location of Stone | Treatment: Noninvasive | | | | | | Invasive |
|---|---|---|---|---|---|---|---|
| Above iliac crest | In-situ ESWL | → | Manipulation of stone back to renal pelvis, ESWL | → | Antegrade or retrograde ureteroscopic lithotripsy | → | Open surgery |
| Below iliac crest | Manipulation of stone up ureter | → | Retrograde ureteroscopic lithotripsy | | | | Open surgery |

## Complications of Ureteral Manipulation

Possible complications of ureteral manipulation include ureteral perforation, avulsion of the ureter secondary to ureteroscopic stone manipulation, and urinary tract infection secondary to unsterile manipulation. Damage to the ureter is a rare occurrence in the hands of the experienced endourologist. With liberal use of ureteral lubrication, the incidence of ureteral perforation is very low (0.5 percent). Late sequelae of ureteral manipulation may be ureteral stricture or refluxing of an orifice that has been dilated to accommodate a ureteroscope. At UCLA, with a mean follow-up of more than 2 years, none of the latter complications have been observed.

## RESULTS

The regimen described in this chapter results in a rate of successful treatment of 96.6 percent (Table 10-5). Patients who have residual stones can usually be rendered stone free by using the ureteroscope.[10,11] Open surgery has become a rarity, and is only used when all other methods have failed, or when it is indicated to correct anatomic alterations existing at the time of stone removal.[6,7,10,11] It can be anticipated that most patients (95 percent) treated by this combined approach will require only one session of ESWL for a successful outcome.[11]

In most cases, stone manipulation is performed in a retrograde fashion. The antegrade route is chosen when the distal ureter is not readily accessible because of supravesicular urinary diversion, or when a percutaneous access already exists and the stone is located in the upper two thirds of the ureter.[11] Whenever the stone(s) cannot be repositioned, an attempt is made to pass a ureteral catheter alongside the stone to create an artificial expansion chamber.[11,15]

At UCLA, the success rate for patients whose stones are repositioned into the renal collecting system does not differ from that for renal stones of similar size treated in situ (Table 10-6).[10]

**Table 10-6. Results of Pre-ESWL Stone Manipulation and Success Rate**

| Procedure | Percent of Patients | Average Stone Size (mm) | Success Rate (percent) | Percent of Patients Stone Free at 2 weeks | Percent of Patients Stone Free at 3 months |
|---|---|---|---|---|---|
| Stone manipulation + ESWL | | | | | |
| Ureteral stones repositioned into renal collecting system | 59 | 10.4 | 100 | 85 | 98 |
| Upper ureter + stent | 23 | 9.6 | 94.6 | 92 | 99 |
| Mid-ureter + stent | 7.5 | 10.2 | 91.2 | 89 | 99 |
| In situ treatment | | | | | |
| Upper ureter, no stent | 8.4 | 9.8 | 72 | 70 | 100[a] |
| Mid-ureter, no stent | 2.1 | 9.7 | 74 | 74 | 100[a] |

[a] Stone-free after auxiliary procedures.

**Table 10-7. Energy and Number of Shock Waves Administered in Relation to Stone Position at Treatment**

| Procedure | Successful Disintegration (percent) | Average No. of Shock Waves | Energy (kV) | |
|---|---|---|---|---|
| | | | Average | Range |
| Stone manipulation + ESWL | | | | |
| Ureteral stones repositioned in the RCS | 100 | 1,360 | 20.0 | 19–21 |
| Upper ureter + stent | 94.6 | 1,620 | 22.4 | 21–23 |
| Mid ureter + stent | 91.2 | 1,710 | 22.2 | 21–23 |
| In-situ treatment | | | | |
| Upper ureter, no stent | 72 | 1,900 | 23.4 | 22–24 |
| Mid-ureter, no stent | 74 | 1,880 | 23.0 | 22–24 |

The same amount of energy is needed to disintegrate them as for native stones in the kidney. This seems to indicate that an expansion space is the most important variable in the ESWL treatment of ureteral stones, and that the other factors previously suggested, such as mucoid lining, play only a minor role.[11]

When treated in the upper ureter, or in the mid-third of the ureter above the iliac crest, a success rate almost as high as for stones relo-

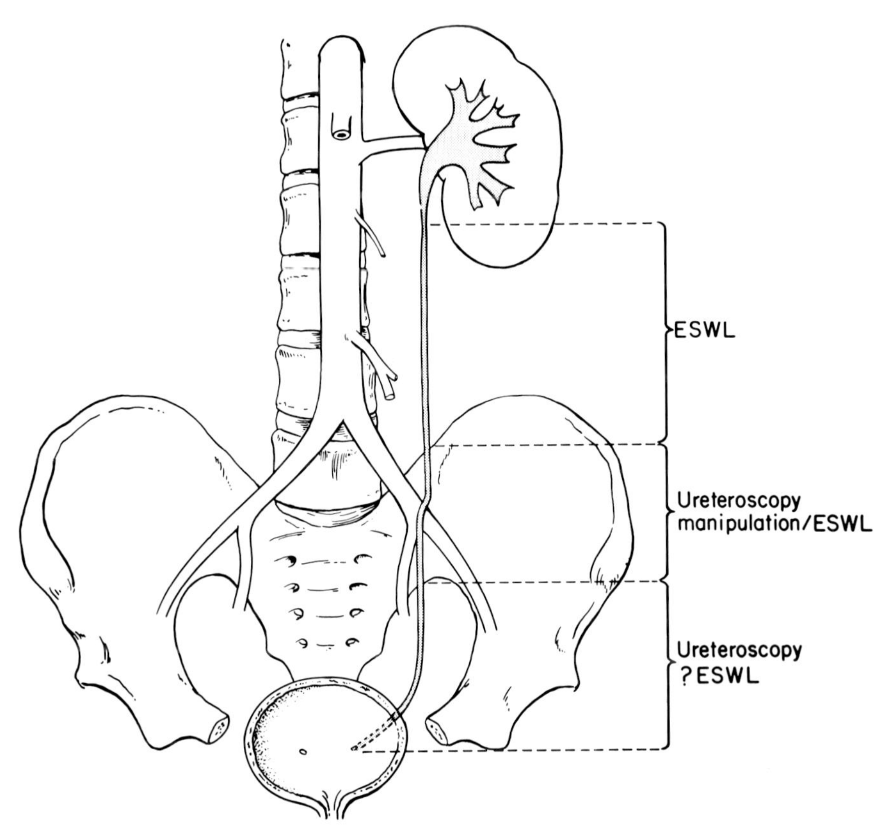


Fig. 10-11. Treatment of ureteral stones at New York Hospital–Cornell Medical Center.

cated into the renal collecting system can be achieved once a stent is passed about the stone—94.6 percent success and 91.2 percent success, respectively (Tables 10-2, 10-7).[11] Greater energies are required for successful disintegration when stones are treated in the ureter, even in the presence of an adequate expansion chamber. For stones in this location, 23 to 24 kV are routinely used.[11]

In the selected cases at UCLA in which ESWL was performed in situ, the success rates were 72 percent for upper and 74 percent for mid-ureteral stones. Substantially more shock waves were needed for disintegration than for repositioned stones. The success rate was much lower for totally obstructing stones with a history of more than 6 weeks that were treated in situ in the ureter. A single ESWL treatment is successful in only approximately 50 to 60 percent of such cases.[1–3,6,7]

At New York Hospital–Cornell Medical Center, similar results have been reported.[12,16] In a series of 100 symptomatic primary upper ureteral stones (9.8 mm average diameter), 72 percent of the patients underwent preprocedural cystoscopy for stone manipulation, and 39 percent (28/72) of these stones were impacted and could not be dislodged proximally to the kidney or upper ureter. The stone-free success rate at 3 months for dislodged stones (9.1 mm average) was 93 percent. Of 56 ureteral stones treated in situ (impacted or not manipulated), 47 patients (84%) were rendered stone free at 3 months; four stones (13.6 mm average) treated in situ failed to disintegrate.

Thus, the convincingly high success rates have made the combined approach the preferred procedure for stones located in the mid or upper ureter, or for any ureteral stone associated with ipsilateral upper collecting system stones.[11]

Lithotripsy of small renal stones prior to ureteral migration and of proximal ureteral stones early in their symptomatic course may significantly alter the incidence of distal ureteral calculi requiring hospitalization, cystoscopy, or ureteroscopy (Fig. 10-11).

## REFERENCES

1. Chaussy C, Schmiedt E, Jocham D et al: First clinical experience with extracorporeally induced destruction of kidney stones by shock waves. J Urol 417, 1981
2. Chaussy C, Schmiedt E: Shock wave treatment for stones in the upper urinary tract. Urol Clin N Am 10:743, 1983
3. Chaussy C, Schmiedt E, Jocham D et al: Extracorporeal shock wave lithotripsy (ESWL) for treatment of urolithiasis. Urology 5:59, 1984
4. Fuchs G, Miller K, Rassweiler J: Alternatives to open surgery for renal calculi: Percutaneous nephrolithotomy and extracorporeal shock wave lithotripsy. In Schilling W (ed): Klinische und Experimentelle Urologie. Zuckschwerdt, Munich, 1984
5. Fuchs G, Miller K, Rassweiler J, Eisenberger F: Extracorporeal shock wave lithotripsy: One year experience with the Dornier lithotripter. Eur Urol 11:145, 1985
6. Lupu AN, Fuchs GJ, Chaussy C: Treatment of ureteral calculi by extracorporeal shock wave lithotripsy. Urology (in press)
7. Miller K, Fuchs G, Rassweiler J, Eisenberger F: Treatment of ureteral stone disease: The role of ESWL and endourology. World J Surg 3:445, 1985
8. Mueller SC, Wilbert D, Alken P, Thueroff JW: Model for extracorporeal shock wave lithotripsy (ESWL) of ureteral stones. Abstract in Second World Congress on Percutaneous Renal Surgery, June 14–15, 1984
9. Chaussy C, Fuchs G: World experience with extracorporeal shock wave lithotripsy for the treatment of urinary stones: Assessment of its role after 5 years of clinic use. Endourology Newsletter 1:7, 1986
10. Chaussy C, Fuchs G: Extracorporeal shock wave lithotripsy (ESWL) for the treatment of urinary stones. In Gillenwater J (ed): Textbook on Adult and Pediatric Urology, Year Book, Chicago, 1986
11. Chaussy C, Schmiedt E, Jocham D et al: Extracorporeal shock wave lithotripsy, 2nd Ed. S Karger Verlag, Munich, 1986
12. Riehle R, Naslund E: Treatment of upper ureteral stones with extracorporeal shock wave lithotripsy. Surg Gynecol Obstet (in press)
13. Clayman RV, Castaneda-Zuniga WF: Tech-

niques in endourology: A guide to the percutaneous removal of renal and ureteral calculi. University of Minnesota, Minneapolis, MN, 1984
14. Kahn RI: Endourological treatment of ureteral calculi. J Urol 135:239, 1986
15. Lupu A, Fuchs G, Chaussy C: A new approach to ureteral stone manipulation for ESWL. Endourology Newsletter, 1:13, 1986
16. Riehle R: Extracorporeal shock wave lithotripsy of ureteral stones. Semin Urol 4:175, 1986

# 11

# ESWL of Distal Ureteral Calculi

Jens Rassweiler
Ferdinand Eisenberger

In the past, distal ureteral stones seemed to be unsuitable for extracorporeal shock wave lithotripsy (ESWL) because the iliac brim marked the caudal limit for shock wave exposure.[1–7] Therefore, the vast majority of pelvic stones—if not spontaneously passable—were treated endoscopically using a Zeiss loop, a basket, or retrograde ureteroscopy (Dormia, forceps, ultrasound lithotripsy, laser)[5,7–12] (Fig. 11-1). More recently, however, successful positioning, location, and disintegration of calculi distal to the sacral outlet have been reported using either the Dornier lithotripter[13,14] (personal communication, K. Miller and R. Hautmann, Ulm, F. R. G.) or the Siemens device.[15] These promising results may lead to symptomatic lower ureteral stone disease being treated regularly by the noninvasive ESWL technique.

## POSITIONING

For the treatment of distal ureteral calculi, the shock wave can be applied either via the gluteal region or perineally using the birth canal[13] (personal communication, K. Miller and R. Hautmann). In case of a gluteal shock wave exposure, the patient has to be positioned completely flat with a horizontally adjusted backrest and a very distal arrangement of the leg support leaving the gluteal region free. Additionally, the patient lies on two crossed belts to stabilize his position. Finally, if the stone is close to the sacral bone, the patient is rotated to the stone-bearing side as in the case of upper ureteral calculi (Fig. 11-2).

If the patient is not positioned flat, he tends to slide caudally, which may lead to two problems. First, the patient cannot be lowered sufficiently along the Z axis, and thus the lowering of the intensifiers may be impossible because of the elevated thighs. Second, the patient cannot be sufficiently moved up along the X axis due to the limitations of the stretcher's movement within the tub (X coordinate = −200 mm).

For perineal shock wave lithotripsy, the patient is placed in a sitting position on the stretcher, with crossed belts below. Thus the patient actually sits on the ellipsoid and the birth canal is used for shock wave exposure[13,14] (personal communication, K. Miller and R. Hautmann).

With supine flat positioning of the patient, ureteric stones at the level of the sacroiliac joint cannot be treated by ESWL because the shock wave energy is attenuated by the sacral and iliac bones. However, there are promising initial reports of treating such calculi with the patient lying prone on a modified stretcher.[14]

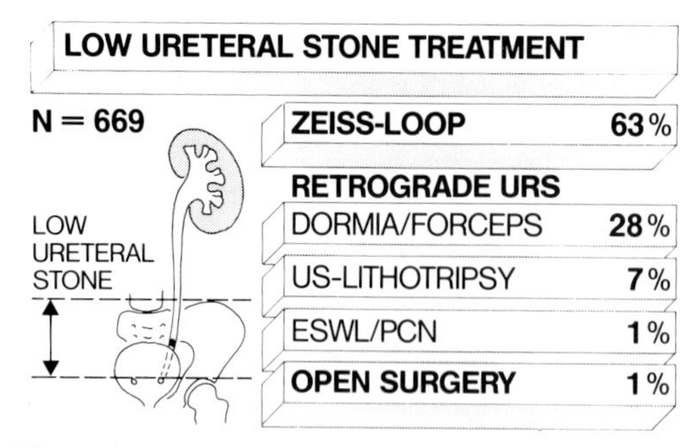


Fig. 11-1. Treatment for distal ureteral calculi at the Urologic Department of Katharinenhospital, Stuttgart (October 1983 to August 1985). URS = ureteroscopy.

## LOCATION

In case of distal ureteral calculi, optimal localization should be achieved with minimal x-ray exposure of the patient. Based on optimal positioning, this may be achieved by the following approach. First, flat positioning of the patient has the advantage of allowing easy orientation on the screen, as it corresponds to the actual kidney–ureter–bladder (KUB) film (Fig. 11-3). Therefore we start by prelocating the stone without fluoroscopy using the patient's anatomic landmarks (vertebral column, coccyx, symphysis, greater trochanter, and anterior superior iliac spine). Subsequently the stone is localized by intermittent use of fluoroscopy with early collimation on the screen once the stone is near to the crosshair; localization begins with the stone-sided anteroposterior (AP) monitor. For a better orientation in the pelvic region, it is useful to look on the KUB film at an oblique angle according to the direction of the central beam of the two radiographic systems.

Short time, high-current fluoroscopy (snapshots or quick pics) is only used if normal fluoroscopy does not provide enough information for stone localization and disintegration; however, it is used routinely at the end of the treatment (Fig. 11-4). Consequently, long time, high-current fluoroscopy with increase of exposure time (usually to about 1.3 sec) should only be used if snapshots do not provide sufficient information, i.e., in case of superposition of air-filled intestine or if slightly opaque calculi are treated. Naturally, optimal adjustment of the intensifiers towards the abdomen or the thighs for distal calculi as well as the adjustment of the monitors is required. However, overinflation of the balloons at bottom of the tub should be avoided because this may elevate the patient

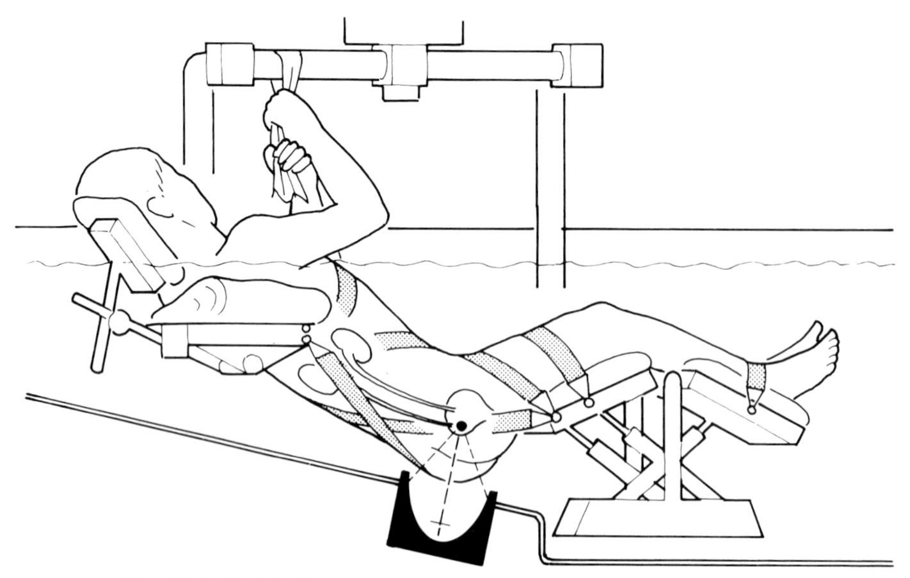

Fig. 11-2. Positioning technique for ESWL of distal ureteral calculi.

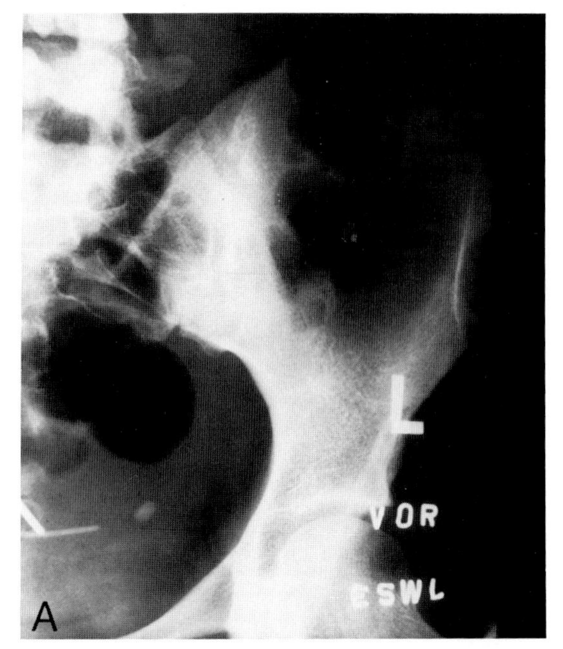


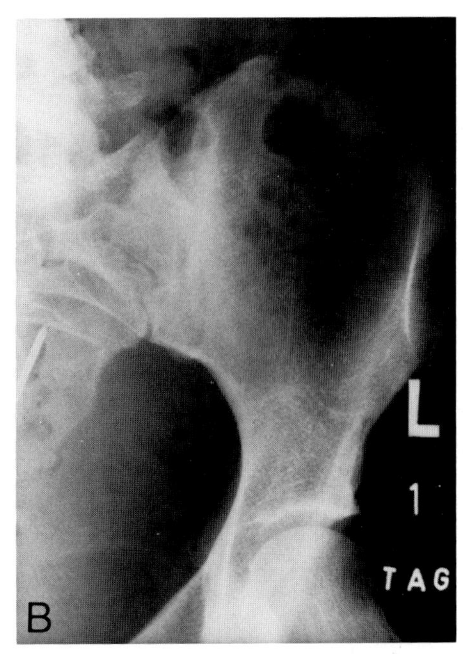


Fig. 11-3. In-situ ESWL for distal ureteral calculi; KUB films before **(A)** and after **(B)** gluteal shock wave lithotripsy.

during screening and thus lead to movement of the stone out of F2 once the balloons are deflated again.

If these imaging techniques cannot resolve the locating difficulties, there are further possibilities. First, the use of contrast dye may be helpful, particularly in the case of small stones, where further information of stone disintegration is of minor importance once the stone can be focused. Another possibility is the insertion of a ureteral catheter below the stone to mark its position.[3,5,13] However, once an endourologic procedure is necessary, one should simply remove the distal ureteral calculus using either a Zeiss loop, a basket, or ureteroscopy.

With this locating technique, our mean fluoroscopy time amounted to 108 (48 to 259) sec. An average of eight snapshots (0.333 msec, 55 to 65 kV, 160 to 200 mA), ranging from 6 to 20, were necessary, which corresponded to an overall x-ray exposure of 7 Gy (collimation included) per session. However, this may be reduced considerably with increasing experience.

## SIMULATION

In obese patients, patient positioning and stone localization should be simulated beforehand. If the distance from the loin skin to the stone is more than 12 cm it exceeds the distance from the upper rim of the ellipsoid to F2, and thus the stone cannot be focused on the monitors. If simulation is performed without water in the tub, the buoyancy of the body once the tub is filled must be taken into account. The body will float slightly even if the belts are well tightened. The distance lost in this way amounts to about 1.5 cm on the Z axis.

## FOCUSING

For large, obstructive ureteral calculi, cranial focusing is recommended because the upper rim of the stone represents the optimal interface between urine and calculus, thus minimizing the attenuation of shock wave energy due to mucosal edema around the stone.[6] With increasing disin-

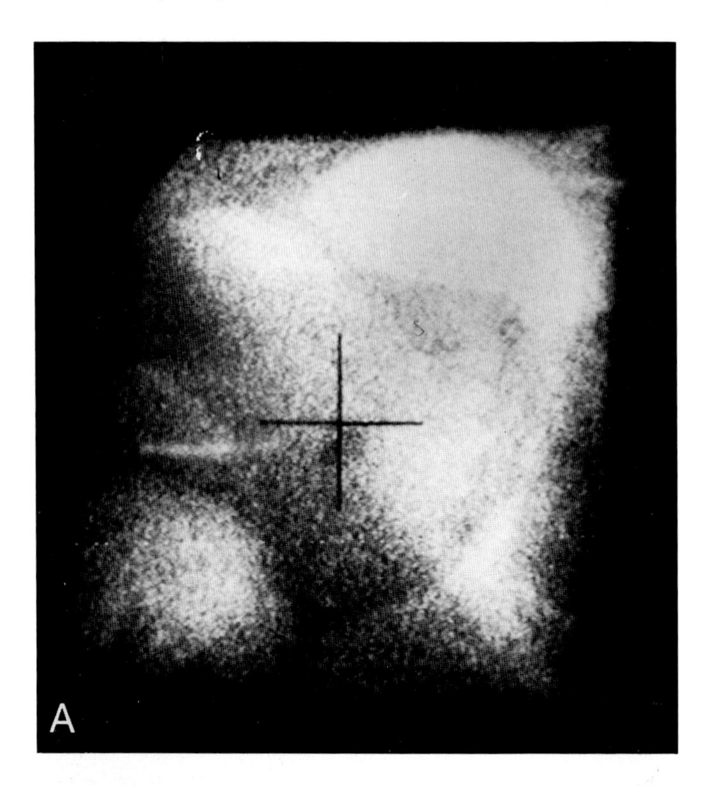


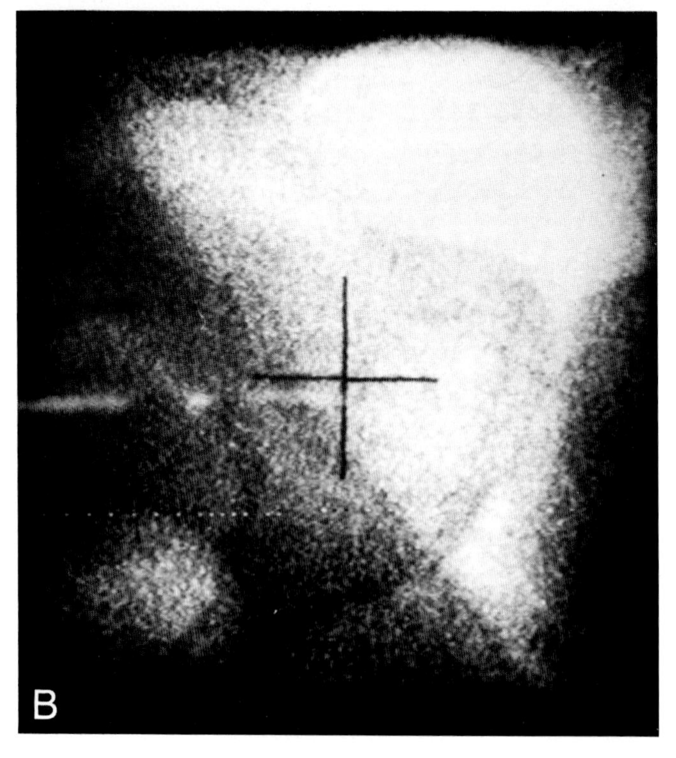


Fig. 11-4. Fluoroscopic control (short-time, high-current technique) during in-situ ESWL for distal ureteral calculi demonstrating progressive disintegration of the stone (AP monitor).

tegration of the calculi, urine permeates the upper fragments and increases the areas of interfaces as it decreases the tension of the ureteral wall; both effects improve shock wave efficiency.

### Shock Wave Energy

Renal calculi are treated first with low-energy shock waves (15 to 16 kV) to cause fine fragmentation and to prevent early scattering of the large fragments. Impacted ureteral calculi, on the other hand, should be treated with higher energies (22 kV) (Table 11-1). There is no risk of the fragments scattering, and some of the energy will be attenuated by the edematous ureteral wall.[5,6] In consequence, more as well as stronger shock waves are required to disintegrate ureteral than renal stones. In our series we used an average of 1,650 shock waves (700 to 2,000) with a mean energy of 20 kV (18 to 22 kV) (Table 11-1). The maximum number of shock waves, however, should not exceed 2,500 to 3,000.

### Auxiliary Measures

Our stone center does not recommend any ancillary measures prior to ESWL for distal ureteral stones. Our treatment policy is based on minimizing invasiveness and avoiding any unnecessary manipulations. Once the stone has to be approached endoscopically, it should be removed by that route, using a Zeiss loop, a basket, or retrograde ureteroscopy.

**Table 11-1. Results of In-Situ ESWL for Distal Ureteral Calculi ($n$ = 20)**

| | |
|---|---|
| Number of shock waves | 1,650 (700–2,000) |
| Generator voltage | 20 kv (18–22) |
| Fluoroscopic time | 108 sec (48–259) |
| Number of snapshots | 8 (6–20) |
| Colic | 4 (20%) |
| Auxiliary measures | 2 (10%) |
| Hospital stay | 4 days (2–7) |

## CONTRAINDICATIONS

Distal ureteral stones in obese patients may not be treatable by ESWL because it may not be possible to lower the patient enough to bring the stone to F2. Also, even if the stone can be positioned at F2, large thighs may prevent the image intensifiers from being lowered. Patients weighing more than 90 kg are usually not candidates for prone-position ESWL because of the even greater distance from the abdominal skin to the stone. In cases of spinal deformity, adequate positioning of the patient may not be possible.

Another contraindication for shock wave lithotripsy is ureteral stricture distal to the stone. Such cases should be treated by retrograde ureteroscopy with endoscopic incision of the stenosis, balloon dilatation, or even ureterocystoneostomy.

Most stones at the level of the sacroiliac joint seem not to be suited to ESWL, despite initial promising experience with prone-position shock wave lithotripsy.[14] Such calculi should be extracted using a ureteroscope.[5,7,10–12]

## SUMMARY

A few preliminary experimental studies have detected no mutagenic or carcinogenic effects of shock waves.[16,17] Nevertheless, our center still excludes fertile women from pelvic ESWL. It should be noted that exposure to x-rays during treatment is not significantly greater than for endourologic procedures such as retrieval of stones by a basket or Zeiss loop, or retrograde ureteroscopy.

Finally, it must be emphasized that the introduction of new technology has not changed the general indications for interventional therapy for ureteral stones.[18–22] Interventional treatment is only indicated when there is:

Persistent hydronephrosis with the risk of loss of renal function

Persistent fever with the risk of septicemia

Persistent colic or pain with the risk of intestinal ileus

A stone larger than 8 mm (unlikely to pass spontaneously)

## REFERENCES

1. Alken P, Hardeman S, Wilbert D et al: Extracorporeal shockwave lithotripsy (ESWL): Alternatives and adjuvant procedures. World J Urol 3:48, 1985
2. Chaussy C, Schmiedt E, Jocham D et al: First clinical experience with extracorporeally induced destruction of kidney stones by shockwaves. J Urol 127:417, 1982
3. Chaussy C, Griffith D, Dretler S: Extracorporeal shockwave lithotripsy/percutaneous lithotripsy—case reports (abstract #133). J Urol 135:137A, 1986
4. Fuchs G, Miller K, Rassweiler J, Eisenberger F: Extracorporeal shockwave lithotripsy, one-year experience with the Dornier lithotriper. Eur Urol 11:145, 1985
5. Miller K, Fuchs G, Rassweiler J, Eisenberger F: Treatment of ureteral stone disease: The role of ESWL and endourology. World J Urol 3:53, 1985
6. Müller SC, van Haverbeke J, Seweifi A, Alken P: Der hohe Harnleiterstein—ein Problem trotz extracorporaler Stosswellenlithotripsie. Akt Urol 16:294, 1985
7. Rassweiler J, Miller K, Fuchs G, Eisenberger F: Ureterorenoscopic stone extraction (URS)—its role in the management of urolithiasis. Proceedings of the 2nd Congress of Endourology, Mainz, June 1984
8. Dretler SP, Watson GM, Murray S, Parrish, JA: Laser fragmentation of ureteral calculi (abstract #425). J Urol 135:210A, 1986
9. Gumpinger R, Miller K, Fuchs G, Eisenberger F: Antegrade ureteroscopy for stone removal. Eur Urol 11:199, 1985
10. Huffman JL, Bagley DH, Schönberg HW, Lyon ES: Transurethral removal of large ureteral and renal pelvic calculi using ureteroscopic ultrasonic lithotripsy. J Urol 130:31, 1983
11. Lyon ES, Kyker JS, Schönberg HW: Transurethral ureteral ureteroscopy in women: A ready addition to the urological armamentarium. J Urol 119:35, 1978
12. Perez-Castro E, Martinez-Pineiro JA: Ureteral and renal endoscopy: A new approach. Eur Urol 8:117, 1982
13. Jenkins A, Lippert MC, Wyker AW, Gillenwater JY: ESWL treatment of distal ureteral stones. Abstracts of Third Congress on Endourology, New York, 20–22, 1985
14. Jenkins A: ESWL treatment of ureteral calculi (abstract #315). J Urol 135:182A, 1986
15. Wilbert DM, Reichenberger M, Noske E et al: New generation multifunctional shockwave lithotripter (abstract #226). J Urol 135:160A, 1986
16. Carrol PR, Shi RY: Genetic toxicity of high energy shockwaves (abstract #749). J Urol 135:292A, 1986
17. Eisenberger F, Chaussy C, Wanner K: Extracorporale Anwendung von hochenergetischen Stosswellen—Ein neuer Aspekt in der Behandlung des Harnsteinleidens (Alken-award 1976). Akt Urol 8:3, 1977
18. Banner MP, Pollack HM: Percutaneous extraction of renal and ureteral calculi. Radiology 144:753, 1982
19. Das S, Harris CJ, Amar AD, Egan RM: Dorsovertical lumbotomy approach for surgery of upper urinary tract calculi. J Urol 129:266, 1983
20. Eisenberger F, Fuchs G, Miller K, Rassweiler J: Non-invasive renal stone therapy with extracorporeal shockwave lithotripsy (ESWL). In Heuck FA, Donner W (eds): Radiology Today. Vol. 3. Springer Verlag, Berlin–Heidelberg–New York, 1985, p. 161
21. Hulbert JC, Lange BH: The percutaneous removal of difficult upper urinary tract calculi. World J Urol 3:19, 1985
22. Reddy PK, Hulbert JC, Lange PH et al: Percutaneous removal of renal and ureteral calculi: Experience with 400 cases. J Urol 134:662, 1985

# 12

# Staghorn Calculi

Although partial and complete staghorn calculi constitute less than 10 percent of symptomatic stones, they seem to generate 90 percent of the discussion concerning the efficacy of ESWL. Clearly, the larger the stone burden, the higher the likelihood of residual renal fragments after ESWL. And while the fate of residual calcium oxalate fragments is not yet known, history would suggest that incomplete removal of staghorn infection stones presages early and rapid recurrent stone growth.

In an effort to utilize new noninvasive technology and maximize stone-free rates, the combination of percutaneous nephrostolithotomy and ESWL is now advocated for large-volume stones, especially in dilated intrarenal collecting systems. Nevertheless, urologists are not in agreement as to which complex stones require which combinations of techniques. Can internal ureteral stenting with repeated ESWL achieve the same success as percutaneous "debulking" prior to shock wave lithotripsy? Would a large-caliber nephrostomy tube placed before ESWL and irrigated during the treatment make percutaneous nephrostolithotomy unnecessary? Probably, as experience is gained, several effective routes to a successful result and a stone-free patient will be found.

Certainly, as both authors outline in this chapter, decisions must be individualized depending on the variables of stone size (burden), intrarenal anatomy, presence of obstruction, presence of infection, and patient habitus. Treatment plans may need to be altered depending on the success of the initial procedures.

One must not forget that each staghorn renal calculus is surrounded by a patient whose personal agenda and needs may affect the choice of treatment. A motivated 32-year-old woman executive can be advised differently from a 68-year-old retired banker with moderately symptomatic bladder outlet obstruction. ESWL monotherapy with repeat treatments, suitable for the younger patient, may lead to prolonged hospitalization, ureteral obstruction, secondary procedures, and resistant bacterial colonization of nephrostomy tubes in the older patient; for him, percutaneous combined therapy might be the most effective, safest, and most conservative approach.

In addition, a physician must realize that a sole owner/proprietor of a small business, or a single mother with three children, may still wisely see surgery as the most time-efficient course of action to avoid multiple ESWL treatments, tubes, irrigations, and hospitalizations.

*R.A.R.*

# Combined Percutaneous and ESWL Treatment

James E. Lingeman

The treatment of staghorn calculi has always represented a special challenge for the urologist. Traditionally, these stones have been removed by anatrophic nephrolithotomy, a procedure popularized and performed with excellent results by Boyce.[1,2] With the advent of extracorporeal shock wave lithotripsy (ESWL), a new therapy potentially applicable to staghorn calculi became available. The noninvasive nature of this technology was very appealing, and initial reports from Germany suggested that ESWL monotherapy for staghorn calculi was feasible. Following the initial phase of Food and Drug Administration (FDA) trials in the United States (limited to the treatment of small solitary stones), staghorn calculi were treated with ESWL as primary therapy. Unfortunately, early results were often unsatisfactory. The large stone bulk in these patients usually necessitated multiple ESWL sessions and resulted in a burdensome volume of gravel to be passed by the patient. Most early cases treated with ESWL monotherapy developed sepsis related to the inevitable obstruction during passage of stone material.[3] Furthermore, the high rate of retained fragments in these early cases was dismaying.

Evaluation of data in patients treated with ESWL alone for stones greater than 3 cm in diameter revealed that further procedures (repeat ESWL, stone manipulation, or percutaneous nephrostomy) were required in 77 percent of cases.[4] Significant residual fragments remained in the upper urinary tract in 57 percent of these cases. Since most staghorn calculi are of an infectious etiology (struvite and/or carbonate apatite), the rate of recurrent stone growth in this situation is probably quite high.[5] Additionally, Chaussy and Fuchs report that 50 percent of patients treated with ESWL monotherapy for small staghorn calculi have residual stone material at three months (personal communication). Therefore, considering the high price to be paid in terms of retreatments and ancillary procedures, experience at our center with ESWL alone suggests that this is not adequate therapy for most staghorn calculi.

Additionally, although Krieger et al,[6] Elder et al,[7] and Clayman et al[8] have reported the removal of staghorn calculi percutaneously, removal of the large staghorn with multiple dendritic branches remains an extremely challenging endourologic procedure. Multiple punctures are commonly required, and even in the most capable hands, the residual stone rate is significant when these stones are treated by endourologic means alone.[9,10]

The advantages of percutaneous and shock wave lithotripsy are complementary (Table 12-1). Experience at the Methodist Hospital of Indiana suggests that a combination of these new

**Table 12-1. Comparison of ESWL and Percutaneous Techniques for Treating Staghorn Stones**

| Technical Considerations | ESWL | Percutaneous |
|---|---|---|
| Large stones | Multiple treatments in most cases | Single treatment in most cases |
| Multiple stones | Easy | Harder |
| Stone in lower pole | Residual gravel common | Residual stones uncommon |
| Stone in upper pole | Easy | Harder |
| Morbidity relative to surgery | Very low | Low |

techniques is appropriate for most large and/or complex renal calculi. By adding ESWL to percutaneous nephrostolithotomy, all areas of the kidney can be accessed for stone removal, using a single percutaneous puncture in most cases.

A partial staghorn calculus is defined as a renal pelvic stone with extension into at least two infundibula. A complete staghorn calculus has extensions into most or all caliceal groups. While simple branched stones or large renal pelvic stones are not encompassed by this definition, the concepts presented here are applicable to them as well.

## TECHNIQUE

Preparation of the patient is similar to that for more traditional surgical procedures.[11] Adequate radiographs are essential. Oblique views of the kidney and, occasionally, nephrotomography are necessary to adequately define the extent of the stone material within the kidney. Bacteriological evaluation of the urine is essential, especially in patients with neurogenic bladders and/or urinary diversions where resistant organisms are commonly encountered. Parenteral antibiotic coverage for 48 hours preceding percutaneous debulking is appropriate whenever struvite is suspected, even if urine cultures are negative. Contrast studies and renal scans should be done to demonstrate anatomy and/or function.

Percutaneous debulking is performed as a single-stage procedure in our institution, as previously described.[12,13,14,15] Because these procedures are often complicated and lengthy (average duration 202 min for the initial procedure in the author's series), general anesthesia is preferred. Retrograde ureteral balloon occlusion catheters are used in all cases and facilitate opacification and distension of the collecting system, which can be invaluable in gaining access in these difficult procedures. In addition, the occlusion balloon is inflated during ultrasonic lithotripsy to prevent migration of fragments into the ureter.

Another advantage of the single stage procedure is the opportunity for the radiologist and urologist to combine forces in gaining satisfactory access. Without a doubt, safely achieving a foothold in the collecting system is the single most difficult aspect of this procedure. Typically, the collecting system is filled entirely with stone material, and considerable calicectasis may be present. Frequently, these patients have had previous surgical procedures, again complicating the radiologist's task. Because of these factors, passing the guide wire into the ureter, although desirable, is not always possible. In this circumstance, as much of the guide wire as possible is coiled into the desired calyx. A safety wire is always used. The urologist should realize that the radiologist may not be able to place him in the most desired location because of the aforementioned factors. Occasionally, satisfactory access cannot be achieved. In a situation where the collecting system is too tightly filled with stone material, ESWL immediately followed by percutaneous debulking can be considered. Sometimes, open surgery is necessary in such instances.

In general, access into the lower pole is preferred for debulking procedures. The renal pelvic and lower pole stone material should be removed percutaneously, as completely as possible. The reason for this is that stone material residing in dependent areas of the lower pole is unlikely to pass spontaneously following treatment with shock wave lithotripsy. Therefore, even after adequate treatment with shock wave lithotripsy, a secondary percutaneous procedure is usually necessary to remove this stone material if it is not removed at the initial session. Occasionally, multiple punctures are necessary. At our center, as many as five tracts have been used during removal of a staghorn calculus (another strong argument for general anesthesia). Balloon dilation is used almost exclusively because there is rarely enough room intrarenally for passing dilators. An Amplatz working sheath (Cook, Vantec) is always used. Such an open system easily allows use of a variety of nephroscopes and minimizes intrarenal pressure from the irrigant. Significant fluid absorption during nephrostolithotomy is unusual unless gross extravasation occurs.

Debulking a staghorn calculus is a bloodier procedure than "simple" nephrostolithotomy, and about 20 percent of patients require transfusion. Ultrasonic lithotripsy is generally preferred, with electrohydraulic lithotripsy occasionally being required for extremely hard stones (uric acid and cystine). At the time of the initial percutaneous procedure, all stone material accessible with the rigid instrument is removed. No attempt to remove substantial amounts of caliceal or remote stone material with flexible nephroscopy is made at this session. Using the combined treatment plan, these stones are disintegrated subsequently with shock wave lithotripsy and then irrigated out later if necessary. A re-entry malecot or simple red rubber catheter is preferred over a Foley catheter for nephrostomy drainage following percutaneous debulking because the Foley balloon may obscure residual stone material during subsequent radiographs and ESWL treatments.

Following a rest period of 2 to 4 days, residual stone material is then fragmented with ESWL. A maximum of 2,000 shock waves at 20 to 24 kV is used in a single session. Occasionally, more than one session will be required to adequately pulverize all of the remaining stone material. If gravel does not pass promptly (within 48 hours) after ESWL, it is unwise to assume that it will do so, and further intervention (ESWL, chemolysis, secondary percutaneous procedures) is usually required. It is important to ensure adequate pulverization of these residual calculi before attempting secondary percutaneous procedures, as gravel is much easier to wash out than solid stone material. A blood administration pump is used to pressurize irrigant with the flexible nephroscope to increase water flow.

The percutaneous debulking procedure should always be performed first, except for partial staghorn calculi confined to the upper pole, which may be successfully treated with ESWL alone. This approach is preferred for several reasons: (1) the percutaneous procedure may be the only procedure required; (2) the nephrostomy tube assures good drainage, minimizes the risk of urosepsis, and provides a route for the rapid egress of gravel after ESWL; and (3) the nephrostomy tube is also useful for maintaining access to the kidney for secondary percutaneous procedures, for instilling contrast to localize stones during ESWL, and for chemolysis. The tube is thus left indwelling until the patient has been rendered stone free radiographically.

Although about 90 percent of staghorn calculi may be treated with this combined approach, not all patients are good candidates. Patients who are not candidates for ESWL must have a good chance of being rendered free of stone with nephrostolithotomy alone; otherwise, open surgery is preferred. Patients considered marginal candidates for ESWL (patients with renal ectopy, kyphoscoliosis, extreme obesity) should be simulated in the lithotripter before deciding on the approach to be used. Patients with extremely complex intrarenal architecture (a judgment dependent, at least in part, on the operator's endourologic skills) or with infundibular stenosis may be better served by anatrophic nephrolithotomy.

The results obtained with this combined approach at the Methodist Hospital of Indiana are comparable to those reported for anatrophic

**Table 12-2. Composition of Residual Fragments from 52 Renal Units**

| Composition | Number of Renal Units | Patients |
|---|---|---|
| Calcium oxalate/brushite | 4/11 (36%) | 4 |
| Struvite | 3/33 (9%) | 2 |
| Uric acid | 0/3 (0%) | 0 |
| Cystine | 1/5 (20%) | 1 |
| Total | 8 (15%) | 7 |

nephrolithotomy.[1–3] Significantly, more than 90 percent of kidneys containing infectious stones (struvite and/or carbonate apatite) can be rendered completely stone free (Table 12-2). This less invasive approach is especially useful in patients with urinary diversion or neurogenic bladders (25 percent of cases in our institution) who are at higher risk for recurrent calculi because of recalcitrant urea-splitting bacteriuria.

About 25 percent of kidneys can be rendered stone free with the initial percutaneous procedure, but most will require a combination of percutaneous and shock wave lithotripsy. Very few kidneys containing staghorn calculi can be rendered stone free with ESWL alone. Those that can generally contain smaller partial staghorn calculi confined to the upper half of the kidney.

An average of 2.6 procedures are required per renal unit in our institution. The typical sequence is percutaneous debulking followed by ESWL. Any gravel not passing promptly is irrigated out with the flexible nephroscope. This procedure can usually be performed under local anesthesia. Renacidin irrigation via the nephrostomy tube is performed for 48 hours if struvite is suspected once the kidney is radiographically free of stones.

One patient with bilateral complete staghorn calculi accounted for two of the three kidneys with residual struvite in our series. The patient was a quadriplegic with ileal conduit urinary diversion. He had highly resistant urea-splitting organisms in his urine, and we felt he was likely incurable from a bacteriologic standpoint, whether he was rendered stone free or not. Therefore, we elected to treat him in the least invasive fashion possible and performed only solitary punctures on each kidney. About 95 percent of his stone material was removed, leaving only small, potentially passable fragments. He has been followed closely and any fragments exhibiting growth are now periodically treated easily with ESWL before his stone burden reaches a critical point. This approach allows for control of his stone disease while minimizing risk to renal function and morbidity. Patients such as this, with chronic urea-splitting bacteriuria, or those with difficult-to-control metabolic problems (some cases of cystinuria, intestinal hyperoxaluria, etc.) are the most likely to benefit from this less invasive approach. Thus, while most of these challenging patients can be rendered stone free with a single percutaneous access and ESWL, multiple punctures are avoided in situations where a long-term cure is deemed unlikely.

In our institution we have not experienced greater complications with this approach than with less complicated percutaneous procedures. Temperature elevation is common, which is not surprising given that most of these stones are of an infectious etiology. This fact underscores the need for close attention to cultures and antibiotic coverage. Serious complications related to the percutaneous puncture (arteriovenous fistula, pneumothorax, hydrothorax, colon perforation) are uncommon, but can be troublesome when they do occur. The management of staghorn calculi with a combined approach is technically demanding. A definite learning curve for these complex cases will be evident, even for those facile in endourologic techniques.

## CASE STUDIES

### CASE 1

A 38 year old woman presented with a complete staghorn calculus (Fig. 12-1A,B). She was neurologically normal. The referring physician had placed an indwelling stent because of pain. This case illustrates why the percutaneous procedure is done first, as it was possible to clean this relatively simple collecting system in a single session (Fig. 12-1C,D). Although urine cultures were negative, a stone culture grew *Proteus mirabilis* and the stone was 100 percent carbonate apatite. It is extremely important that no residual stone material be allowed to remain because this category of patient has a high probability of long-term cure if rendered stone free.[16] The nephrostomy tube allowed access for hemiacridin irrigation for 48 hours prior to its removal. This may be important after ultrasonic lithotripsy or ESWL as radiographically undetectable minute particles invariably remain and could provide a site for future stone formation.

If ESWL had been attempted in this patient as primary therapy, more than one session would likely have been required given the volume of stone present. In addition, the volume of gravel presented to the ureter would doubtless have been troublesome, the ureteral stent notwithstanding.

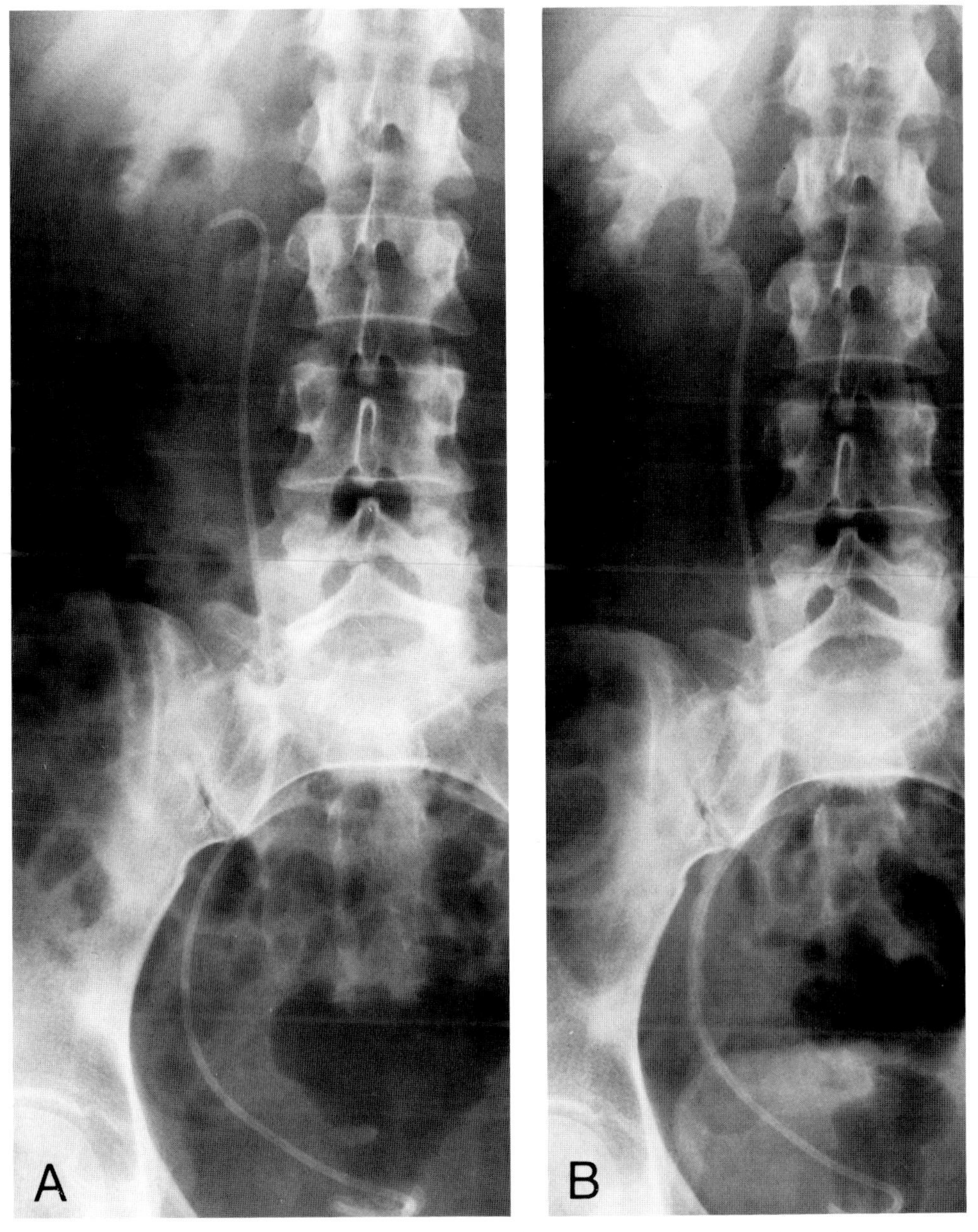


Fig. 12-1. **(A)** KUB film. **(B)** 15-min film from IVP. (*Figure continues.*)

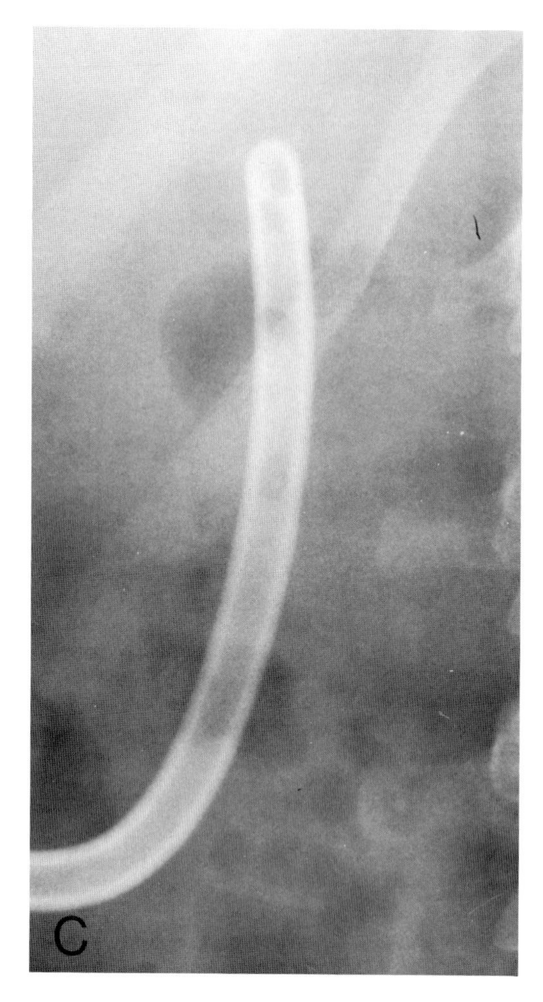


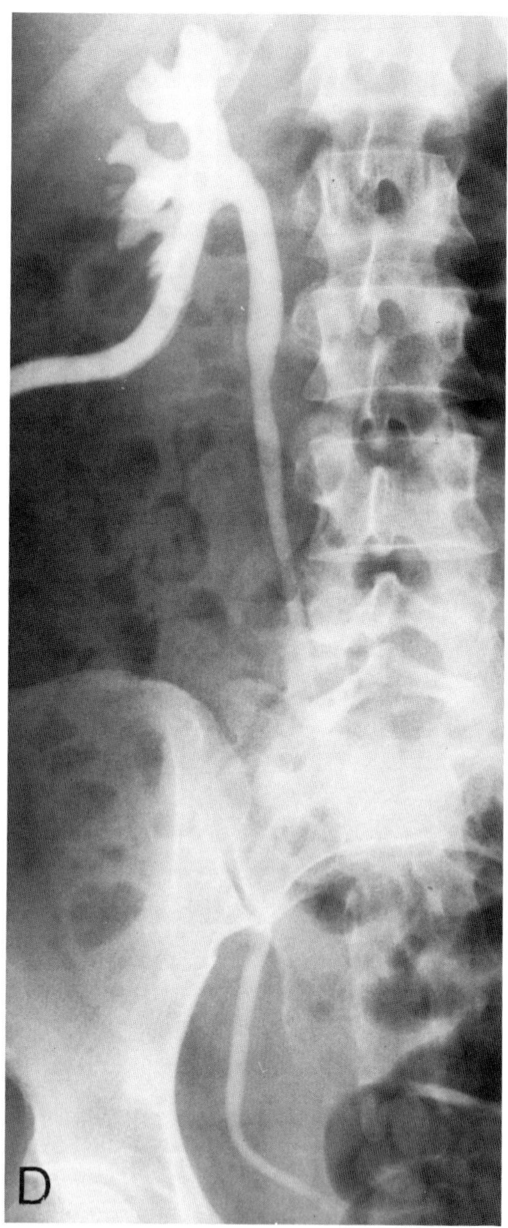


Fig. 12-1 (*continued*). **(C)** KUB film 48 hours after nephrostolithotomy. **(D)** Nephrostogram.

## CASE 2

This 56-year-old woman had bilateral complete staghorn calculi associated with normal renal function (Fig. 12-2A,B). These stones could certainly have been removed completely by percutaneous means although multiple sessions[4,17] and/or multiple punctures would likely be required. By approaching the kidneys via the most accessible route (i.e., lower pole posteriorly), debulking of most of the stone material was easily accomplished (Fig. 12-2C,E). As noted in Figure 12-2D,E,F), the stone material in the mid and upper half of the kidneys passed rapidly (mostly via the nephrostomy tubes) after

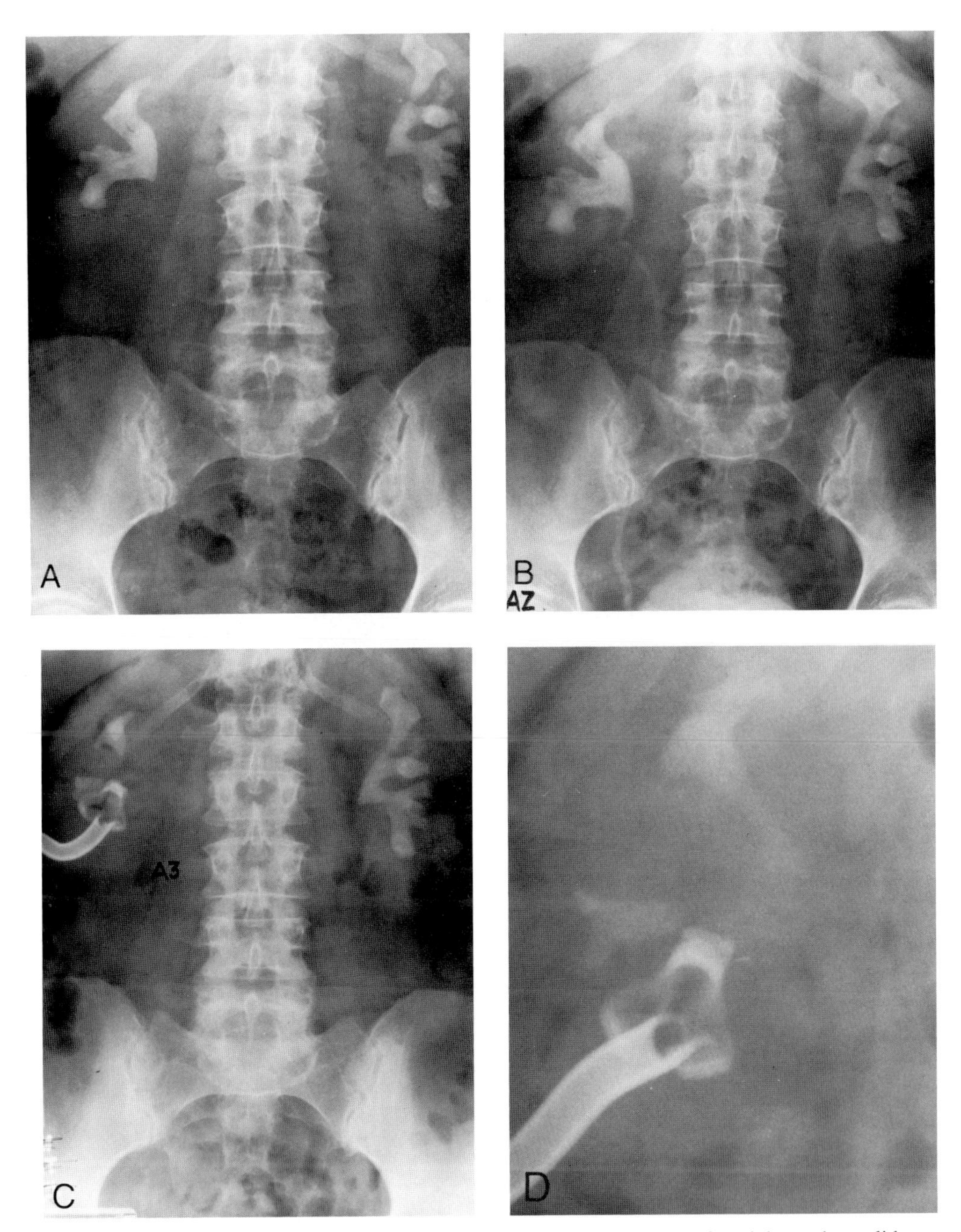


Fig. 12-2. **(A,B)** KUB and 15-minute IVP films. **(C)** KUB film 24 hours after right nephrostolithotomy. **(D)** KUB film 24 hours after right ESWL; note projection of gravel down into upper pole infundibulum. All the gravel passed within 48 hours. (*Figure continues.*)

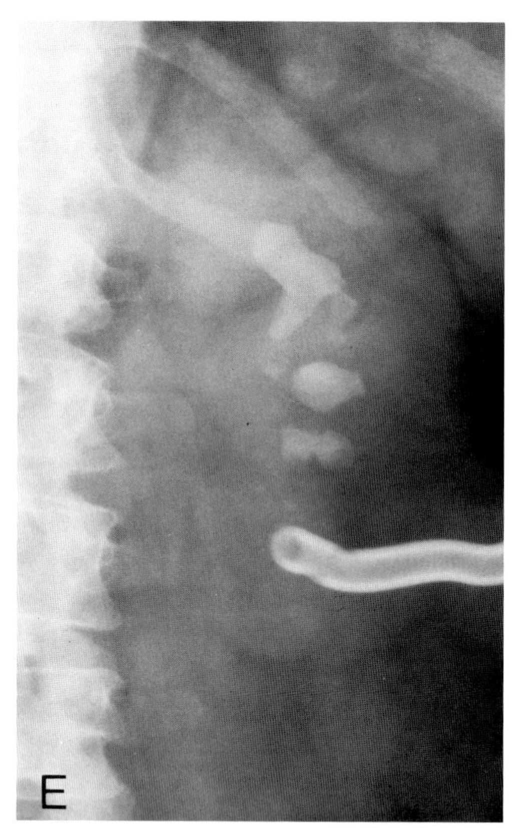


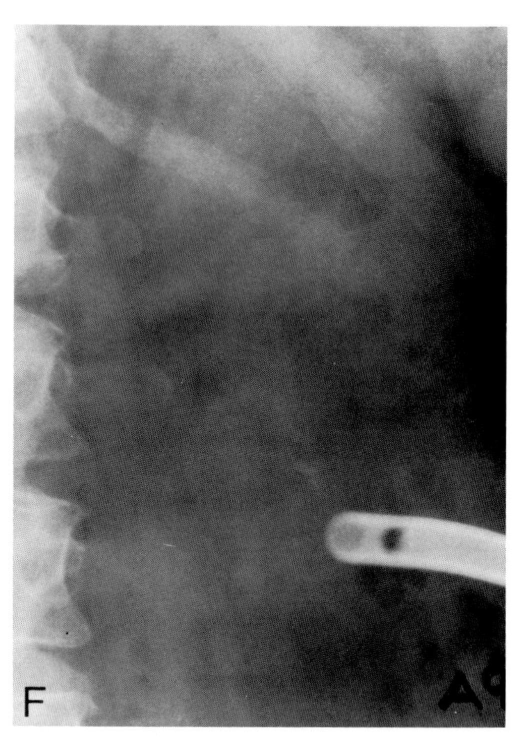


Fig. 12-2 (*continued*). **(E)** KUB film after left nephrostolithotomy. **(F)** KUB film 24 hours after left ESWL, revealing only scant gravel remaining in the lower pole. KUB film 6 weeks after treatment was negative.

ESWL. A secondary percutaneous procedure was therefore not required in this instance. Hospital stay was 17 days total and she remains stone and infection free 15 months following her procedures.

## CASE 3

The large and complex stone (Fig. 12-3A) in the left kidney of this 36-year-old woman required multiple ESWL sessions to adequately pulverize the remaining caliceal stone material following percutaneous debulking (Fig. 12-3C,D). Nephrotomograms are routinely performed the day after ESWL, as assessing the adequacy of stone fragmentation may be difficult. All stone material must be thoroughly broken up as this facilitates its passage and/or irrigation out at the time of a secondary percutaneous procedure, if required. In this case, as is not uncommon, the primary puncture did not allow access to all caliceal groups with the flexible nephroscope, and a second puncture into the middle pole was required to render her stone free (Fig. 12-3E,F). Discharging patients in the hope that residual stone material will pass spontaneously is usually unwarranted. Making maximum use of the nephrostomy tube for flexible nephroscopy and irrigation before removing it is critical to achieving good results. This patient was free of stone and infection at 1 year follow-up (Fig. 12-3G,H).

## CASE 4

This 27-year-old man (Fig. 12-4) had undergone a left nephrectomy several years earlier when excessive bleeding occurred at the time of attempted staghorn calculus removal. His serum creatinine was 2.7 mg percent when he presented for treatment of a large right staghorn calculus associated with *Proteus*

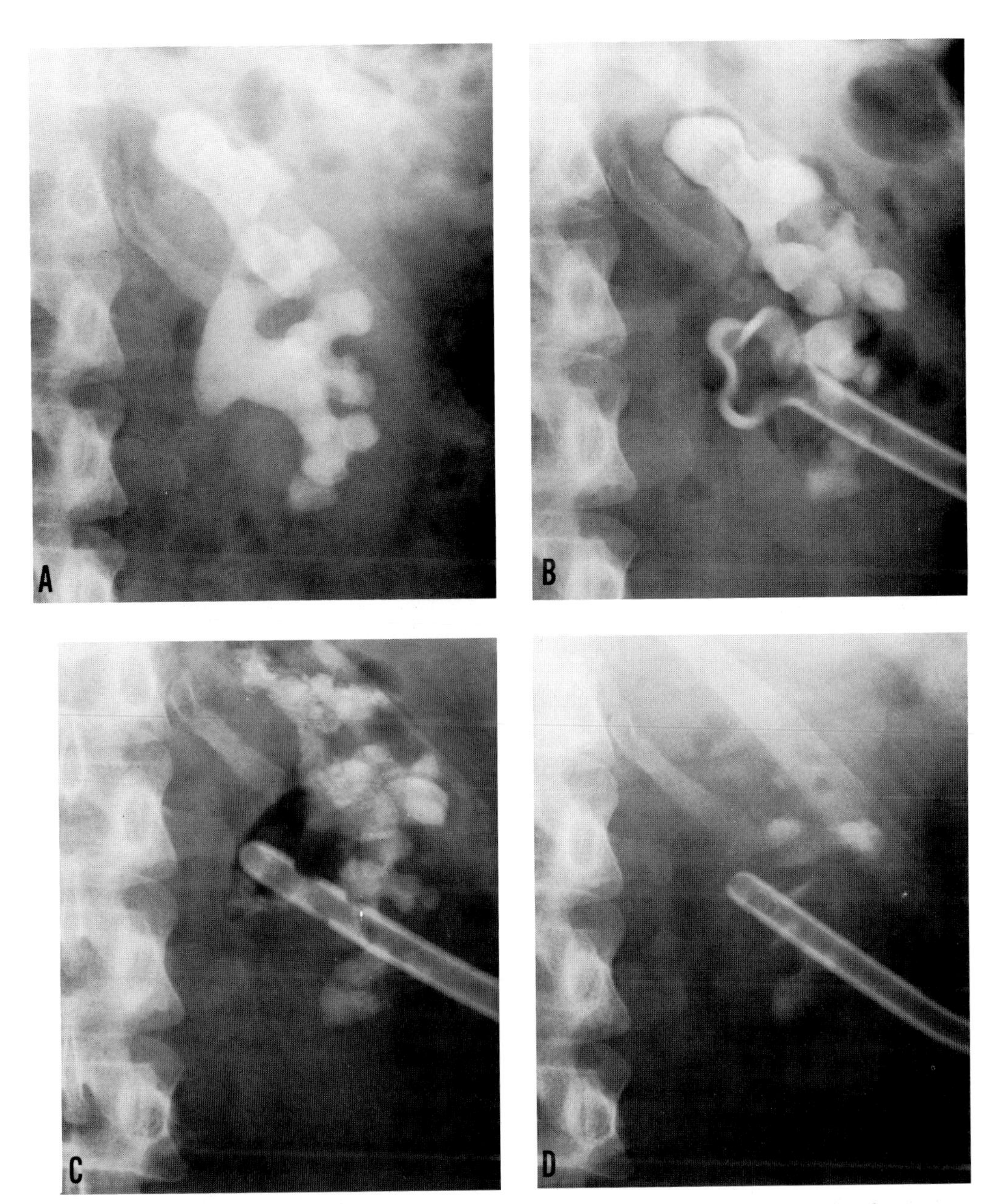


Fig. 12-3. **(A)** KUB film demonstrating stone with multiple dendritic branches. **(B)** KUB film after percutaneous debulking. **(C)** KUB film 24 hours after first ESWL. **(D)** KUB film 24 hours after third ESWL. The gravel appears well pulverized but much remains. (*Figure continues.*)

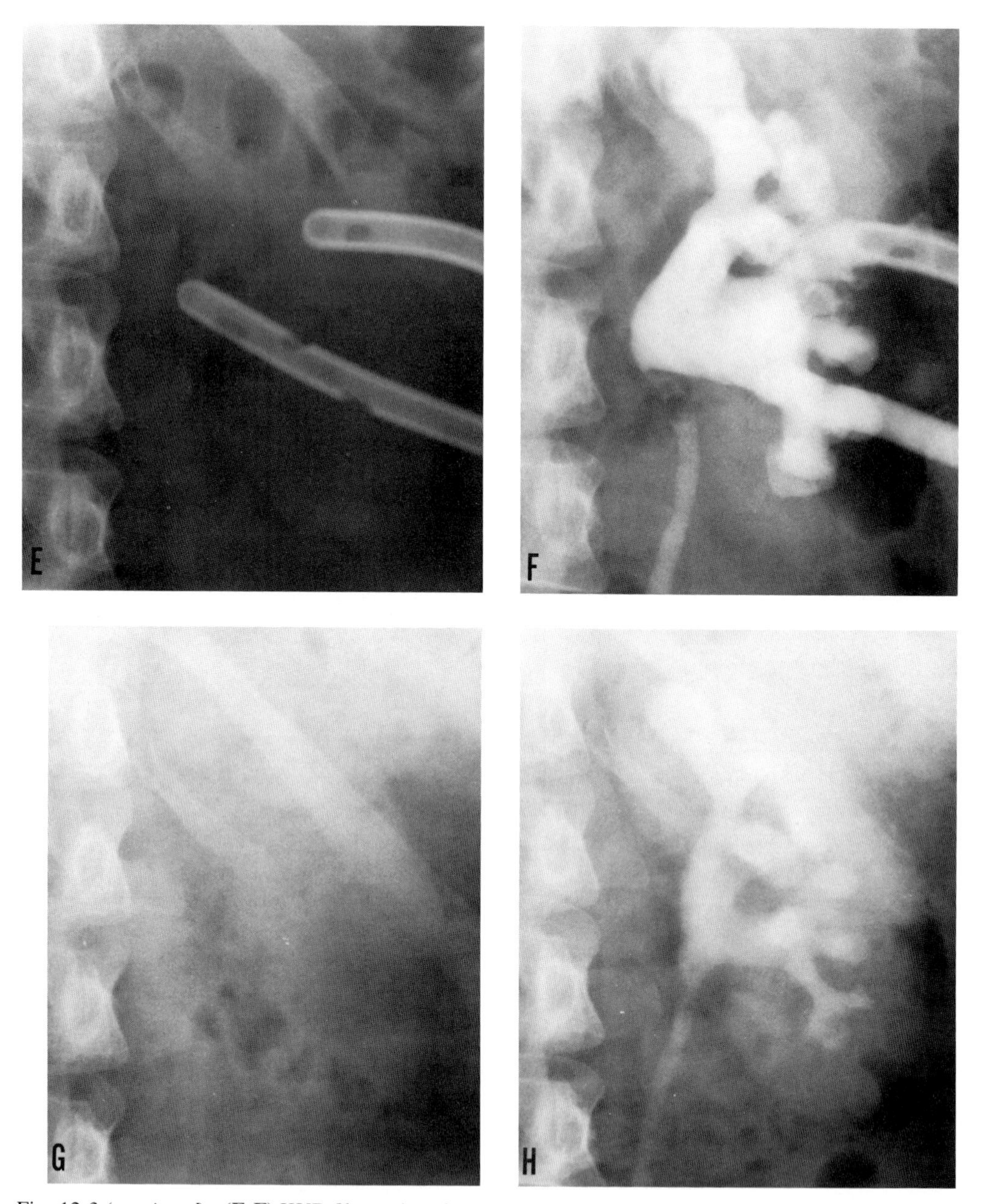


Fig. 12-3 (*continued*). **(E,F)** KUB film and nephrostogram after secondary percutaneous procedure requiring a second puncture into the middle pole. **(G,H)** Follow-up after one year.

bacteriuria (Fig. 12-4A). Retrograde pyelograms (Fig. 12-4B) suggested a complex cluster of calices in the lower pole. Approaching this stone through the lower pole would likely require multiple punctures, a situation to be avoided if possible in a solitary kidney. In addition, the upper pole contained a huge caliceal stone that would have been difficult (if not impossible) to access from a lower pole puncture. A stone of this size in the upper pole presents problems for ESWL also as multiple sessions would doubtlessly be required. Therefore, although supracostal puncture carries increased risk,[18] this approach was elected. In addition to allowing removal of the upper pole and renal pelvic stone material, good vision with the rigid nephroscope was provided into the most dependent calices (Fig. 12-4C). The small stone remaining after percutaneous debulking was then treated with ESWL. Not surprisingly, gravel persisted in the mid and lower pole (Fig. 12-4D), but could be easily visualized with the flexible nephroscope and irrigated out through the upper pole nephrostomy tract.

The choice of upper pole access in this complex situation allowed the patient to be rendered stone free (Fig. 12-4E) with a single puncture. A small (10%) pneumothorax did occur, but did not require any specific treatment.

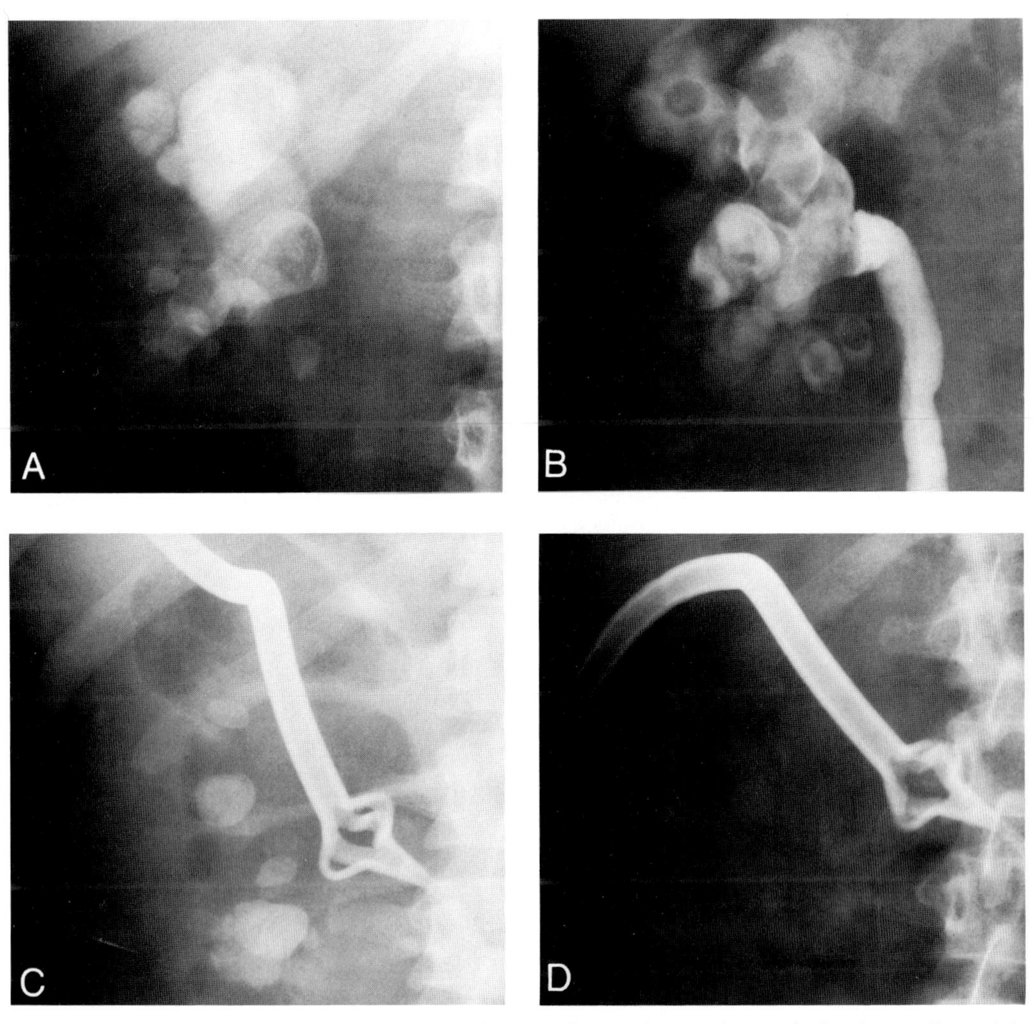


Fig. 12-4. **(A,B)** KUB film and retrograde pyelogram of extensive staghorn calculus in a solitary right kidney. **(C)** KUB film after percutaneous debulking via a supracostal approach. **(D)** KUB film 24 hours after a single ESWL treatment. (*Figure continues.*)

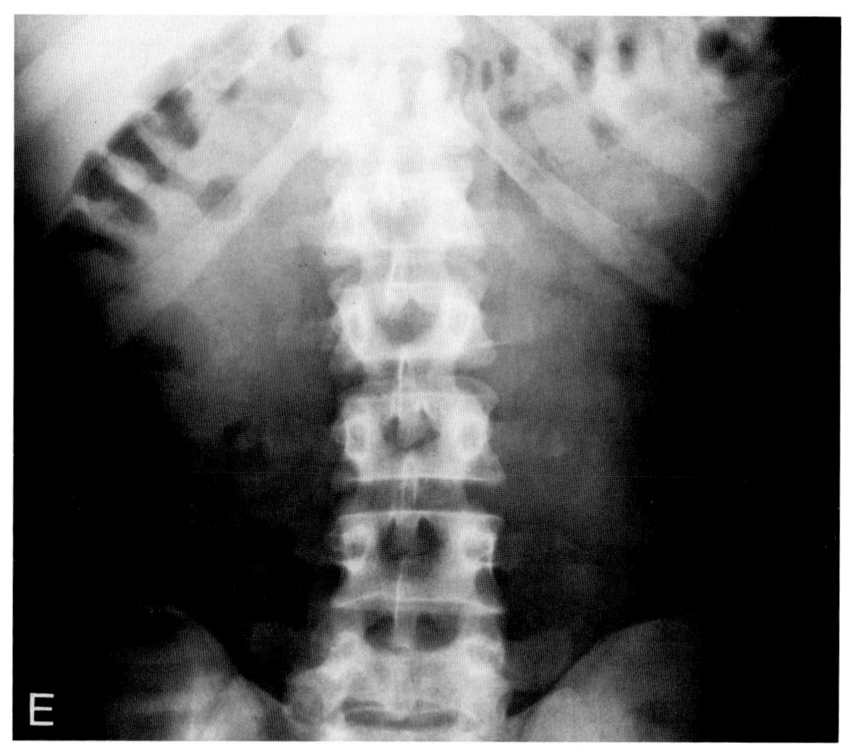


Fig. 12-4 (*continued*). **(E)** KUB film 2 months after discharge.

## COMPARISON TO ANATROPHIC NEPHROLITHOTOMY

The combination of percutaneous debulking and ESWL in our institution represents a reasonable alternative to anatrophic nephrolithotomy for various reasons. Convalescence and hospital stay are shorter when compared to anatrophic nephrolithotomy[19] (Table 12-3). Blood loss is less than with anatrophic nephrolithotomy. Stubbs, Resnick, and Boyce[20] reported an average transfusion rate of 2.7 units per case for anatrophic nephrolithotomy, compared to 0.4 units at the Methodist Hospital of Indiana for our approach. Preservation of renal function is comparable to that with anatrophic nephrolithotomy.[19,20] Complications associated with open surgery such as atelectasis, pneumothorax, and wound infection are less common using a combined approach. Since patients have little incisional pain, early ambulation is the rule, thereby minimizing the risk of deep venous thrombosis. There were no instances of delayed hemorrhage requiring transfusion or open operation in our series. Hospital and physician costs were slightly higher in our institution when the

**Table 12-3. Length in Days of Stay (LOS) and of Postoperative Stay (POS) of Patients after Treatments for Staghorn Stones**

| | Percutaneous Alone | ESWL Alone | Percutaneous + ESWL | Total | Comparison Anatrophic[a] |
|---|---|---|---|---|---|
| No. of patients | 14 | 2 | 36 | 52 | 22 |
| LOS | 8.1 | 17 | 14.2 | 12.6 | 16 |
| POS | 5.2 | 13 | 10.8 | 9.5 | 12.9 |

[a] Performed at the Methodist Hospital of Indiana, 1984–85.

combined approach was compared to anatrophic nephrolithotomy,[21] primarily owing to the need for 2.6 procedures per renal unit. The rate of residual fragments (15%) is similar to published results with open surgery.[6,16,19,22,23] Although these techniques are less invasive, they do not make management of metabolic factors less important[24] nor should the tenet of leaving no residual struvite stones be discarded.[5,16,25] Significantly, only 9 percent of renal units containing struvite were left with residual stones.

## CONCLUSIONS

Combined percutaneous debulking and ESWL are highly effective in completely removing most staghorn calculi. Advanced endourologic skills are required for good results. Nephrostolithotomy should be performed as the initial procedure, except when dealing with smaller upper-pole partial staghorn calculi. Morbidity is lower than with comparable open surgical techniques, and convalescence is shorter, although in-hospital costs are probably slightly higher.

Since percutaneous and extracorporeal shock wave lithotripsy may be used repetitively without increasing technical difficulty, this is an obvious advantage over repeated open operations for staghorn calculi. Patients at high risk for recurrence, such as those with cystinuria, intestinal hyperoxaluria, or recalcitrant urea-splitting bacteriuria, would benefit most from this less invasive approach.

## REFERENCES


1. Boyce WH: Surgery of urinary calculi in perspective. Urol Clin N Am 10:585, 1983
2. Boyce WH: Nephrolithotomy. In Glen JF (ed): Urologic Surgery. 3rd Ed. JB Lippincott, Philadelphia, 1983
3. Kahnoski RJ, Lingeman JE, Coury TA et al: Combined percutaneous and extracorporeal shock wave lithotripsy for staghorn calculi: An alternative to anatrophic nephrolithotomy. J Urol 135:679, 1986
4. Lingeman JE, Coury TA, Newman DN et al: Comparison of results and morbidity of percutaneous nephrostolithotomy and extracorporeal shock wave lithotripsy. J Urol (in press)
5. Griffith DP: Infection induced stones. In Coe FL (ed). Nephrolithiasis, Pathogenesis and Treatment. Year Book, Chicago, 1978
6. Krieger JN, Rudd TG, Mayd ME: Current treatment of infectious stones in high-risk patients. J Urol 132:874, 1984
7. Elder JS, Gibbons RP, Bush WH: Ultrasonic lithotripsy of a large staghorn calculus. J Urol 131:1152, 1984
8. Clayman RV, Surgy V, Miller RP et al: Percutaneous nephrolithotomy: An approach to branched and staghorn renal calculi. JAMA 250:73, 1983
9. Segura JW, Patterson DE, LeRoy AJ et al: Percutaneous removal of kidney stones: Review of 1,000 cases. J Urol 134:1077, 1985
10. Segura JW, Patterson DE, LeRoy AJ, Williams MJ: Percutaneous removal of staghorn calculi: Review of 86 cases. J Urol: 133:182A, 1985
11. Resnick MI: Evaluation and management of infectious stones. Urol Clin N Am 8:265, 1981
12. Brannen GE, Bush WH, Correa RJ et al: Kidney stone removal: Percutaneous versus surgical lithotomy. J Urol 133:6, 1985
13. Segura JW, Patterson DE, LeRoy AJ et al: Percutaneous lithotripsy. J Urol 130:1051, 1983
14. Wickham JA, Kellett MJ: Percutaneous nephrolithotomy. Br J Urol 53:297, 1981
15. Alken P, Hutchenreiter G, Gunther R, Marberger M: Percutaneous stone manipulation. J Urol 125:463, 1981
16. Silverman DE, Stamey TA: Management of infectious stones: The Stanford experience. Medicine 62:44, 1983
17. Preminger GM, Clayman RV, Hardeman SW et al: Percutaneous nephrostolithotomy vs. open surgery for renal calculi. JAMA 254:1054, 1985
18. Young AT, Hunter DW, Castaneda-Zuniga WR et al: Percutaneous extraction of urinary calculi: Use of the intercostal approach. Radiology 154:633, 1985
19. Boyce WH, Elkins IB: Reconstructive renal surgery following anatrophic nephrolithotomy: Follow-up of 100 consecutive cases. J Urol 111:307, 1974
20. Stubbs AJ, Resnick MI, Boyce WH: Anatrophic nephrolithotomy in the solitary kidney. J Urol 119:457, 1978
21. Mosbaugh PG, Newman DM, Lingeman JE et al: Cost comparisons of the options currently

available in the treatment of upper urinary tract stone disease. Report of American Urologic Association Ad Hoc Committee to study the safety and clinical efficacy of current technology of percutaneous lithotripsy and non-invasive lithotripsy. American Urologic Association, May 16, 1985

22. Wickham JEA, Coe N, Ward JP: One hundred cases of nephrolithotomy under hypothermia. J Urol 112:702, 1974
23. Vargas AD, Bragin SD, Mendez R: Staghorn calculus: Its clinical presentation, complications, and management. J Urol 127:860, 1982
24. Preminger GM, Peterson R, Peters PC, Pak CYC: Current role of medical treatment of nephrolithiasis: Impact of improved techniques of stone removal. J Urol 134:6, 1985
25. Sant GR, Blaivas JG, Meares EM: Hemacidrin irrigation in the management of struvite calculi: Long-term results. J Urol 130:1048, 1983

# Treatment of Large Renal Calculi by ESWL

Herbert Brandl

While small renal pelvic stones and renal caliceal stones are handled with relatively simplicity by extracorporeal shock wave lithotripsy (ESWL), the treatment of staghorn stones by ESWL requires an expertise on the part of the urologist which exceeds the mere knowledge of ancillary procedures. With increasing stone mass, additional therapeutic pathways must be explored—nephrolithotomy, percutaneous lithotripsy, and ureteroscopic relief of the steinstrasse.

## PREPARATIVE METHODS

In addition to standard preoperative diagnostic studies (coagulation studies, intravenous pyelogram (IVP), urinalysis, urine culture, bowel prep), an oblique abdominal radiographic view should be prepared, especially with staghorn stones, to estimate in two dimensions the size and approximate volume of the stone. If the stone mass is found to be large, it is advisable to remove the greater part of the stone with the help of percutaneous lithotripsy, and then disintegrate the rest of the calculus with ESWL.

If a urinary tract infection is present, antibiotics must be administered for 48 hours before ESWL; they should be continued after ESWL and after discharge from the hospital.

## TREATMENT OF PARTIAL STAGHORN STONES

### ESWL Monotherapy—Single Treatment

Smaller partial staghorn stones (< 5 mm thick) can be pulverized with one shock wave session (Fig. 12-5). However, complete disintegration of a stone is not assured at the beginning of the treatment, and it is sensible to focus first on the stone mass at the entrance to the ureter, second on stone in the pelvis, and last on stone in individual calices. Since complications in the form of subcapsular and perirenal hematomas increase with increasing shock energy and shock number, in our clinic we do not exceed a shock energy of 18 to 23 kV and 1,500 to 2,000 individual shocks per treatment.

### ESWL Monotherapy—Multiple Treatments

If partial staghorn stones are not completely reduced to spontaneously dischargeable fragments in one 2,000 shock treatment, additional treatments may be necessary. After all disintegrated material and small fragments have been spontaneously discharged, the destruction of

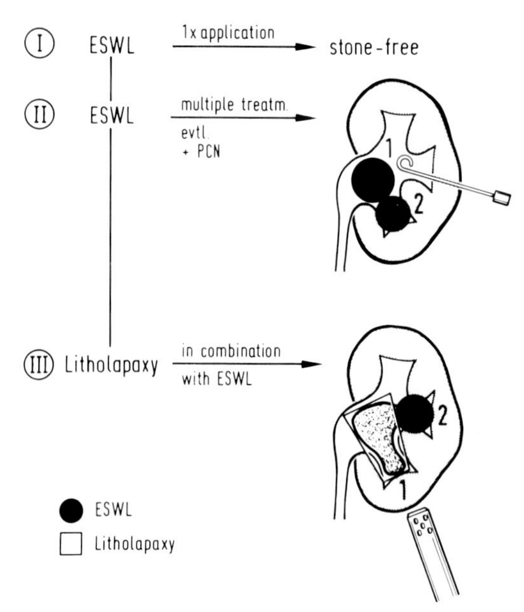


Fig. 12-5. Strategies for the treatment of partial staghorn stones.

residual caliceal remnants may be completed in a second or third treatment (Fig. 12-5). However, consecutive steinstrassen can lead to prolonged obstruction which must be relieved through percutaneous nephrostomy. Percutaneous relief is especially indicated in the case of infection stones receiving antibiotic treatment. Decompression of the obstructed kidney avoids pyelonephrosis with possible urosepsis.

Distal ureteral steinstrassen can result in ureteral obstruction and a failure to discharge fragments for several days. Placing a percutaneous nephrostomy tube frequently reestablishes spontaneous discharge of the stone fragments, and the patient is made stone free in a shorter amount of time. This phenomenon is due to reestablishment of ureteral peristalsis after relief of obstruction.

Percutaneous nephrostomy is a genuine alternative to transurethral or ureteroscopic stone removal. Should larger stone fragments (so-called "pilot stones") lodge at the distal end of a steinstrasse and block stone discharge for several days, they must be removed with a basket or ureteroscope before a second or third ESWL treatment can be attempted.

## Combination Therapy: Percutaneous Nephrolithotomy and ESWL

Partial staghorn stones with larger stone burdens (more than 5 mm thickness in oblique abdominal projection) should be treated with percutaneous nephrolithotomy and ESWL in combination. This combined therapy minimizes the stone burden to be passed and reduces the chance of prolonged obstruction which may lead to degradation of the kidney function.

In the past, stone portions in the lower pole calices were treated with ESWL to facilitate placement of the nephrostomy tube. However, this method is no longer needed because of increased experience and improved nephrolithotomy technique. At present, the primary treatment for stones in lower pole calices, other accessible calices, and the renal pelvis is percutaneous removal; any remaining concretion is then treated with ESWL (Fig. 12-6). This technique largely avoids difficult and lengthy percutaneous manipulations such as the use of the

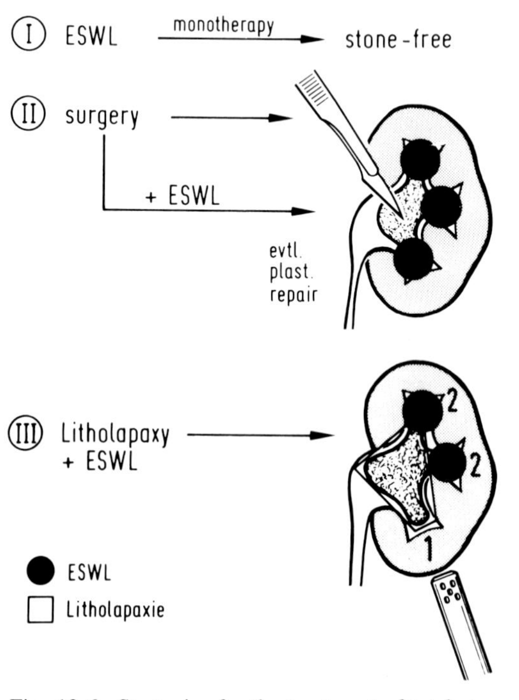


Fig. 12-6. Strategies for the treatment of total staghorn stones.

flexible nephroscope on branched calices and the insertion of numerous nephrostomy tubes. These techniques are now indicated only in special cases, for example where there is caliceal stenosis.

Between the percutaneous nephrolithotomy and the ESWL treatment, there should be a pause of 5 to 8 days. ESWL done too soon after percutaneous nephrolithotomy can lead to perirenal or intrarenal hematomas which can even require open surgery. However, in our experience of more than 2,800 patients, this problem was observed only once.

The nephrostomy tube is left in place until after ESWL to facilitate fragment discharge; stone particles are also discharged through a nephrostomy tube. Stone fragments can be washed out by irrigation of the lumen, and chemolysis can be conducted through the nephrostomy tube (Fig. 12-7).

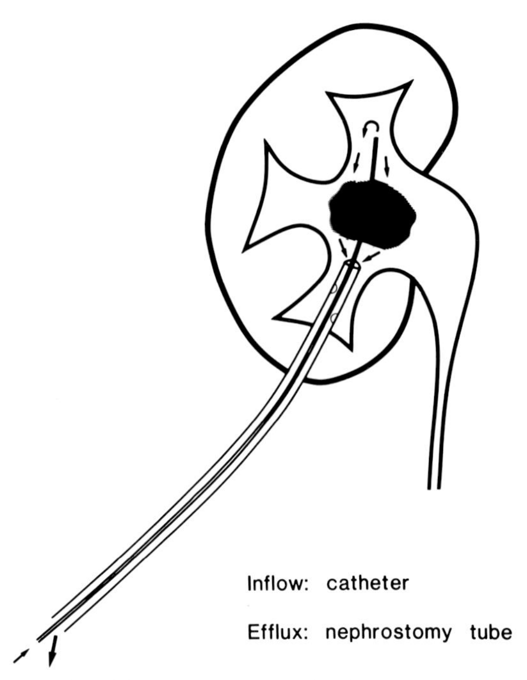


Fig. 12-7. Arrangement of irrigation setup.

## TREATMENT OF TOTAL STAGHORN STONES

### Monotherapy

Total staghorn stones with smaller masses (up to 5 mm thick in the oblique abdominal projection) can be eliminated through ESWL monotherapy in one or more treatments, just as can partial staghorn stones (Fig. 12-8). Note that first the pelvic stone portions must be destroyed before those in the calices can be treated, especially the medial portion of the stone near the ureteropelvic junction.

### Combination Therapy: Open Surgery and ESWL

Open stone removal is currently indicated only if the patient refuses the time-consuming multiple ESWL and percutaneous therapies, or if anatomical abnormalities of the urinary tract require surgical correction. Infundibular stenosis or severe stenosis of the ureteropelvic junction, unapproachable by percutaneous endopyelotomy, would be examples.

Should it be discovered during surgery that certain portions of the staghorn stone are too difficult to reach through a pyelotomy and would necessitate additional nephrostomies, these stones may be left in situ and after recovery from surgery (about 8 to 10 days after operation) may be disintegrated with ESWL. The option of using ESWL shortens the operation and spares the parenchyma from numerous nephrotomies. Nephrostomy tubes inserted during surgery may be left in place during ESWL treatment.

### Combination Therapy: Percutaneous Nephrolithotomy and ESWL

As with partial staghorn stones, total staghorn stones with larger stone masses should first be removed from the lower pole calices, the renal pelvis, and the accessible calices by means of percutaneous nephrolithotomy, and after 5 to 8 days the remaining stone mass may be treated with ESWL (Fig. 12-9).

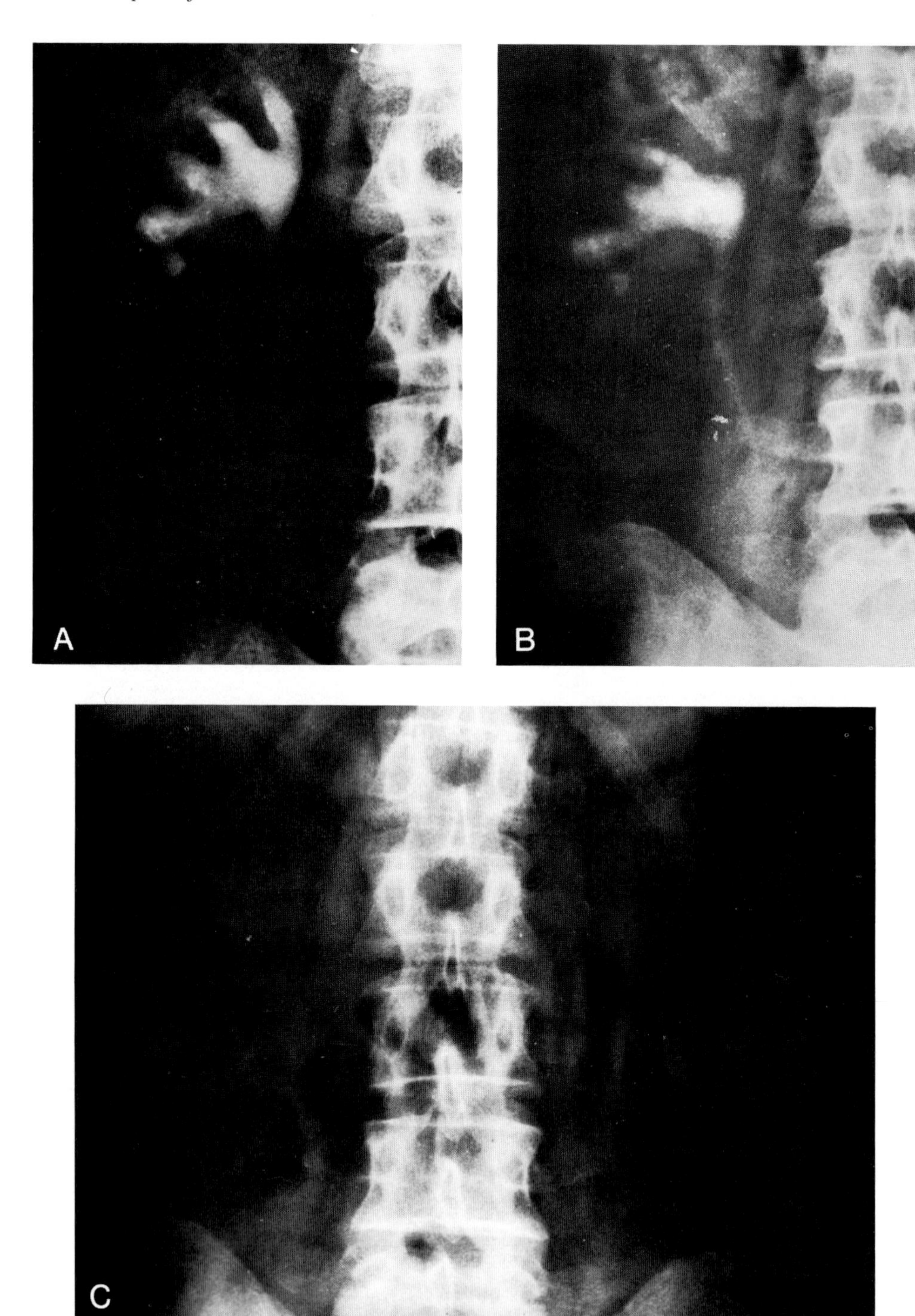


Fig. 12-8. ESWL monotherapy of a total right staghorn calculus (57-year-old woman). **(A)** Stone before ESWL. **(B)** Shortly after ESWL, the stone is largely disintegrated, and the upper third of the ureter is already filled with stone particles. **(C)** Five days after ESWL, patient is stone free with no auxiliary treatments.

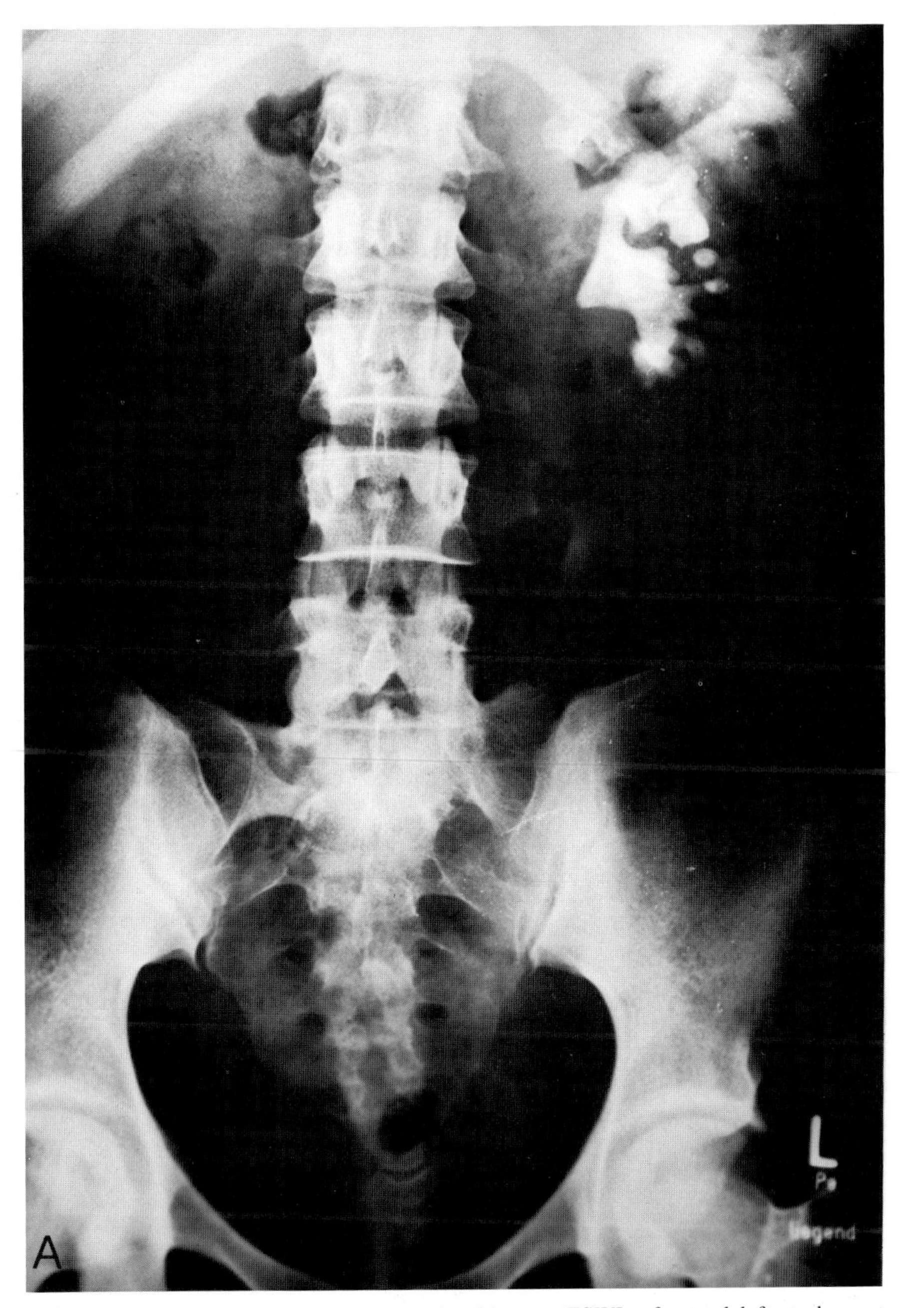


Fig. 12-9. Combination therapy: percutaneous nephrolithotomy/ESWL of a total left staghorn stone in a 26-year-old woman. (**A**) Stone before treatment. (*Figure continues.*)

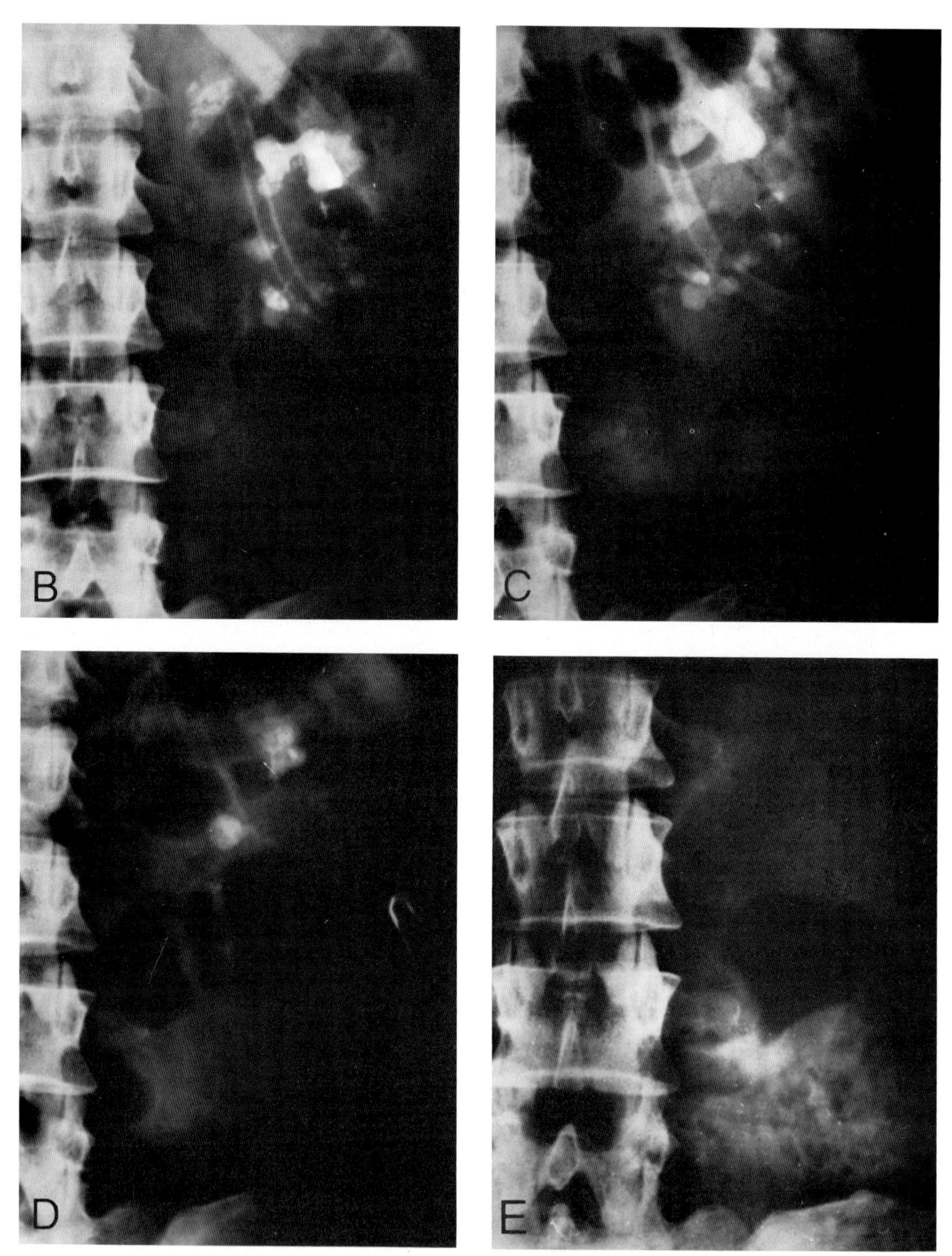


Fig. 12-9 (*continued*). **(B)** Portions of the stone after percutaneous nephrolithotomy with installed nephrostomy tube. **(C)** Disintegrated stone particles after ESWL (8 days after percutaneous nephrolithotomy). **(D)** Stone fragments are noticeably absent after spontaneous discharge through the ureter and the nephrostomy tube (12 days after percutaneous nephrolithotomy). **(E)** Patient is stone free 19 days after percutaneous nephrolithotomy and 11 days after ESWL. The nephrostomy tube was removed.

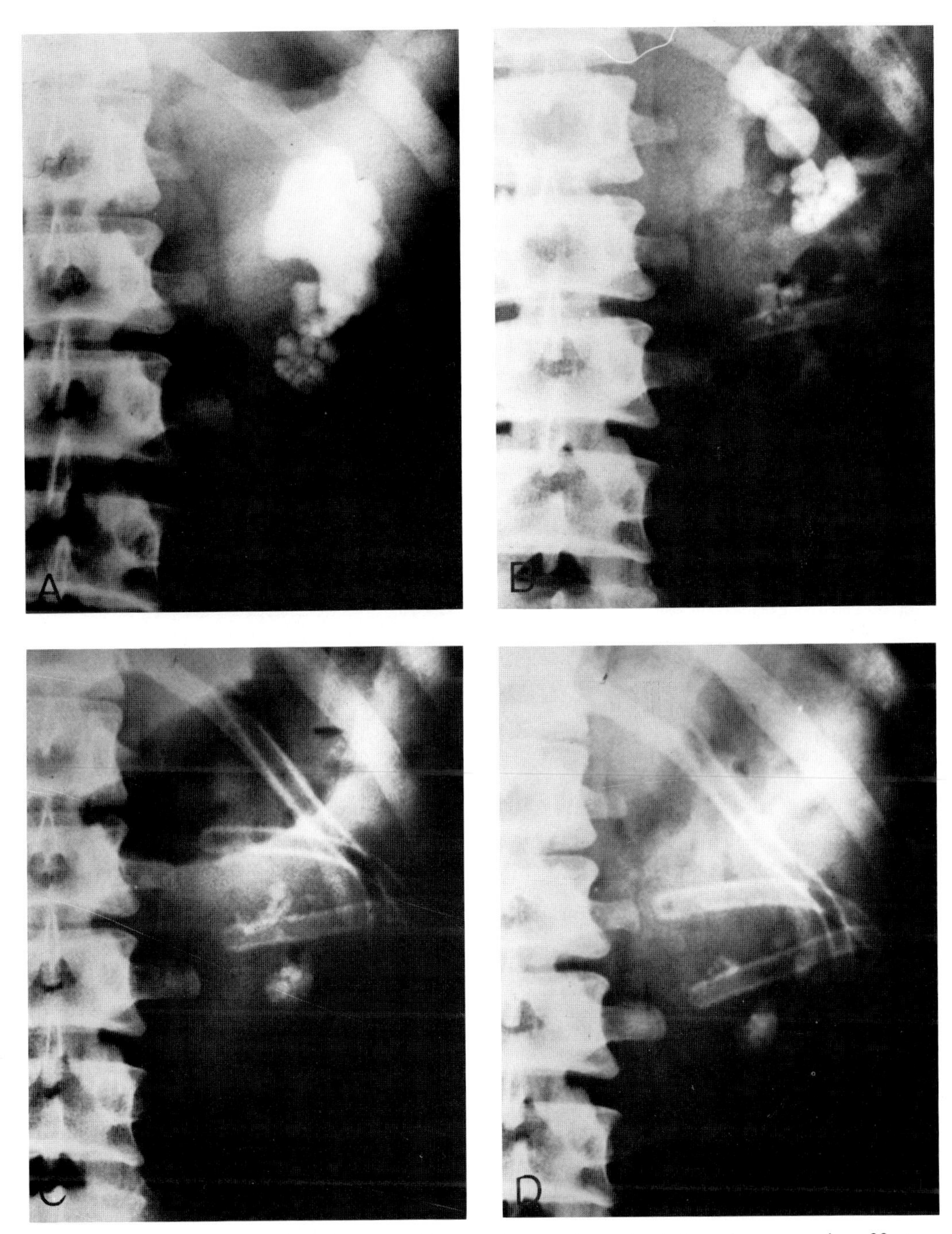


Fig. 12-10. Percutaneous nephrolithotomy, ESWL, and chemolysis of a left staghorn stone in a 22-year-old man. **(A)** Stone before treatment. **(B)** Stone remaining after the first percutaneous nephrolithotomy. **(C)** Stone remaining after the second percutaneous nephrolithotomy (three nephrostomy tubes were placed). **(D)** Stone after ESWL treatment. The remaining concretions are well disintegrated. (*Figure continues.*)

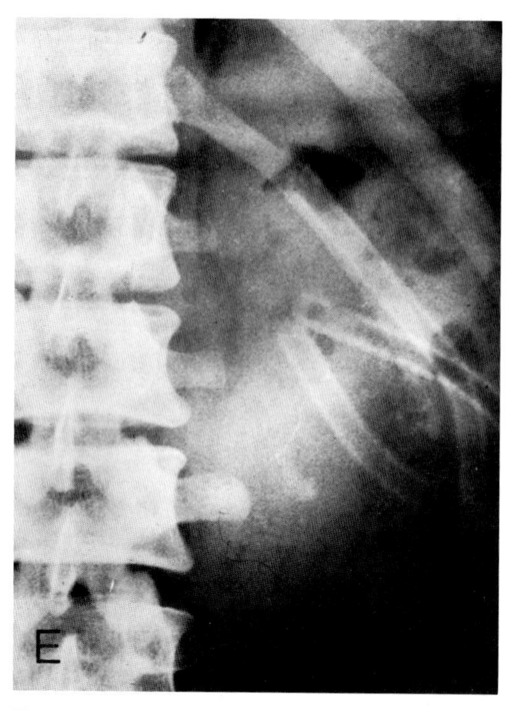


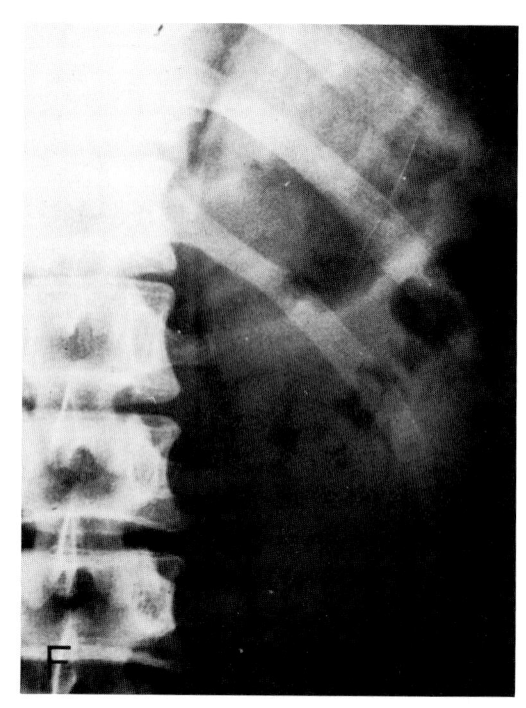


Fig. 12-10 (*continued*). **(E)** The stone particles that were not spontaneously discharged from the lower pole calices were irrigated out through the nephrostomy tube. **(F)** The patient is stone free after 10 days of percutaneous chemolysis. The nephrostomy tube was removed (27 days after the first percutaneous nephrolithotomy, 22 days after the second percutaneous nephrolithotomy, and 15 days after ESWL).

## Combination Therapy: Percutaneous Nephrolithotomy, ESWL, and Chemolysis

Residual fragments of infection stones or cystine stones not expelled after percutaneous nephrolithotomy and ESWL may be washed out by irrigation or dissolved by percutaneous chemolysis. Chemolytic agents may be introduced through a percutaneous nephrostomy tube; for infection stones, this agent is Renacidin (magnesium citrate); for cystine stones, either THAM-E (a buffer solution with pH 10.2) or a 2.5 percent solution of *N*-acetyl-L-cysteine can be used. Figure 12-7 shows the catheter arrangement for local irrigation therapy; the inflow is through a thin ureteral catheter which is pushed through a percutaneous nephrostomy tube. The irrigation solution flows around the concretion particles and travels out through the nephrostomy tube.

By means of percutaneous chemolysis, stone particles can be dissolved within 5 to 14 days, depending on their size (Fig. 12-10).

## SUMMARY

When ESWL is used in combination with ancillary methods, 85 to 90 percent of all kidney stones can be treated successfully. Open operative procedures for stone disease are therefore indicated only in special cases, such as obstruction requiring surgical correction. Patients must be cooperative and be informed that the modern techniques of stone treatment are occasionally time consuming, especially with staghorn stones, and that numerous manipulation and extraction procedures may be required. However, combination therapy spares the patient open surgery and is therefore usually the best approach, especially for patients with recurrent stones.

# 13

# Outpatient Shock Wave Lithotripsy

Martin L. Madorsky

Since the first kidney stone in a human was disintegrated by extracorporeal shock wave lithotripsy (ESWL), more than 100,000 patients with renal or ureteral lithiasis have been treated by this method worldwide.[1] Certainly, the lithotripter has revolutionized the treatment of renal calculus disease.

However, because of the high capitalization required, not every hospital will have a lithotripter and, furthermore, not every urologist will have access to ESWL to treat his patients. A urologist may not have admitting privileges at a hospital where a machine has been installed, or may not have a practice volume that warrants the training period at a specified training center required to become proficient in the operation of the ESWL machine.[2] Urologists with small practices or those specializing in uro-oncology or pediatrics may not want to make the commitment of time or expense to gain this expertise.

With the above in mind, a geographically centralized outpatient stone treatment center was developed in Fort Lauderdale, Florida. This concept gives patients of regional urologists access to ESWL. Soon, urologists in Los Gatos, California and Winston Salem, North Carolina opened ESWL centers to be used primarily for treatment of outpatients.

The safety of outpatient ESWL has been a subject of controversy. Although the Food and Drug Administration (FDA) investigational protocol required patient hospitalization, several of the original FDA investigators felt that outpatient ESWL would be safe for selected patients. Griffith analyzed the hospital stay of ESWL patients at the Klinikum Grosshädern in Munich, West Germany, and reported that the average stay after ESWL (6 days, range 1 to 19 days, $n = 47$) was less than after percutaneous nephrostolithotomy (7 days, range 11 to 22 days; $n = 98$).[3] However, during the early phase of the Munich clinical trials, patients were observed carefully as inpatients, and numerous social and demographic considerations (worldwide referrals, government-supported health care system for German citizens, and specific logistic considerations) often precluded earlier discharge. Since 65 percent of the ESWL patients in this study required no postprocedural analgesia, it seemed logical to perform the procedure in an outpatient setting on selected patients with small stones and normal medical histories.

While currently a common and accepted procedure, outpatient surgery is a relatively new concept. Not until 1970 were free standing surgical centers established.[4] Currently it is estimated that between 20 and 40 percent of all hospital inpatient surgery can be done safely in an ambulatory setting.[4] In 1978, Natof showed that complications resulting from a lack of postoperative observation in the hospital were rare and did not outweigh the benefits of outpatient surgery.[5]

In Munich, all patients were admitted to Grosshädern hospital prior to lithotripsy. In the United States, because the initial FDA protocol required patient hospitalization, only 3 percent of the patients undergoing ESWL during the investigational period were treated as outpatients.[2] In most centers, the minimum stay was 48 hours, with a range of 2 to 14 days depending on stone size and type, presence of urinary infection, or postprocedural complications.[6,7]

Most of the original FDA investigational facilities were rapidly converted for lithotripter installation and were not designed to treat outpatients. After FDA approval was received in December, 1984, many new centers across the United States were designed to facilitate outpatient ESWL treatments. Considering the demand by both insurance companies and the government for increased outpatient medical care, this seemed to be a prudent and practical approach.

## THE OUTPATIENT FACILITY

The report of the National Kidney Foundation on Extracorporeal Shock Wave Lithotripsy outlines the physical characteristics of an ESWL facility.[8] An outpatient facility must be a complete medical unit equipped to handle all of the anticipated medical problems associated with the procedure. It should be designed solely for ESWL, which will result in an effective patient flow and treatment schedule. The outpatient ESWL unit should be near a full-service hospital for treating all medical and surgical emergencies that may occur during and following treatment. Staff requirements and equipment are the same as for an outpatient ambulatory surgical center. By operating the center as an outpatient facility and referring post-treatment patients back to their own urologist for follow-up, all of the hospitals used by urologists in the service area can participate in the medical and surgical treatment of patients treated by ESWL.

The unit itself should be contained in an area of 2,500 to 4,000 square feet. The Dornier lithotripter requires a dedicated room of approximately 20 ft × 20 ft with a room height of 10.25 ft. A room containing the water treatment center and hydraulic system of the lithotripter is also necessary.

A cystoscopy suite should be near if not next to the ESWL machine, enabling the urologist-operator to place ureteral catheters and stents, perform retrograde urographic studies, and occasionally manipulate stones. Ideally, hard-copy radiographic and fluoroscopic capabilities should be available in the cystoscopy suite. Having the cystoscopy suite close to the ESWL machine increases the efficiency and safety of treatment. A hard-copy radiographic film is taken immediately before treatment and is essential in determining stone location. Radiographs obtained just before discharge are used to advise the patient and the referring urologist of the results of treatment.

Full anesthesia facilities are essential, including a complete pre- and postoperative area with cardiac monitoring, suction, oxygen, medications, and a qualified nursing staff (Fig. 13-1). Bioengineering support services must be rapidly available, and a waiting area, business areas, dressing rooms, film viewing area, and conference room are desirable options.

Most centers have also found that the lithotripter unit requires a rather large support staff. The minimal staffing in an outpatient center should include a pre- and post-treatment nurse as well as a nurse to assist in patient care during cystoscopy and lithotripsy. Assistants to help move and position that patient are also essential. A minimum of two clerk/receptionists as well as an x-ray technician and treating urologist are also required. The 3,500 ft$^2$ facility of the Kidney Stone Center of South Florida requires a minimum of seven full time staff to maintain efficient treatment schedules.

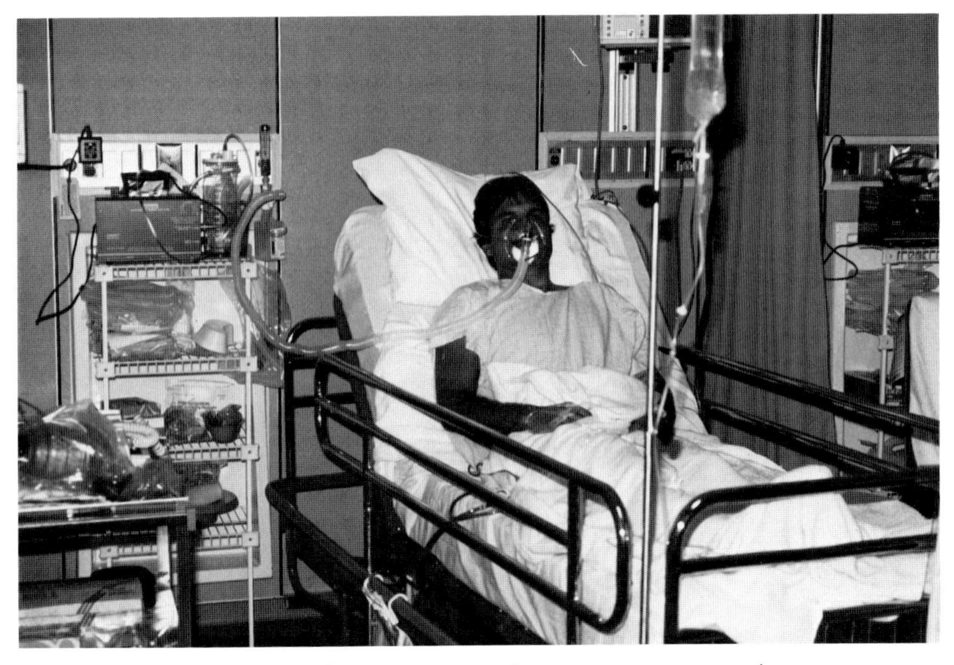

Fig. 13-1. Patient in recovery room after treatment as an outpatient.

## INVOLVEMENT OF THE UROLOGIST

An outpatient facility should be designed to accommodate the needs of both patients and their referring or treating urologists. Rapid, easy access to the technique and close communication between operator and referring urologists are essential.

In agreement with the recommendations of the ad hoc committee, all ESWL operators must be urologists and must have completed a mini-residency, participating in the care of a minimum of 30 ESWL cases at one of the designated training centers.[2] Each newly trained urologist should be carefully observed for 2 or 3 treatment days as he assumes the primary responsibility of the treatment at the outpatient ESWL center.

Urologists not trained or not interested in performing ESWL treatments still remain the primary urologists for their patient. In addition to evaluating stone patients for referral to the lithotripsy center, they follow their patients after stone fragmentation, and perform secondary procedures as indicated.

Since a stone treatment center is not a complete urologic center, all patients referred for treatment should be referred by a urologist. This minimizes the chance for misdiagnosis of associated diseases, such as tumors, bleeding disorders, or congenital abnormalities. Since 7 to 10 percent of patients treated will require a secondary procedure (cystoscopy with stents, percutaneous nephrostomy, ureteroscopy, or rarely ureterolithotomy)[7,9] the patient must be followed by a urologist after treatment to ensure satisfactory post-ESWL care.

## PATIENT SCREENING

All patients referred for outpatient treatment should have prior evaluation by a urologist. The patient questionnaire and radiographs (which include a recent kidney–ureter–bladder (KUB) film and representative films from a contrast study) are then sent to the ESWL-trained urologist at the outpatient center for patient screening prior to treatment (Fig. 13-2). At the time of

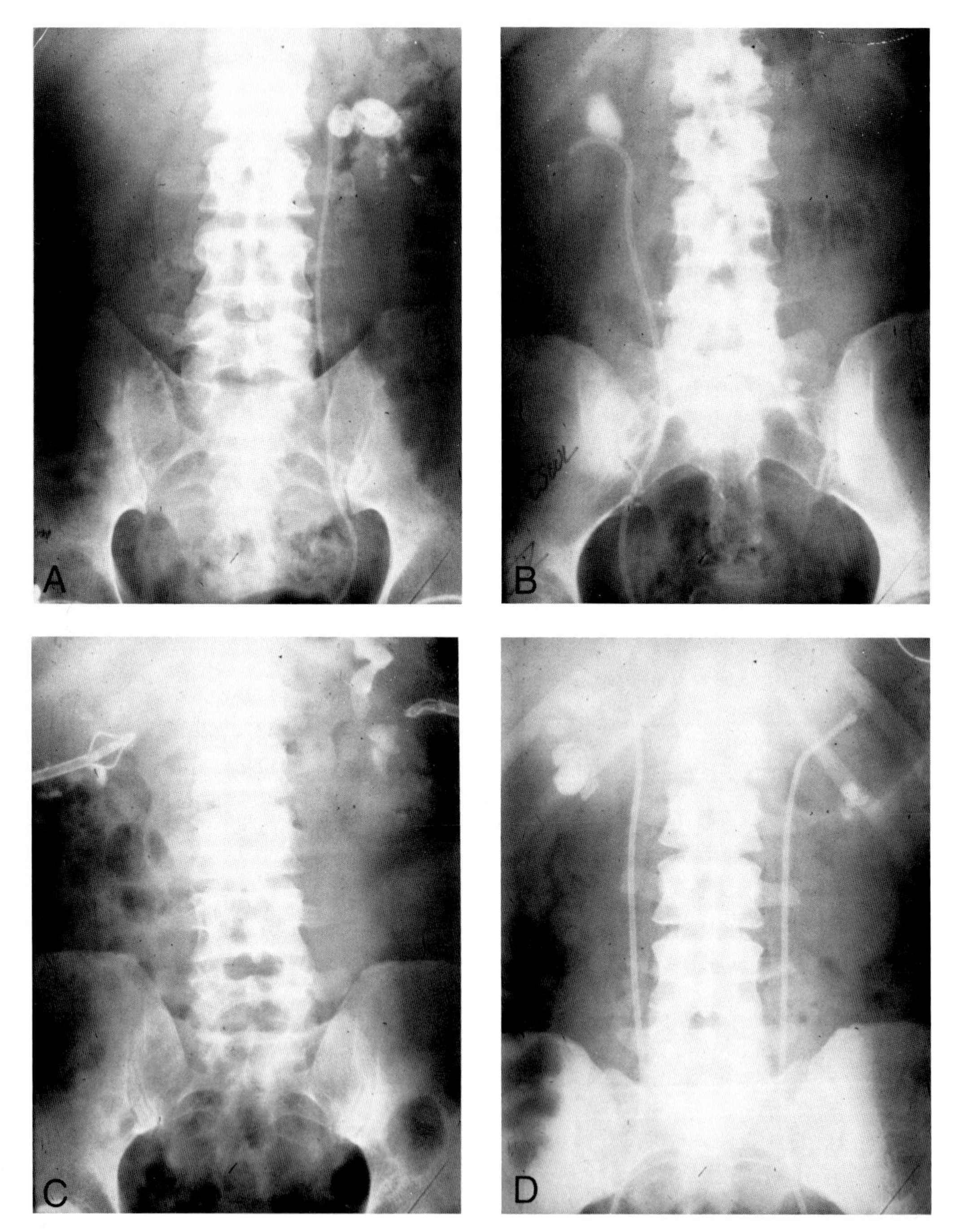


Fig. 13-2. Radiographs of typical stones suitable for outpatient treatment.

review, plans for preprocedural stenting, perioperative antibiotics, medical clearance, and type of anesthesia are made by the reviewing urologist.

Once approved for treatment, the patient is asked to come to the outpatient lithotripter facility 1 to 2 days before treatment. At that time the patient is given a chest x-ray, electrocardiogram, hemogram with bleeding profile (prothrombin time (PT), partial thromboplastin time (PTT)), electrolytes, and other necessary laboratory procedures. The patient and interested family members then view an educational film about ESWL and talk with a treating urologist and anesthesiologist. The patient is given a tour of the facility by a staff nurse to view the pre- and post-treatment areas, as well as the lithotripter. This tour alleviates many of the patient's anxieties regarding the procedure. Careful pretreatment evaluation eliminates patients with significant cardiopulmonary risks and identifies those patients with bleeding diatheses.

Patient selection adheres to the FDA classification criteria for ESWL.[6,7,9] After careful review and planning, all types of category A and category B patients have been accepted by our unit, but any patient with a cardiac pacemaker, renal artery calcifications, or severe cardiopulmonary compromise has been rejected. Patients whose habitus precluded proper patient positioning on the ESWL gantry have also been rejected.

## PATIENT TREATMENT

Upon arrival at the center on the day of treatment, the patient is again questioned regarding his medical history and whether he has been without food or water since midnight.

Intravenous fluids are started and the EKG electrodes are positioned. Both the IV site and electrodes are covered with waterproof sealing (Steri-Drape, 3M). The patient is then ready for treatment.

At our center, the vast majority of patients have been treated under general inhalation anesthesia, most commonly high frequency jet ventilation (HFJV). HFJV was first used or investigated for ESWL at the University of Florida.[10,11] This technique helps keep the stone at F2 by decreasing diaphragmatic movement.

Epidural anesthesia is not desirable for outpatient ESWL because patients may take up to 3 hours to recover lower extremity motor function which may delay discharge from the outpatient center. Also, the possibility of a "wet" epidural attempt (subdural injection of anesthetic) always exists; patients with dural penetration should be kept at bed rest for 6 to 10 hours. At the Kidney Stone Center of South Florida, two patients have been treated with local infiltration and paravertebral blocks with xylocaine. This treatment, however, requires a very cooperative patient and some intravenous sedation.

All patients with ureteral calculi are treated with either an indwelling double-J stent or ureteral catheter. Patients with a stone burden over 20 mm have double-J stents inserted prior to ESWL. In most cases, this cystoscopic procedure is done under general anesthesia just before placing the patient on the gantry.

During the procedure, most patients are treated with intravenous antibiotics. Small caliceal stones (<6 mm) are treated without antibiotics if there is no evidence of acute or chronic infection. All patients with ureteral stones or stones greater than 8 mm in diameter are given antibiotics to prevent obstructive pyelonephritis. If cystoscopic manipulation has been performed, our choice is either a long-acting cephalosporin or an aminoglycoside.

Patients receiving more than 1,500 shocks are usually treated with 10 to 20 mg of furosemide. This treatment has two major benefits. First, it helps clear the hematuria and thereby reduces post-treatment colic caused by blood clots. This, in turn, facilitates the early removal of catheters and subsequent discharge of the patient from the treatment center. Second, furosemide helps restore to normal the elevations in right atrial and pulmonary capillary wedge pressures and further eliminates the possibility of post-ESWL pulmonary edema that can occur with high frequency jet ventilation anesthesia.[12]

At the completion of the procedure, the patient is sent to the recovery room where he is monitored as after any surgical procedure. If a Foley catheter has been inserted, it is usually removed once the patient is alert and gross hematuria has cleared. Standard ureteral catheters are removed before discharge. Double-J stents can be removed after a few days in the referring urologist's office. This can be done without cystoscopy simply by pulling on a monofilament suture attached to the stent and secured outside the urethra. Intravenous fluid therapy is maintained until the patient is alert, stable, and tolerating oral fluids without difficulty. For immediate post-treatment nausea, patients are treated with 10 mg IV metoclopramide HCl (Reglan). This drug is safe, efficacious, and has minimal side effects.

Analgesia immediately after ESWL treatment is provided by 1 to 2 ml of fentanyl (Sublimaze) injected intravenously, when needed. Fentanyl takes effect almost immediately, and its analgesic effect lasts 30 to 60 minutes, so it is ideal for outpatient use. Fentanyl minimizes nausea and the prolonged narcotizing effect that often prolongs recovery time after any type of surgery. Once the patient is ready for discharge from the center, prescriptions are given for oral antibiotics, analgesics, and antiemetic suppositories.

From this point on, the patient's care is assumed by the referring urologist. ESWL-treated patients should be followed on an outpatient basis in the same way as any patient with a passable distal ureteral calculus. Each patient is instructed to contact his physician for signs of significant fever, severe uncontrolled pain, nausea, or any other unexpected reaction. Patients are advised to visit their aftercare urologist 24 to 72 hours after treatment. In most uncomplicated cases, double-J catheters are removed 3 to 5 days after ESWL treatments. In patients with evidence of infection or severe pain, the urologist may elect to leave the stents in place for a few more days.

At 30 to 45 days after ESWL treatment, questionnaires are sent to the patient's aftercare physician. These questionnaires, designed to be completed in just a few minutes, record the post-treatment course, including pain, complications, hospitalizations, and final stone-free results.

## RESULTS

Although U.S. data are preliminary, it appears that outpatient ESWL from selected patients can be safely performed without a greater complication rate than previously reported for inpatients.[6,7] Each outpatient center has developed an individualized protocol for patient management, and certainly the definition of outpatient varies from unit to unit.

At the Piedmont Stone Center, any patient with a stone larger than 23 to 25 mm in maximum diameter will have a 6 mm double-J stent placed prior to the procedure to minimize the risk of post-treatment colic. In many cases, the stent is inserted by the referring urologist 1 week prior to treatment. After 1 week of internal stent drainage, if the stone treated is 3 cm or less in diameter, the double-J stent will be removed immediately after ESWL. For stones larger than 3 cm, indwelling ureteral stents are left in place. A 2–0 prolene suture attached to the distal end of the double-J stent allows easy subsequent removal by the referring urologist (personal communication, P. Coughlin and C. Reid, High Point, NC, 1986).

Urologists at the Piedmont Stone Center report that 20 percent of stone fragments from large stones (4 to 6 cm in diameter) pass around the stent within a week. After 2 weeks, usually 50 percent of fragments have passed, and after removal of the double-J stent, most of the remaining fragments pass through the dilated ureter. The incidence of secondary procedures, hospitalizations, and retreatments after stent removal is not reported, and the overall stone-free rate is not yet available (Table 13-1).

Of note, at the Northern California Kidney Stone Center at Los Gatos, patients requiring admission after ESWL were usually admitted the same day of treatment, and most went home the day after admission. The most common reason for admission was pain, followed by nausea and vomiting.[13]

**Table 13-1. ESWL at the Piedmont Stone Center (July 10, 1985 through January 21, 1986)**

| Treatment | Number |
|---|---|
| Outpatient ESWL treatments | 662 |
| Retreatments | 17 |
| Admission after ESWL | |
| Immediate | 4 |
| Within 1 month | 17 |

At the University of Alabama in Birmingham, the in-hospital location of the lithotripter allows inpatient or outpatient treatments. In an analysis of 543 patients treated with ESWL, less than 10 percent were admitted to the hospital the evening before ESWL.[14] Double-J ureteral stents were placed in 260 patients several days to several weeks before ESWL to relieve urinary obstruction or for large stones greater than 20 mm unaccompanied by obstruction. Whistle-tip ureteral catheters were placed in another 158 patients just before ESWL for poorly calcified or ureteral calculi. These were removed upon completion of ESWL.

After ESWL, 149 patients were sent home from the holding area within several hours, and 162 were admitted into a motel that is an integral part of the hospital and offers skilled nursing care. Most motel patients were able to return home on the next day. The 149 discharged patients and the 136 motel patients who required no further care constituted 55 percent of all patients treated. The other 26 patients placed in the motel required some form of nursing care, usually intramuscular injections for pain or nausea. Of these, 24 were able to return home the day after ESWL, and two required hospital admission for persistent pain.[14]

The remaining 231 of the 543 patients were admitted to the hospital either before or after ESWL. Reasons for admission varied; sometimes they were not based on medical necessity. Several insurance carriers required admission as a condition for reimbursement. Some referring physicians were uncomfortable with the idea of outpatient ESWL, and therefore requested admission. The mean postoperative stay for admitted patients was 1.8 days. Of the 231 patients admitted, 170 were discharged on the first postoperative day.

The type of anesthesia administered did not influence patient disposition. Of the 143 patients given general anesthesia, only 55 were admitted to the hospital, whereas 205 of the 395 patients given epidural anesthesia were admitted.[14]

Of the 543 patients, 81 (14.9%) required a secondary procedure, most often cystoscopy and manipulation of ureteral stone fragments. Ureteroscopy was performed in nine patients, percutaneous nephrostomy in 10, and open surgical stone removal in two. ESWL retreatment was necessary in 35 patients.[14]

At the Methodist Hospital in Houston, Texas, 600 outpatients treatments have been performed successfully, and currently, 40 percent of daily treatments are performed as outpatients (personal communication, B. Finlayson, March, 1986).

In a combined series from three outpatient centers, 550 patients were treated from June 20, 1985 to October 1, 1985 (Table 13-2). The post-treatment hospitalization rate was 7.8 percent; most patients were admitted for control of post-treatment pain.

Of note, these three outpatient centers treated more than 50 percent of their patients with either ureteral catheters or double-J stents. Double-J stents were used preoperatively for ESWL treatments where there was a large stone burden. (A stone burden is defined as the sum, in millimeters, of the length and width of all the stones; Fig. 13-3). A stone burden greater than 20

**Table 13-2. Combined Outpatient Study at Three Centers[a]**

| | Number | Percent |
|---|---|---|
| Total patients | 550 | |
| Post-ESWL hospitalization | 41 | 7.8 |
| Retreatments | 17 | 3.1 |
| Age range | 14–84 | |
| Ureteral catheters | 139 | 25.3 |
| Stents (double-J) | 137 | 24.9 |
| Anesthesia | | |
| General | 548 | 99.5 |
| Epidural | 2 | 0.5 |

[a] Los Gatos, California, Winston Salem, North Carolina, and Fort Lauderdale, Florida.

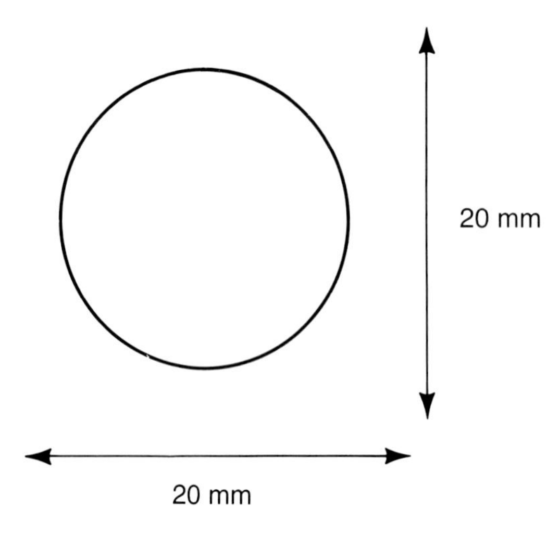


Fig. 13-3. Technique for measuring stone burden: if total stone measures 20 mm in both dimensions, the stone burden is 40 mm.

mm was felt to require pretreatment stenting (Table 13-3). Ureteral catheters were used to facilitate the treatment of ureteral calculi, nonopaque calculi, or kidneys with multiple stones. The catheters help visualize the stones during treatment by enabling the operator to use contrast material if necessary.

Patients were selected for treatment at these centers according to the previously discussed criteria. Patients with large staghorn calculi had nephrostomy tubes place 1 to 7 days before treatment. Patients with active infections were placed on therapeutic levels of antibiotics before treatment and maintained on antibiotics until stone free.

**Table 13-3. ESWL at Fort Lauderdale**

| | Number | Percent |
|---|---|---|
| Total patients | 223 | |
| Inpatients | 0 | 0 |
| Outpatients | 223 | 100 |
| Post-ESWL hospitalization | 22 | 10 |
| Age range | 11–84 | |
| Ureteral catheters | 73 | 32.7 |
| Stents (double-J) | 100 | 44.8 |
| Anesthesia | | |
| General | 221 | 99 |
| Local | 2 | 1 |
| Retreatments | 6 | 2.7 |

The patients averaged 2 hours in the recovery room before discharge from the center. Twenty-two patients were admitted to the hospital after treatment. Two patients required immediate hospitalization for pulmonary edema. One was an 82-year-old man with severe diabetes mellitus and arteriosclerotic cardiovascular disease who developed acute pulmonary edema immediately after treatment. He was admitted to the adjacent hospital, and was discharged 7 days later. The other patient was a 42-year-old male, without history of any cardiopulmonary disease. After thorough evaluation in the hospital, no underlying cause could be found for the pulmonary edema. However, post-ESWL pulmonary edema is a recognized complication.[10] Although the mechanisms are not fully known, it has been shown that submersion in water may act as a gravity suit, increasing right atrial and pulmonary capillary wedge pressures. Lower water levels in the bath and increased use of furosemide during the procedure should help prevent pulmonary edema.

Two patients developed urinary retention within 12 hours after treatment and were hospitalized by their own urologists. Two patients developed sepsis 2 to 7 days after treatment, and subsequently required admission for parenteral antibiotics. Three patients were admitted for obstruction; percutaneous nephrostomy was required in two of these and ureteral catheterization in one. Thirteen patients were admitted for renal colic 12 to 72 hours after treatment. Some of the earlier admissions were patients who had a stone burden of greater than 20 mm, and were treated without double-J stents.

Six patients considered to have retained stones required retreatment; of these, four had staghorn calculi, and two had impacted ureteral stones. After retreatment, the two patients with ureteral stones are stone free. As the length of follow-up increases, more patients may be discovered to have retained stones.

At the Kidney Stone Center of Fort Lauderdale, the patient's post-treatment course is monitored by calling the patient 24 to 48 hours after treatment, and reviewing the physician post-treatment questionnaires. Successful outpatient ESWL has been achieved by (1) careful pretreatment evaluation, (2) meticulous attention to the outlined details during treatment, (3) innovative postoperative pain control, and (4) increased use of ureteral catheters and double-J stents.

The usefulness of ureteral catheters and double-J stents cannot be overemphasized (Table 13-3). These devices facilitate safe treatment as well as permit outpatient treatment of patients with larger stones. Double-J stents help dilate the ureter, and secondarily facilitate passage of stone particles. Thus, in selected patients, larger stones, between 5.0 and 6.0 cm in size, can be treated with minimal complications and diminished need for analgesics. These maneuvers greatly reduce the number of post-ESWL secondary procedures. Only 4 of our first 223 required additional intervention after treatment. One patient required cystoscopy and bilateral ureteral catheterizations for obstruction, and three patients with symptomatic hydronephrosis needed a percutaneous nephrostomy after treatment.

## COST

One of the prime reasons to perform ESWL in an outpatient setting is to reduce medical cost. Current length of hospitalization stay for inpatient ESWL is 2 to 3 days with a range of 0 to 7 days and a mean of 2.5 days.[3] The hospitalization expense, including pretreatment studies, is projected to be approximately $3,500. Outpatient lithotripsy should reduce this cost by at least one half. The National Kidney Foundation report states that 80,000 treatments per year will save $160 million in costs over open surgical procedures. By treating 80 percent of all these patients as outpatients, it is our opinion that another $100 million in medical costs can be saved.

## SUMMARY

The benefits and results of outpatient ESWL must be compared to the known results of inpatient treatments[6,7] and weighed against the disadvantages of outpatient treatment. Outpatient treatment requires a greater patient participation. The patient is responsible for the total pretreatment preparation, including bathing and refraining from eating or drinking. Each patient is also responsible for post-treatment monitoring of temperature. Patients treated in an inpatient setting have less responsbility for their own well-being, and are constantly reminded of the importance of hydration, ambulation, and antibiotics.

For patients with large staghorn calculi, compromised cardiopulmonary states, and/or difficult socioeconomic situations, inpatient treatment is required.

Outpatient ESWL is a safe, cost effective, and viable form of treatment for many carefully selected patients (Table 13-4). Patient acceptance of outpatient ESWL has been better than anticipated, and third-party payers and employers are rapidly accepting this form of ESWL treatment.

**Table 13-4. Pointers for Successful Outpatient Management**

1. Careful medical screening of patients
2. Judicious use of catheters, double-J stents, and percutaneous nephrostomy tubes
3. Anesthesia technique carefully designed for outpatient ESWL
4. Enthusiastic, knowledgeable nurses to educate, encourage, and support the patients
5. Proper use of antibiotics
6. Scheduling of the more difficult cases earlier in the day
7. If the patient lives 3 hours or more from the center, it is wise to recommend that he stay in a local motel the night of the procedure
8. Careful urologic follow-up after ESWL by the patient's hometown urologist

## REFERENCES

1. Press Release, Dornier Medical Systems, Inc., Atlanta, GA, October 23, 1985
2. Report of American Urological Association Ad Hoc Committee to Study the Safety and Clinical Efficacy of Current Technology of Percutaneous Lithotripsy and Non Invasive Lithotripsy, May 16, 1985. American Urological Association, Inc., Baltimore, MD, p.6,16
3. Griffith DP, Chaussy C, Meacham R, Seale C: New techniques for the removal of urinary stones: A preliminary descriptive comparison of percutaneous lithotripsy and extracorporeal shock wave lithotripsy. (In preparation)
4. Davis JE, Detman DE: The ambulatory surgical unit. Ann Surg 175:856, 1972
5. Natof HE: Complications associated with ambulatory surgery. JAMA 244:1116, 1980
6. Drach G: Summary of American clinical trials of ESWL. Third World Congress on Endourology. New York, 1985
7. Riehle R, Fax W, Vargha E: Extracorporeal shock-wave lithotripsy for upper urinary tract calculi. JAMA 255:2043, 1986
8. Resnick M: Report to National Kidney Foundation—Extracorporeal Shock-Wave Lithotripsy Committee, February 26, 1985
9. Riehle RA: Extracorporeal shock-wave lithotripsy. Urology Grand Rounds. Marion Laboratories, Kansas City, MO, 1985
10. Boysen PG, Carlson CA, Banner MJ, Gravenstein JS: Ventilation during anesthesia for ESWL. In Gravenstein JS, Peter K (eds): Extracorporeal Shockwave Lithotripsy—Technical & Clinical Aspects. Butterworths, Stoneham, MA, 1986
11. Finlayson B, Newman R, Hunter PT et al: Efficacy of ESWL for stone fracture. In Gravenstein JS, Peter K (eds): Extracorporeal Shock Wave Lithotripsy—Technical & Clinical Aspects. Butterworths, Stoneham, MA, 1986
12. Health and Public Policy Committee, American College of Physicians. Lithotripsy. Ann Intern Med 103:626, 1985
13. Dale R: Outpatient Treatment. Paper presented at the 2nd ESWL symposium, Indianapolis, IN, February, 1986
14. Burns J, Crow A, Breaux E: Outpatient extracorporeal shock wave lithotripsy. Urol Clin N Am (in press, 1987)

# 14

# Financial and Hospital Management Issues

Henry C. Alder

Hospitals are encountering increasing pressures from the public and private sectors to control health care costs. The federal government and private industry have begun to influence how hospitals and physicians conduct their affairs. For example, with the onset of the federal government's prospective payment system under diagnostic related groups (DRGs), financial resources for hospitals are no longer unlimited. In addition, hospitals are now facing new forms of competition from the private sector such as freestanding emergi-centers and outpatient diagnostic/treatment centers. These changes have resulted in hospitals assuming greater risks for their decisions. The allocation of hospital financial resources for new clinical services such as extracorporeal shock wave lithotripsy (ESWL) requires that issues in planning and implementing this capability be clearly identified. Once these managment issues have been identified, hospitals can compare ESWL with other opportunities and decide which one to pursue. The purpose of this chapter is to address the management issues associated with ESWL so that, together, physicians and hospital management can make better decisions regarding whether or not to pursue bringing this new service to their community.

## ESWL MANAGEMENT ISSUES

There are six issues that physicians and hospital management should consider before deciding whether or not to establish an ESWL capability. These six issues are as follows:

1. Volume of stone patients
2. Investment and operating costs
3. Payback and return on investment
4. Reimbursement policies
5. Patient referrals
6. Competition for the treatment of stone patients

For a comprehensive treatment of these issues, see ref. 1.

ESWL patient volume projections are important in identifying the potential demand and utilization of an ESWL facility for both the short and the long term. Patient volume can vary significantly from one region of the country to another. If the volume of patients does not justify bringing ESWL into a community or region, alternatives may need to be considered.

The investment and operating costs of an ESWL service are not trivial. The purchase price of the Dornier lithotripter Model HM-3 (Dornier

Medical Systems, Inc., Marietta, GA) is approximately $1.7 million. Investment costs reach $2 million or more when the cost for facility construction or renovation to install the lithotripter is included. In addition, the annual operating costs are a very significant concern and depend largely upon the annual number of treatments.

In today's highly competitive and constrained health care environment, the concept of financial return or payback assumes a broader significance. It must be determined to evaluate the merit of any clinical service opportunity. ESWL payback is based upon the traditional financial concept of return on investment. Return on investment draws together the factors of patient volume, operating costs, and treatment charges, with surplus or loss.

ESWL reimbursement relates directly to payback, and is therefore treated separately. There are many payers that reimburse providers, and their rules can vary widely. The status of each payer's reimbursement policy, especially within a specific region or community, must be determined before offering a new service.

Patient referral patterns are an important issue to hospital executive management. Because of the large up-front costs in acquiring a lithotripter, not every hospital can afford to offer ESWL capability. Since ESWL will initially be provided by a select group of urologists, the question arises whether urologists who do not have access to a lithotripter will refer their patients to urologists with ESWL privileges.

Finally, competition for stone patients will likely occur as other treatment modalities such as second generation lithotripters and drug therapy become available. Potential competition must be assessed to project the utilization of an ESWL capability in both the short and long term. Let us examine each of these issues.

## PATIENT VOLUME

According to expert opinion, approximately 2 to 3 percent of the adult population of industrialized Western nations suffers from urolithiasis or kidney stone disease.[2] If we apply this assumption to the United States (where many patients suffer from stones in the upper urinary tract and kidney), it can be calculated that more than 2.2 million Americans have kidney stone disease. However, not all sufferers of kidney stone disease are hospitalized or seek medical attention. In fact, according to 1984 hospital discharge statistics from the National Center for Health Statistics (NCHS), approximately 423,000 Americans were hospitalized for stones in the kidney and ureter, but only about 140,000 patients actually underwent stone surgery.[3] In other words, only 19 percent of the potential stone patient population was hospitalized. Of those hospitalized, only 33 percent underwent surgery in 1984. The nonsurgical patients evidently underwent medical (drug) treatment. However, not all of the 140,000 patients who underwent stone surgery are eligible for ESWL. Approximately 77 percent of patients who undergo traditional stone surgery are ESWL candidates, according to an independent study of ESWL operators conducted by this author. By 1990, and as operators gain more experience with ESWL, about 90 percent of stone surgery patients may be eligible for shock wave lithotripsy.

Another issue pertaining to patient volume is the inconsistent distribution of stone patients in the United States. By examining 1983 hospital discharges in the four major regions of the United States, one observes a higher prevalence of stone disease in the South and lower prevalence in the West (Fig. 14-1). This variance may be caused by local or regional physician/hospital practice patterns, as well as regional climate, diet, and other predisposing factors. Whatever the reason, the stone patient population variation in a specific area has to be predetermined in order to estimate more precisely the number of candidates for ESWL treatment. Failure to do so may result in inappropriate utilization projections.

The national kidney stone patient volume is not expected to change appreciably over the next several years. Based on a preliminary analysis and forecasting by this author, it appears that stone patient hospitalization is closely re-

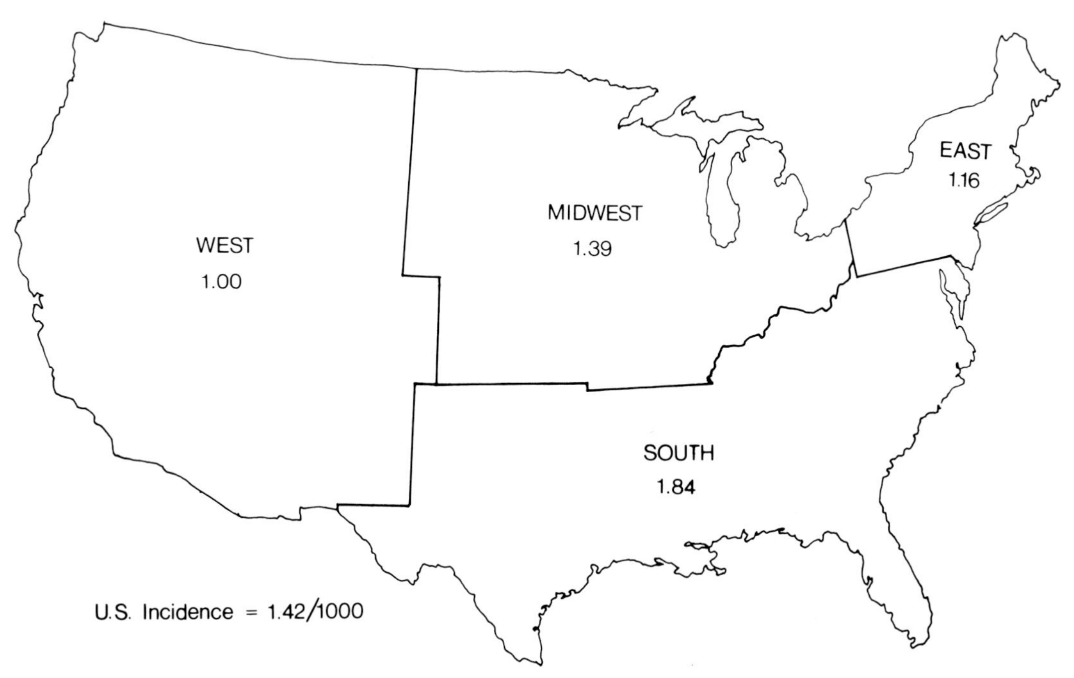


Fig. 14-1. Regional incidence of kidney and ureter stones per 1,000 population. Based on primary diagnosis of discharged hospitalized patients, from the 1983 CPHA data for ICD-9-CM codes 592.0 and 592.1 (Alder HC: Lithotripters: Noninvasive Devices for the Treatment of Kidney Stones. AHA Hospital Technology Series, AHA-012828. American Hospital Association, Chicago, IL, 1985.)

lated to population growth and aging. For example, it is expected that the hospitalization rate of stone patients nationally will rise very gradually from 1.42 admissions per 1,000 population in 1985 to approximately 1.50 in 1990—provided physician practice patterns and therapeutic procedures do not change. Assuming a U.S. population of 249 million in 1990, about 373,000 people are expected to be hospitalized for kidney and ureter stones. If, as in 1984, 33 percent of those admitted require stone surgery and 90 percent are eligible for ESWL, only 111,000 patients will be treated.

## INVESTMENT AND OPERATING COSTS

The investment and operating costs of an ESWL service are very significant. For many hospitals and urologists, these costs can be a "barrier to entry" in providing ESWL services. A successful ESWL capability requires a precise identification of these costs and a strategic plan to obtain the fastest financial return, or return on investment. For illustrative purposes, the Dornier lithotripter will be used as a model to define ESWL investment and operating costs. The Dornier model will then be compared with the potential investment and operating costs of second-generation lithotripters which are expected to become commercially available in 1987.

The investment cost of the Dornier lithotripter ranges from \$1.85 million to \$2.7 million. This cost includes not only the \$1.7 million price of the lithotripter and related hardware, but also the cost to renovate an existing area or to construct a facility to house the device and its components in a dedicated environment. A basic dedicated facility requires from 400 to 600 $ft^2$ of space, special utilities such as water conditioning equipment, and room for support services such as patient preparation, recovery, and radiographic capabilites. The average cost to renovate an existing facility ranges from $150,000 to $250,000. However, if additional nonessential facilities, such as cystoscopy suites, patient waiting areas, dressing rooms, and offices are added, the maximum facility renovation and/or construction cost can exceed $400,000 and easily reach $1 million.

In comparison, the total investment cost for second-generation lithotripter is expected to be less than $1 million, including the cost for equipment and site preparation. Second-generation lithotripter developers such as Medstone International, Inc., Costa Mesa, CA; International Biomedics, Inc., Bothell, WA; Edap, Marne-la-Vallee, France; and Northgate Research Division of Monaghan Medical Corporation, Plattsburgh, NY, are predicting purchase prices from $400,000 to $850,000. Surely, the anticipated lower investment cost of second-generation lithotripters will attract buyers who could not afford the price of a Dornier lithotripter or those who were unable to realize a rapid return on investment from the more expensive technology. It is expected that second-generation machines will be especially attractive to providers whose treatment volumes do not support a $1 million-plus annual operating cost.

## METHODS OF FINANCING

The large capital expenditure for ESWL technology has motivated physicians and hospital executive management to seek creative ways to finance the acquisition of a lithotripter. Financing strategies are based on the best estimate of how quickly return on investment can be achieved. Earlier purchasers of the lithotripter financed the technology in a number of different ways. Many hospitals purchased the hardware outright, using operating revenue and funded depreciation as a source of capital. Those without sufficient funds acquired bank loans or obligated themselves to long-term debt financing through the issuance of low-interest tax-exempt bonds. Still other providers, in an attempt to reduce or spread the risk associated with a major technology purchase, aligned themselves with financial partners. For example, many urologists have formed foundations, associations, and co-operatives in an effort to collectively acquire a lithotripter. From these collective groups a number of joint-venture relationships have emerged, such as between urologists and hospitals, urologists and independent third parties, and urologists, hospital, and third party. These types of relationships have fostered a great deal of interest among health care policy analysts who see physician collectivism and its resulting bargaining power as the modus operandi for future acquisition of new technology.

## OPERATING COSTS

ESWL operating costs are not trivial. They can vary dramatically from one operator to another, and depend upon the fixed costs of setting up the capability, the projected number of patients to be treated, and the variable cost of treating a single patient. Fixed costs include the salaries of urology support staff, clerical costs, facility costs, and depreciation and interest costs (Table 14-1). The first year total fixed costs (which are the same whether 100 or 1,000 patients are treated annually) are approximately $980,000/year for the Dornier lithotripter using 1985 figures.

**Table 14-1. Annual ESWL Fixed Costs (First Year of Operation for the Dornier Lithotripter)**

| | |
|---|---|
| Salaries, wages, and benefits (Assumption: 7 full time equivalents added to urology department staff, one RN/nurse manager, one ESWL technician, two clerical staff, two practical nurses/orderlies, 1 nurse anesthetist) | $208,000 |
| Insurance, liability | $ 75,000 |
| Telephone | $ 7,500 |
| Office supplies, postage | $ 9,500 |
| Service contract | $125,000 |
| Facility rental or mortgage | $ 25,000 |
| Facility maintenance, utilities | $ 10,000 |
| Depreciation (5-year straight-line depreciation of $2,000,000 investment) | $400,000 |
| Interest expense (annual cost for borrowing $1,000,000 at 12% for 5 years) | $120,000 |
| Annual total fixed costs | $980,000 |

**Table 14-2. ESWL Variable Costs (Dornier Lithotripter)**

| | |
|---|---|
| Electrodes (assumption: two electrodes per treatment at $200 per electrode) | $400 |
| Medical supplies (syringes, towels, disposable bathing suits, etc.) | $ 50 |
| Billing and/or collection costs | $200 |
| Total variable costs | $650 |

The variable cost, or cost for each treatment, is $650. This includes the cost of two electrodes, medical supplies, and facility billing/collection fees (see Table 14-2). If a facility plans to provide 1,000 treatments each year, the total variable cost to the operator would be $650,000. Added to the aforementioned fixed costs, the total annual cost to operate an ESWL facility would be $1,630,000, or $1,630 per treatment. Note that this $1,630 treatment cost is only the technical component of providing ESWL services. It excludes other treatment costs, such as urologist and anesthesiologist professional fees, x-ray and laboratory test charges, recovery room charges, and hospitalization (room and board) costs.

The second generation lithotripters are expected to have lower fixed and variable operating costs because of technological enhancements such as the elimination of the water bath, improved imaging capability, and advances in stone localization techniques. Initial estimates by this author indicate that fixed costs would be less than $600,000 during the first year of operation. Reduced operating costs may also occur as a result of the need for smaller facilities and less support staff than presently used for the Dornier lithotripter, lower salary expenses, reduced facility upkeep and utility expenses, and lower depreciation and interest costs.

Anticipated costs would be approximately $300 per treatment rather than the $650 per treatment using the Dornier unit. However, these costs could be greater if one or more of the second-generation manufacturers adopt a usage fee or royalty arrangement for each treatment. If, for instance, 1,000 patients are to be treated annually using a second-generation lithotripter, the total annual operating costs would be $900,000 per year or $900 per treatment (excluding vendor's usage fee, professional fees, and other treatment costs such as x-ray, lab, and hospitalization costs) (see Table 14-3).

## PAYBACK AND RETURN ON INVESTMENT

Since the onset of the prospective payment system, hospitals no longer are reimbursed based on their cost for the care of Medicare patients. Instead, a system of diagnostic related groups (DRGs) has been devised to provide a fixed payment for each patient's diagnosis. This cost-containment system has motivated other health care payers to adopt similar payment structures for hospital services. By knowing in advance the specific payment to be received for inpatient services such as ESWL, hospitals have become more acutely aware of the need to precisely assess costs, to project the return on investment (ROI) for the service provided, and to estimate the time for "payback."

**Table 14-3. Comparison of Dornier Lithotripter and Second Generation Lithotripter Costs**

| | Dornier Lithotripter | Second-Generation Lithotripter |
|---|---|---|
| Annual fixed costs | $980,000 | $600,000 |
| Unit variable cost | $650 | $300 + usage fee |
| Total operating costs for 1,000 treatments per year (cost per treatment)[a] | $1,630,000 ($1,630/treatment) | $900,000 + usage fee ($900/treatment + usage fee) |

[a] The total operating costs and cost per treatment is the technical component of providing ESWL. It does not include other costs that are incurred as a result of treatment, such as radiology and clinical laboratory costs, hospitalization costs, and urologist and anesthesiologist professional fees.

Identifying payback time requires analyzing expected patient volume, potential income from direct treatment charges, income from fixed payments negotiated with various health care payers, and all operating costs associated with providing an ESWL service. One method to identify when a service begins to generate a positive net income is the break-even analysis. Simply stated, a break-even analysis identifies the revenue and expenses that can be expected at discrete levels of patient volume. Figure 14-2 illustrates a break-even analysis for the Dornier lithotripter and a hypothetical second-generation lithotripter. The dotted, short dashed, and long dashed lines illustrate the revenue an ESWL service will realize assuming an assessment per treatment of $4,000, $3,000, or $2,000 respectively. The light solid and dark solid lines identify the expenditures at various treatment volumes for the Dornier lithotripter and a hypothetical second-generation lithotripter, respectively.

In this example, the expenditure lines only indicate the fixed and variable costs for supporting the lithotripter and its facilities. They do not include professional fees and other hospital ancillary expenditures such as x-ray, clinical lab, and hospitalization. The intersection points of the solid expenditure lines with the hyphenated revenue lines are called the break-even points. At the break-even point, patient revenue equals treatment costs, and the net income derived from ESWL services would be zero. Payback begins to occur after break-even is achieved. Based on this analysis, the break-even point for the Dornier lithotripter, at a fee of $2,000 per procedure, would be at 726 procedures a year. At $3,000 and $4,000 per procedure, the break-even points would be 417 and 293 procedures a year, respectively. Likewise for the second generation lithotripter (dark solid line), the break-even points at $2,000, $3,000, and $4,000 per procedure would be 347, 219,

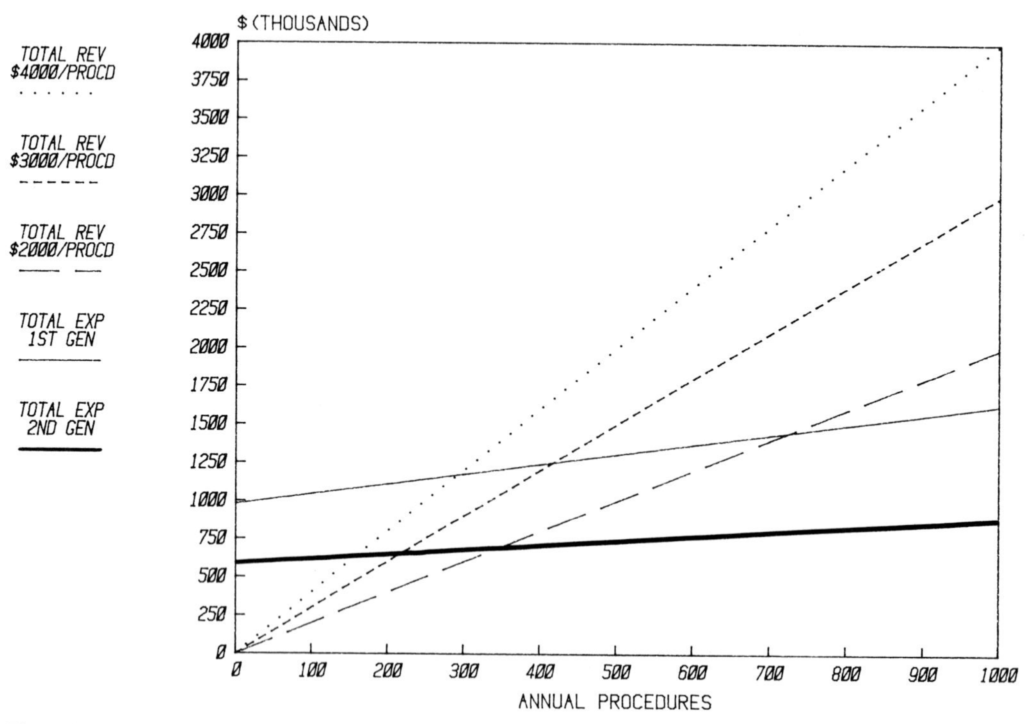


Fig. 14-2. Break-even analysis for first- and second-generation ESWL units. (Adapted from Alder HC: Lithotripters: Noninvasive Devices for the Treatment of Kidney Stones. AHA Hospital Technology Series, AHA-012828. American Hospital Association, Chicago, IL, 1985.)

and 160 treatments a year, respectively. Clearly, one can see that once the second-generation lithotripters become available, they may offer a much more rapid return on investment than the first-generation machine.

## REIMBURSEMENT POLICIES

Reimbursement is the fourth management issue. ESWL reimbursement policies vary from one payer to another. Generally speaking, however, most health care payers reimburse without reluctance. In fact, many commercial insurers pay for ESWL quite readily, and have not imposed significant restrictions on the use of ESWL. Evidently, most commercial payers have been convinced of the added safety, cost effectiveness, and rapid patient convalescence of shock wave lithotripsy when compared to other traditional stone surgery procedures. Blue Cross-Blue Shield plans also reimburse hospitals and physicians for ESWL treatment. However, individual group plans negotiate the technical (the basic treatment cost for machine use and electrodes) and professional fees with hospitals and physicians.[4] The Blue Cross-Blue Shield Association has recommended to the group plans that the professional component be reimbursed as a surgical procedure.

The major reimbursement issue, as of this writing, is Medicare policy for ESWL payment. Medicare, which is administered by the Health Care Financing Administration (HCFA) of the U.S. Department of Health and Human Services (USHHS), established a policy in 1985 that classified ESWL as a medical rather than a surgical procedure. Although the classification has little or no immediate impact on the professional component (Part B) of Medicare reimbursement, its impact on hospitals is significant. Medicare reimbursement for ESWL (classified as DRG 324) averages about $1,550 for a patient without complications. According to an independent study conducted by FLR Health Resources (Atlanta, GA) in Fall 1985, the average loss incurred by hospitals for each Medicare patient ranged from $750 to $1,000. Moreover, in some regions of the United States, Medicare does not reimburse for outpatient ESWL.

## COMPETITION FOR STONE PATIENTS

The final ESWL management issue is the competition for stone patients that is expected to emerge as more lithotripters are distributed in the United States. By the end of 1986, there are expected to be up to 150 lithotripter installations. The emergence of one or more second-generation lithotripter models in 1987 or 1988 will most likely increase this number to over 300 units by 1990. However, as described above, the number of candidates for ESWL treatment admitted in hospitals in the United States in 1990 is not expected to exceed 111,000. Assuming 300 lithotripters are available, each facility will treat about 370 patients a year.

It is also anticipated that new drugs will effectively compete with ESWL by preventing the formation of new stones and slowing the growth of existing stones in recurrent stone-forming patients. Drugs such as sodium cellulose phosphate and potassium citrate are just now emerging as potential medical treatments for kidney stones.

Undoubtedly, in the future there will be keen competition among ESWL facilities for the treatment of a limited number of stone patients, unless the indications for shock wave lithotripsy are expanded to include lower ureteral stones, gallstones, and common duct stones.

## PATIENT REFERRALS

One of the most sensitive issues associated with ESWL is the referral of stone patients by urologists to a lithotripter operator. Since ESWL has been dubbed the most significant advancement in urology in 20 years, most urologists who treat stone patients want to become affiliated with a lithotripter facility. However, the cost of the new technology and its initial diffusion to major tertiary care medical centers have somewhat limited urologist and patient access.

Moreover, many ESWL facilities have limited privileges to a few urologists. This "closed staff" arrangement has created tensions among urologists who do not have privileges and those who do. As a result, urologist referral networks have not always been easy to establish. A number of urologic lithotripsy groups and hospitals have attempted to circumvent this issue by placing the lithotripter in a "neutral" location, and transporting patients from the referring urologist's hospital to the facility on the day of treatment. After treatment and a brief recovery period at the ESWL facility, the patient is returned to the hospital for continuing care by the referring physician. Freestanding outpatient facilities have used the neutral site concept to encourage urologists to refer the patients to the facility with the explicit understanding that stone patients would be returned to the referring urologist after treatment for follow-up care.

## CONCLUSION

Despite these challenging management issues, many hospitals are prepared to take risks to establish an ESWL clinical service. They see ESWL as an opportunity to expand their market share, or to penetrate an existing market. Other hospitals see ESWL as a necessary extension of their existing urologic/stone service. As of this writing, the interest and enthusiasm generated by ESWL has resulted in a demand for lithotripters that far exceeds their present supply. Nevertheless, hospital managements need to carefully examine the issues described above in relation to their own institution before acquiring an ESWL capability.

## REFERENCES

1. Alder HC: Lithotripters: Noninvasive Devices for the Treatment of Kidney Stones. AHA Hospital Technology Series, AHA-012828. American Hospital Association, Chicago, IL, 1985
2. Chaussy C et al: First clinical experience with extracorporeally induced destruction of kidney stones by shock waves. J Urol 1982
3. Preliminary data, National Hospital Discharge Survey, National Center for Health Statistics, Hyattsville, MD, 1984
4. Extracorporeal Shock Wave Lithotripsy: Clinical Assessment, Utilization and Cost Projections. Blue Cross-Blue Shield Association, Chicago, IL, May 1985, pp. 30–44

# 15

# World Experience with Shock Wave Lithotripsy

Gerhard J. Fuchs
Christian G. Chaussy

Since the first experimental efforts to use extracorporeally induced shock waves to disintegrate human kidney stones were undertaken in Munich, West Germany, only a little more than 10 years has elapsed.

In February 1980, after extensive testing in vitro and finally in an animal stone model,[1] the method was introduced into clinical use by Christian Chaussy.[1,2] It quickly proved effective, safe, and reliable, and has dramatically changed the management of stones in the upper urinary tract.[2–15] In 1983, further distribution of the lithotripter began first in West Germany, and subsequently in the USA, Europe, and Asia. In December 1984, the Food and Drug Administration (FDA) approved the method in the United States. At present more than 220 units are operational in 21 countries worldwide, with more than 250,000 treatments successfully performed (see Table 1-1).

At present, less than 1 percent of all stone patients are excluded from consideration for ESWL (Table 15-1). A pathologic bleeding profile continues to be a contraindication, as it places the patient at an increased risk for perirenal bleeding after shock wave treatment. Experience with patients who have received ESWL after a pathologic bleeding profile has been corrected indicates that these patients are at a somewhat higher risk of developing a perirenal hemorrhage; nevertheless, ESWL, has also become the treatment of choice in those patients.[8,9] Patients suffering from cardiovascular diseases, rendering them unsuitable for anesthesia, are of course not eligible for ESWL treatment. Pregnancy also has to be considered a contraindication, as the effect of shock waves on the fetus is not known.

Urologic contraindications are severe obstructions at any level distal to the stone. For technical reasons, patients who are too tall, too small, or too obese to fit on the patient support system must also be excluded (Table 15-1).[2,6,8,9,11,12,14]

The range of indications initially established in Munich[2] was soon confirmed by other centers [10–12,15] (personal communication, T. Saka and H. Tanda, Sapporo, Japan, 1986). Owing to the high success rate of ESWL, the low rate of periprocedural complications, and the absence of long-term adverse effects, the range of indications has been expanded over the years to include most upper-tract urinary stones (Table 15-2).[8–13]

**Table 15-1. Current Contraindications to ESWL**

| |
|---|
| General |
| Untreated bleeding disorder |
| Not fit for anesthesia |
| Pregnancy |
| Technical |
| Gross obesity (>130 kg body weight) |
| Children (<100 cm body length) |
| Patients >200 cm |
| Urologic |
| Obstruction distal to the stone |
| Caliceal neck stenosis |
| Ureteropelvic junction obstruction |
| Ureteral stenosis |
| Significant bladder outlet obstruction |

**Table 15-2. Present Range of Indications for ESWL**

| |
|---|
| Single and multiple stones in the kidney |
| Single and multiple stones in the ureter |
| Infected stones |
| Partial and complete staghorn stones |
| Stones in solitary kidneys |
| Radiolucent stones |
| High-risk patients |
| Children (>100 cm body length) |

## THERAPEUTIC OVERVIEW

Approximately 70 percent of nonselected stone patients are eligible for ESWL monotherapy.[8–10] This group includes patients with single and multiple stones in the kidney (stone mass $\leqslant$ 2.5 cm) and selected patients with ureteral stones located above the iliac crest. The rest of the patient population presents with more complex symptomatic stone disease, and if they are eligible for ESWL, require auxiliary procedures.

### Radiolucent Stones

Although radiolucent stones were considered a contraindication for ESWL in the Munich trials, they can be treated if they are made visible by using contrast medium. To detect the stones fluoroscopically, pre-ESWL cystoscopy is performed and a ureteral stent is placed. During the procedure, contrast dye administered through the stent reveals the stone as a filling defect. This filling defect is centered on the F2 focus of the ellipsoid, and ESWL is carried out in the usual way. However, it is more difficult than usual to decide when to terminate treatment, as there is no positive radiographic image showing a disintegrated stone. The degree of disintegration has to be assessed from the gradual disappearance of the filling defect. After ESWL, the patient must be followed by ultrasound examination and, if necessary, by intravenous pyelography (IVP).[9]

### Staghorn Calculi

There is still considerable controversy as to what strategy is best for large, branched calculi.[8–10,13] Review of 100 consecutive staghorn cases treated at UCLA with ESWL alone revealed certain features characteristic of the patients who became completely stone free. Success with ESWL treatment of staghorn stones depends upon (1) the overall stone burden in the kidney, (2) the shape of the renal collecting system, (3) the architecture of the dependent calices, and (4) the stone composition (Table 15-3). Several ESWL sessions may be necessary to treat large staghorn calculi filling a dilated intrarenal collecting system. The shape of the collecting system is of special importance. The architecture of the dependent calices of the lower pole region is particularly critical. Gross dilation of the dependent calices and/or stenosis of the caliceal neck almost always precludes complete freedom of stone debris. Even after more than 6 months of follow-up, those patients retain small amounts of stone debris (mostly less than 1 percent of the original stone mass). Struvite

**Table 15-3. Selection Criteria for ESWL of Staghorn Stones**

| |
|---|
| Stone burden |
| Shape of renal collecting system |
| Dilation |
| Stenosis |
| Stone in dependent calices |
| Stone composition |

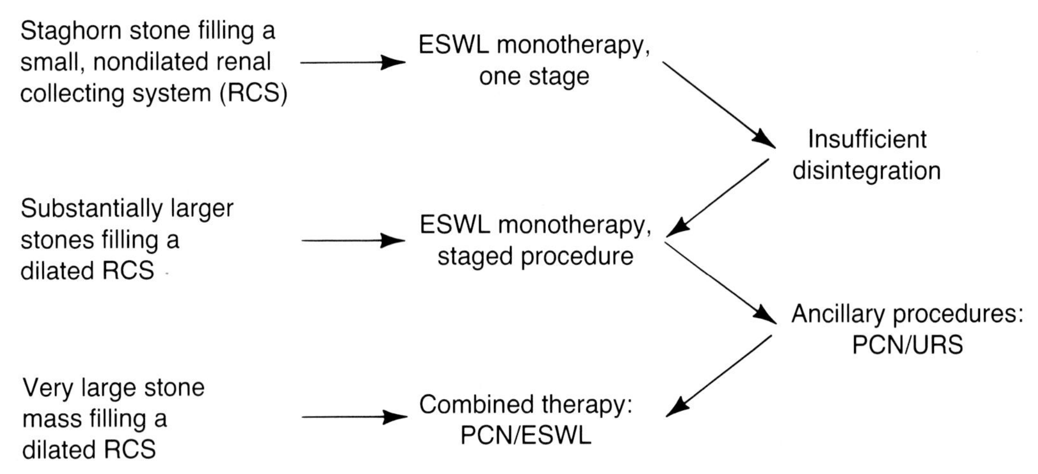


Fig. 15-1. Differential indications in the treatment of staghorn stones.

staghorn stones tend to be particularly responsive to ESWL therapy.

Based on these findings, the indications for ESWL monotherapy of staghorn stones have been revised and follow the selection criteria outlined in Figure 15-1. In partial and complete staghorn stones, ESWL monotherapy is only preferable to other treatment modalities in cases where the stone fills a nondilated collecting system (Figs. 15-1, 15-2). A planned staged ESWL procedure can be performed for staghorn stones that fill a slightly dilated collecting system (Figs. 15-1, 15-3). It has to be pointed out, however, that with increasing stone burden, repeat ESWL and follow-up complications are considerably more common than with smaller stones. Most of the auxiliary procedures—percutaneous nephrostomy tube placement and ureteral manipulations—tube are required after ESWL treatment of complex stones (Table 15-4).[8,9]

In staghorn stones with a very large stone mass, a percutaneous procedure is performed first for debulking the stone. In a second session, ESWL is employed for the disintegration of the remaining caliceal stone parts (Figs. 15-1, 15-4).[8–10] Despite the added invasiveness of the percutaneous procedure, this approach is beneficial for the patient with complicated stone disease as it allows efficient and quick eradication of the stone material.[8–10]

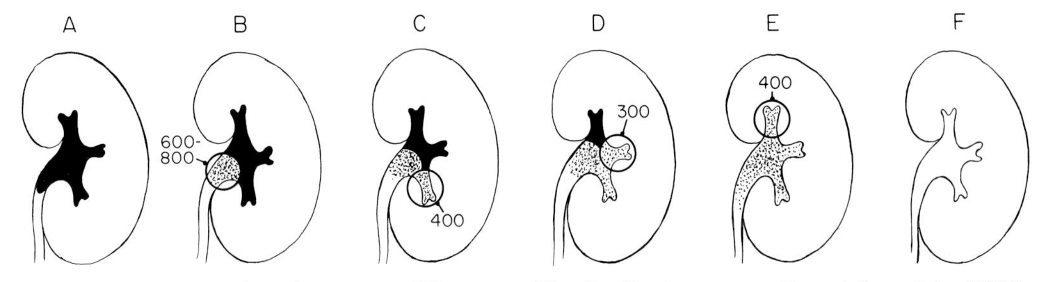


Fig. 15-2. Treatment of staghorn stones filling a nondilated collecting system. From left to right, ESWL commences at the ureteropelvic junction portion of the stone, where 600 to 800 shock waves are delivered. Next, the lower caliceal group is treated with approximately 400 shock waves; then 300 shock waves are administered to the mid-caliceal group and 400 shock waves the upper caliceal group. This leaves a margin of 500 to 700 waves which may be delivered to areas where solid stone material is still detected at this time.

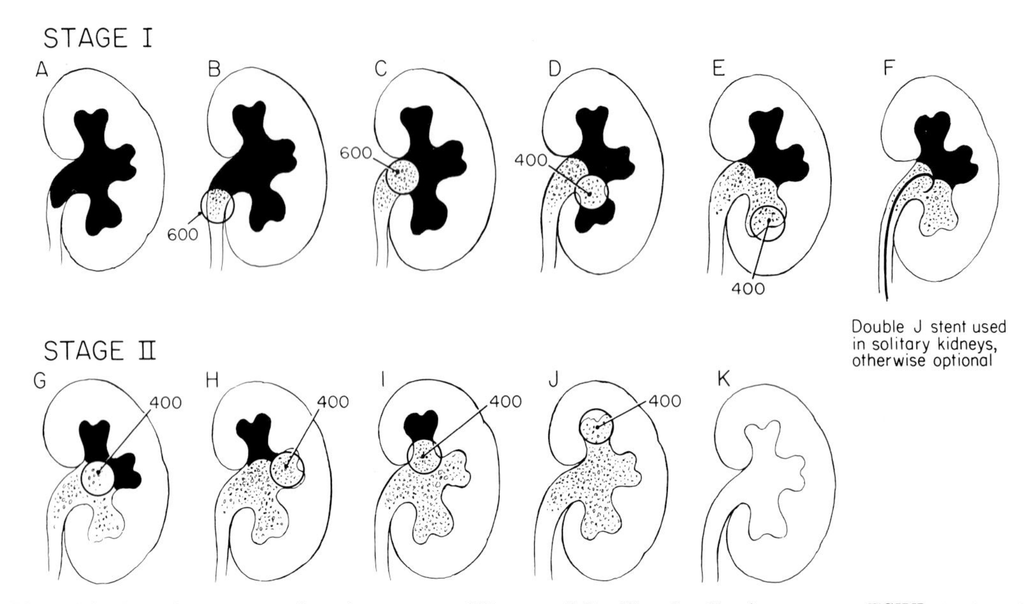


Fig. 15-3. Staged treatment of staghorn stones filling a mildly dilated collecting system. ESWL treatment starts again at the ureteropelvic junction, and in the first session the pelvic stone parts and the lower caliceal group are treated. A double-J stent prevents larger chunks from falling into the ureter between sessions. In the second stage, usually 2 days after the first stage, the remaining stone is treated in a similar fashion, again basically following a predetermined sequence.

## Ureteral Stones

For technical reasons, ureteral stones are usually amenable to primary ESWL only if they are located above the iliac brim, in the mid or upper third of the ureter. Approximately 80 to 85 percent of all ureteral stones that require interventional treatment are located in the upper ureter above the iliac crest. With increasing experience, these stones are being treated effectively with ESWL. It has been found that ureteral stones rarely disintegrate properly if they are impacted and have no room to expand as they disintegrate. To obviate this unfavorable condition, two modalities are available—repositioning the stone(s) into the renal pelvis, and/or maneuvering a ureteral catheter past the calculus. With both procedures the expansion chamber is increased, thereby resulting in a high yield of stone disintegration.[8–9]

Ureteral stones below the iliac crest only become amenable to ESWL when repositioned into the upper ureter or ideally into the renal pelvis in a retrograde or antegrade fashion. Also candidates for pre-ESWL stone manipulation are ureteral stones above the iliac crest in which the radiographic appearance of a narrowing at the stone site indicates absence of a sufficient expansion chamber.

## High-Risk Patients

Patients with hypertension or a heart condition can also receive ESWL treatment without major problems. This understandably necessitates

**Table 15-4. Percent of Patients Undergoing Auxiliary Procedures after ESWL**

| Type of Stone | Ureteral Manipulation | Percutaneous Nephrostomy | Surgery |
|---|---|---|---|
| "Easy" stones (smaller than 2.5 cm) | 5% | 1% | 0 |
| "Problem" (large and staghorn) stones | 22% | 26% | 2% |

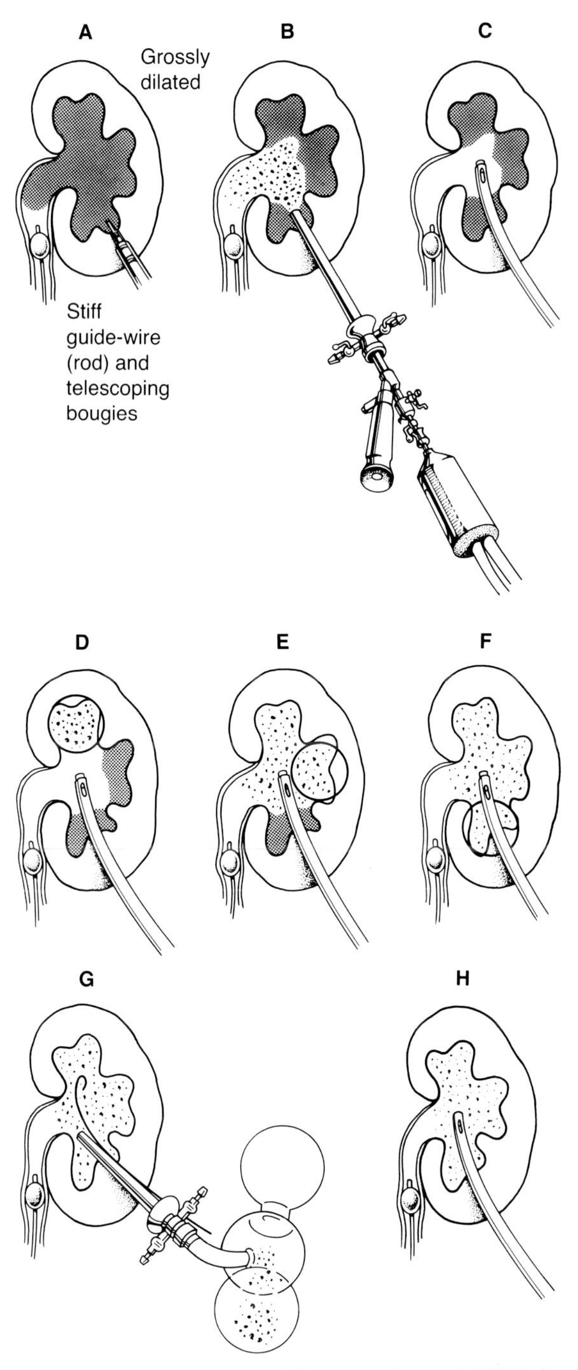


Fig. 15-4. Combined approach of percutaneous debulking procedure and ESWL for the treatment of staghorn stones filling a grossly enlarged collecting system. (**A–C**) First stage: percutaneous debulking procedure to remove most of the stone bulk from the access calyx and the renal pelvis. (**D–F**) Second stage: ESWL of the remaining caliceal stones parts. (**G,H**) At the conclusion of the second session, or when required after additional sessions of ESWL/PCN, the Ellik evacuator is a helpful device to remove large amounts of debris in a relatively short time.

careful evaluation of the patient and appropriate preoperative preparation with medical consultation. These patients are often treated under general anesthesia, which allows close monitoring of the cardiovascular and respiratory parameters (ECG, BP, CVP, PWP).

## Children

Although nephrolithiasis is relatively rare in children, the advent of a noninvasive treatment modality for children has been especially effective. ESWL is particularly advantageous because it can be used repeatedly without detrimental effects. The Dornier lithotripter currently available will accommodate children no smaller than 100 cm. To prevent the lungs from being damaged by the shock waves, a Styrofoam layer is positioned at the back of small patients to cover the lower region of the lungs.

Treatment is performed under general anesthesia to ensure that the child does not move during the procedure and to keep the respiratory excursion low—a further prophylaxis of lung exposure. In general, children get rid of the disintegrated stone material with fewer problems than adults with stones of similar size.[8,9]

## ESWL WORLDWIDE: EXPERIENCE AND RESULTS

The initial results of the first 3 years of treatments by the Munich group revealed a 99 percent rate of successfully treated patients. Ninety percent of patients became completely stone free within a 3 month period, and 9.3 percent of patients had small and spontaneously passable fragments still remaining at the 3-month follow-up. In 0.7 percent of cases, an open surgical or endourologic procedure was indicated to relieve a longer persisting ureteral obstruction (Fig. 15-4, Table 15-5). Evaluation of pre-and post-ESWL renal function showed no evidence of any adverse effects of ESWL on renal function.[2–9] Over a 4 year period, it was found that the renal function of those patients treated with ESWL actually improved with time, which is attributable to successful treatment of obstructing stones.

**Table 15-5. Initial Munich Results (1980–1983)**

| Status | Percent of Patients |
|---|---|
| Stone free | 90 |
| Spontaneously passable | 9.3 |
| Surgery | 0.7 |

As is shown in Table 15-6, the initial results from Munich (1980–1983, selected indications) were convincingly confirmed by the other centers working with the lithotripter. The results in Table 15-6 were compiled from Munich (1980 through 1986); Stuttgart, the second center to work with the kidney lithotripter (1983 to 1986); Sapporo, the first center in Asia (1984 to 1986); and UCLA. Comparison of these results with the overall results from Munich for the last 6 years, including the more complex stones that were later accepted, reiterates the importance of the original stone size as a primary factor in the stone-free rate.

Of note is the fact that UCLA is so far the only center where results have been obtained that do not show a learning curve. The results from the most recent 1,000 patients treated at UCLA are not different from those reported in previous series from Munich and Stuttgart. At UCLA, however, the full range of stones eligible for ESWL was from the beginning accepted for treatment, so that the number of complicated stones in this series is comparably higher than in the other groups.

The incidence of periprocedural complications has been surprisingly low. No perioperative mortality and no loss of a kidney has been observed in our series. Cardiovascular complications during or after treatment have been seen in less than 0.5 percent of cases, and perirenal hematoma in less than 0.2 percent. No open surgical intervention has been required for any reason after ESWL[8,12,14,15] (personal communication, T. Saka and H. Tanda, Sapparo, Japan, 1986).

Post-ESWL complications include insufficient stone disintegration necessitating further procedures, and obstructive symptoms, such as

**Table 15-6. Results of ESWL Treatment (Worldwide)**

| | Munich | Stuttgart | Sapporo | UCLA |
|---|---|---|---|---|
| Date of opening | 2/80 | 10/83 | 9/84 | 3/85 |
| Success rate | 99.0% | 99.1% | 99.0% | 99.5% |
| Status at 3 months follow-up | | | | |
| Stone free | 85% | 85% | 77% | 80% |
| Spontaneously passable fragments | 11% | 11% | 19% | 16% |
| Fragments >4 mm | 2% | 4% | 3% | 4% |
| Open surgery | 1% | 0 | 0.5% | 0 |
| Complications | | | | |
| Fever | 6% | 3% | 5% | 4% |
| Pain/colic | 25% | 28% | 32% | 22% |
| Auxiliary measures | 16.4% | 17.6% | 15.8% | 19.0% |
| Pre-ESWL | 5.1% | 10.3% | 4.4% | 12.2% |
| Post-ESWL | 11.3% | 7.3% | 11.4% | 6.8% |

pain and obstructive pyelonephritis, during the passage of the fragments.

Insufficient stone disintegration mainly depends on the size of the original stone, its hardness, location, and radiographic density. With sufficient experience, the outcome of the ESWL procedure in terms of stone disintegration can be very well predicted. With the exception of a few cystine stones, all calculi can be successfully treated with ESWL, irrespective of their chemical composition.

At present more than 95 percent of all stone patients are eligible for ESWL treatment either as a single therapy (70 to 75 percent), or in combination with endourologic procedures (20 to 25 percent) (Fig. 15-5).

For combined treatment, the approach is carefully planned. Usually the endourologic procedures (PCN-debulking of staghorn stones, stent or ureteroscopic manipulation of ureteral stones) is performed first.[8–10] The efficacy of these endourologic procedures followed by ESWL has

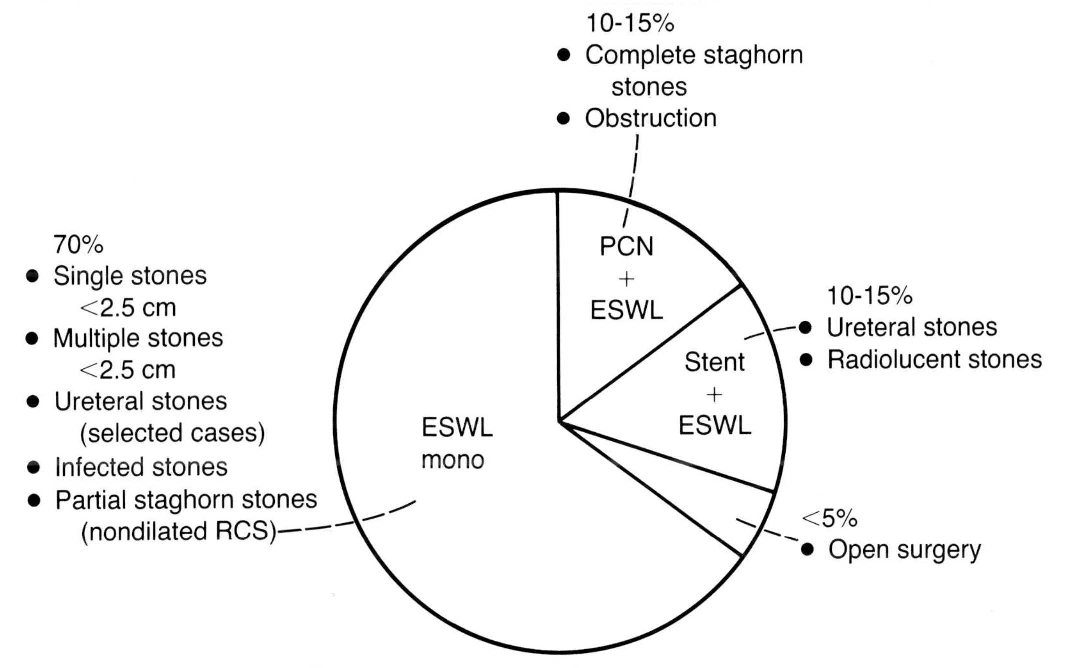


Fig. 15-5. Range of ESWL indications in 1986.

been demonstrated.[8–10] Comparison of the auxiliary procedures performed at various centers shows that the overall frequencies do not differ greatly; the break down into pre- and post-ESWL auxiliary procedures reflects the tactical biases of the different centers[2–10,12,15,16] (personal communication, T. Saka and H. Tanda, Sapporo, Japan, 1986). Further post-procedural complications are mainly secondary to ureteral obstruction. Discomfort or pain is found in approximately 25 percent of all cases, and half of these patients respond well to oral pain medication. Nausea is usually alleviated by rectal antiemetic suppositories.

Auxiliary procedures after ESWL are mainly indicated when obstructive pyelonephritis needs to be relieved to prevent urosepsis. In this situation, the use of a percutaneous renal drainage is advocated. At UCLA percutaneous nephrostomy tube drainage (12 F) is liberally used. Together with appropriate antibiotic treatment this quickly relieves the acute symptoms, and makes urosepsis a very rare complication ($<0.5\%$). Even in the presence of an extended ureteral steinstrasse, the fragments are usually eliminated within a reasonable period of time.[8,9]

Ureteroscopy for treatment of steinstrasse is a tedious, difficult, and highly hazardous undertaking and should only be performed when a larger piece of debris precedes the steinstrasse and blocks the ureter. Most steinstrassen will pass with simple nephrostomy tube drainage and "watchful waiting."

## FUTURE ROLE OF PERCUTANEOUS NEPHROSTOLITHOTOMY

Experience has shown that solitary or multiple kidney stones, less than 2.5 cm in total stone burden, can be considered "easy" stones for ESWL treatment. A disintegration success rate of 99 percent; a stone-free rate between 70 and 90 percent, and a low rate of periprocedural complications can be expected (Table 15-6).

Coincidentally, stones of this size found in approximately 70 to 80 percent of a nonselected patient group are also considered ideal for percutaneous removal. However, when planning a percutaneous nephrolithotripsy, the actual stone location needs special attention, since stones located in middle and upper calices are more difficult to approach via the routine nephrostomy access tract through a lower posterior calyx.

ESWL offers the advantage that all radiopaque renal stones, irrespective of their actual location in the collecting system, can be easily localized and treated. Thus the noninvasiveness, the relative simplicity of treatment, and the lower rate of periprocedural complications explain why percutaneous stone surgery is being increasingly supplanted by ESWL in the treatment of the so-called "easy" stones.

Yet, in renal units with infundibular stenosis preventing proper discharge of the disintegrated fragments, or in kidneys presenting with severe functional alterations of upper urinary tract motility, PCN has the advantage that the underlying anatomic anomality can be corrected at the time of percutaneous stone surgery. Furthermore, in older or immobile patients, it is advantageous to remove rather than to disintegrate the stone, especially in cases with infection stones accompanied by chronie urinary tract infection. For "easy" stones, therefore, stone location and the anatomic configuration of the upper tract influence the choice between extraction and disintegration.

What is the role of ESWL monotherapy in the treatment of solitary or multiple stones 2.5 to 4 cm large? Experience with ESWL of larger stones showed a significant increase in the rate of complications during fragment discharge after successful stone disintegration. This finding is consistent with a relatively high rate of auxiliary procedures needed and a prolongation of the hospitalization time.[8–10] Thus, the choice between a primary percutaneous procedure or a combined shock wave/percutaneous procedure must be made.

In the treatment of approximately 20 percent of large and complex stones, experience with both ESWL and endourology is required to achieve a high success rate in a minimally invasive fashion (Fig. 15-5).[8–13] By no means should a urologist without a strong background in endourology select and treat patients with extracorporeal shock wave lithotripsy.

## IMPROVEMENTS IN THE LITHOTRIPTER

Recently, there has been a multitude of rumors about a "second-generation" lithotripter. From the accessible information to date, it can be concluded that these devices are still in an early experimental phase. Development of such a device is a project of great complexity. Also, it must be reiterated that there is already a device available which does the "stone cracking job" quite well!

Today, the question is what can or should be changed or improved in extracorporeal shock wave lithotripsy. Naturally, the question also arises of what changes are reasonable to anticipate. For the time being, it is surely premature to anticipate fancy gadgets such as a lithotripter for the doctor's carry bag.

All lithotripters (operational models as well as experimental devices) have four main features: the energy source, the focusing device, a coupling medium, and a stone localization system. Generally, the use of shock waves as an appropriate means for contact free disintegration of urinary stones is unquestioned. The same holds true for the elliptical reflector that focuses the energy onto the targeted stone. In addition water has been found to be an ideal coupling medium, and shock waves generated by the use of a spark discharge of a capacitor in a water tub have proven to be safe, reliable and reproducible.

The spark electrode, however, erodes with repeated use, leading to a slight decline in available energy. To maintain a constant output of energy, the spark electrode must be replaced after a specific number of cycles. When this rule is followed the spark electrode can provide a constant energy output. None of the groups working with the lithotripter has yet experienced any technical problems with the spark electrode. Nevertheless, the search for an alternative energy source to generate the shock waves continues.

Laser energy can provide a reliable peak pulse pressure, average pulse pressure, and pulse frequency; however, underwater use is still relatively unexplored. When any energy source is used to generate shock waves, pseudocavitations or minute pollution created in the coupling medium have a relatively strong adverse impact on the propagation of the energy, thus making the system apt to malfunction. Comparison of equipment and the maintenance costs indicates that the more reliable but overpriced spark electrode might be still less expensive in the long run (Table 15-7).

Laser-induced explosion of lead acid pellets as a means of generating shock waves is used in one experimental device. The energy produced is relatively high and inconstant, which results in rather large stone fragments compared to the Dornier lithotripter. Moreover, problems

**Table 15-7. Comparison of Energy Sources Proposed for ESWL**

Objectives:
Longer lasting
Constant energy output
Less expensive

| Energy Source | Long Lifespan | Constant Energy Output | Inexpensive |
|---|---|---|---|
| Spark electrode | − | + | +/− |
| Laser beam | + | +/− | − |
| Lead acid pellets | − | +/− | +/− |
| Piezoelectric | + | (−) | + |

with the handling, transportation, and the storage of the explosive lead acid pellets may preclude distribution of the device.

Generation of shock waves by a piezoelectric discharge has been attempted by another group and is currently utilized in an experimental device. Again, the energy output is inconstant, and the early reported results do not suggest that this might soon become a reliable and reproducible device for routine clinical use.

The second question to be asked is what can be achieved in terms of cost savings and reduced x-ray exposure by substituting an ultrasound localization system for the present x-ray system. The cost of any x-ray equipment employed to reliably localize stones and to monitor the stone disintegration process is much higher than that of ultrasound equipment.

The radiation incurred by the patient during one treatment session of ESWL is in the range of 5 to 20 rads and depends very much on the operator's personal experience. This is slightly less than the radiation delivered to the patient during percutaneous stone surgery and is in the range of radiation delivered during ureteroscopy. Of course, during shock wave lithotripsy there is no radiation exposure to the operator.

In Munich between 1975 and 1977, ultrasound was used as a means of stone localization and for monitoring disintegration. These experiments showed that there was no way to rely on ultrasound as the primary system for sound localization and monitoring of the disintegration process. Similar results were found more recently in Stuttgart with investigations in more than 500 ESWL patients.

Using the most sophisticated ultrasound equipment and imaging techniques now available, the initial localization of the stone is possible for all stones larger than 5 mm when the stone location is unchanged from that on the KUB film taken immediately prior to ESWL treatment.

Proper assessment of the degree of stone disintegration during the process, however, is not possible. Furthermore, particles created during the process of stone disintegration smaller than 5 mm cannot be localized with the ultrasound system when no KUB film is available for comparison (Table 15-8).

**Table 15-8. Experimental Results of Ultrasound Stone Localization**

Objectives:
- Less expensive
- Less x-ray exposure

Experimental results (Munich 1975–1977, Stuttgart 1983–1985):
- Primary localization and focusing possible
- Proper assessment of the degree of disintegration not possible
- Particles <5 mm cannot be localized

Various groups advocating the use of ultrasound systems as a means of reducing hardware expenses and radiation exposure do in fact rely on ultrasound for stone localization. However, to assess the degree of stone disintegration and to determine when to terminate the procedure, they still need to use mobile x-ray equipment. Thus, for the time being, there is no ultrasound device available that would enable the urologist to make the necessary discriminative judgments during treatment.

A tubless lithotripter has also been mentioned as a possible future development that would be advantageous. Patient handling would be easier if the patient did not have to be immersed, and the hypotensive effect of immersion on the cardiovascular system would be eliminated. However, the initial Munich experience with a waterbag system has shown that there are numerous technical problems precluding the routine use of closed systems (waterbag) as opposed to the tub (open system) (Table 15-9). Generation of the shock waves in a closed system results in the formation of air bubbles, which in turn cause problems for effective transmission of the shock waves. The generation of air bubbles must be avoided by all means. Otherwise, the shock wave energy cannot be propagated into the body because it will be absorbed in the coupling medium due to the different acoustic impedances between water and air. Furthermore, the formation of air bubbles in a closed system leads to pseudocavitations, which in turn lead to secondary pressure waves which interfere with the propagation of the original shock wave. Even if all these problems could be

**Table 15-9. Experimental Results with Closed Energy Coupling Devices for Use with the Lithotripter**

Objectives
- Easier to handle
- Needs less space
- Less hypotensic effect on cardiovascular system
- Initially less expensive

Experimental experience (Munich)
- Difficulties in applying the coupling device to the body
- Difficulty in finding appropriate material for the coupling device
- Problems of pressure balance in a closed system
- Problems of achieving air-free coupling of the shock waves
- Problems of secondary pressure waves due to pseudocavitation

solved, there would be still difficulties in applying the coupling device to the body without any air between the device and the body.

In summary, all of these efforts to change the existing device to improve its performance have not yet shown any substantial results. More time, research, and some trial and error must surely be expected before any major equipment changes can be introduced into routine clinical use. As always, the designer of a new lithotripter must give proof of its effectiveness, safety, and clinical reliability.

The Dornier lithotripter, after 8 years of research and development, has proven its clinical efficacy with more than 250,000 successfully treatments in 6 years. During this relatively short period of time the management of urinary stone disease has completely changed, and ESWL has almost completely surplanted open surgical and many endourologic procedures. Urologists using the device, as well as patients undergoing lithotripsy instead of surgery, have learned to appreciate the outstanding complication-free performance of this device.

## REFERENCES


1. Chaussy C, Brendel W, Schmiedt E: Extracorporeally induced destruction of kidney stones by shock waves. Lancet 2:1265, 1980
2. Chaussy C, Schmiedt E, Jocham D et al: First clinical experience with extracorporeally induced destruction of kidney stones by shock waves. J Urol 127:417, 1982
3. Chaussy C, Schmiedt E, Jocham D: Nonsurgical treatment of renal calculi with shock waves. In Roth, RA, Finlayson, B (eds): Stones, Clinical Management of Urolithiasis. Williams & Wilkins, Baltimore, 1982
4. Chaussy C, Schmiedt E: Shock wave treatment for stones in upper urinary tract. Urol Clin N Am 10:743, 1983
5. Chaussy C, Schmiedt E: Extracorporeal shock wave lithotripsy (ESWL): An alternative to open surgery? Urol Radiol 6:80, 1984
6. Chaussy C, Schmiedt E, Jocham D et al: Extracorporeal shock wave lithotripsy (ESWL) for treatment of urolithiasis. Urology 5:59, 1984
7. Chaussy C, Fuchs G: World experience with extracorporeal shock wave lithotripsy for the treatment of urinary stones: Assessment of its role after 5 years of clinical use. Endourology Newsletter 1:5, 1985
8. Chaussy C, Fuchs G: Extracorporeal shock wave lithotripsy (ESWL) for the treatment of urinary stones. In Gillenwater J (ed): Textbook on Adult and Pediatric Urology. Year Book, Chicago, 1986
9. Chaussy C, Schmiedt E, Jocham D et al: Clinical experience with extracorporeal shock wave lithotripsy. p. 95. In Chaussy C (ed): Extracorporeal shock wave lithotripsy, 2nd Ed. Karger, Basel, 1986
10. Eisenberger F, Fuchs G, Miller K: Extracorporeal shock wave lithotripsy and endourology: An ideal combination for the treatment of kidney stones. World J Urol 3:41, 1985
11. Fuchs G, Miller K, Rassweiler J: Alternatives to open surgery for renal calculi : Percutaneous nephrolithotomy and extracorporeal shock wave lithotripsy. p. 216. In Schilling, W (ed): Klinische und Experimentelle Urologie. Zuckschwerdt, Munich, 1984
12. Fuchs G, Miller K, Rassweiler J, Eisenberger F: Extracorporeal shock wave lithotripsy: One-year experience with the Dornier lithotripter. Eur Urol 11:145, 1985
13. Miller K, Fuchs G, Rassweiler J, Eisenberger F: Treatment of ureteral stone disease: The role of ESWL and endourology. World J Urol 3:445, 1985
14. Schmiedt E, Chaussy C: Extracorporeal shock wave lithotripsy of kidney and ureteric stones. Urol Int 39:193, 1984
15. Wickham J, Webb D, Payne S et al: ESWL: The first 50 patients treated in Britain. Br J Urol 290:1188, 1985

# 16

# Thou Shalt Not Cut for Stone

Robert A. Riehle, Jr.

The old man's eyes told me more than he was allowed to. With the collar turned up on his white coat, he watched me slip on the woolen booties to protect the museum floor, and then gestured for me to follow him. In the first room of the Semmelweis Museum in Budapest, he explained in broken English that the dioramas ahead presented the origins of surgery and the men who had perfected the craft. He would introduce me to the famous Hungarian and European surgeons and then place them in proper historic perspective. He shuffled from room to room, where etchings of Risolius, Paré, and Semmelweis shared cabinets with antique surgical instruments, military amputation kits, trephines, and bloodletting knives. These were respected, brave, and studious men, he said. Of course—and he looked at me as a friend for the first time—surgeons have always had a divine and developed gift that patients do not have to understand to appreciate. Each worked in his own way, yet each shared a specific—he stumbled over the word as a new energy reflected from his eyes—a special design to harness a bit of science to assure his patients some relief from suffering and a better quality of life.

A cornucopia of new technology is now available for management of urologic stone disease. New operative procedures for symptomatic urolithiasis have evolved over years rather than decades, and urologists have learned new invasive techniques only to find them rapidly replaced by newer technologic developments.

Anatrophic nephrolithotomy, extended pyelolithotomy, and coagulum pyelolithotomy were surgical techniques that increased surgical success rates and decreased the operative loss of kidneys. Intraoperative nephroscopy with rigid and flexible instruments soon was replaced by the percutaneous approach for extraction and ultrasonic disintegration of renal and ureteral calculi of all sizes and compositions. The rigid ureteroscope has been passed transurethrally for diagnostic examination as well as therapeutic stone manipulations, and prototype flexible ureteroscopes are currently being investigated clinically.

Yet now the nonsurgical and noninvasive technique of ESWL, developed in West Germany, represents the wave of the future for treatment of renal and upper ureteral calculi. How has it affected the way urologists care for their patients? ESWL is probably only the first of many technologies that will replace surgery and consequently alter the traditional role of the surgeon. Both surgeons and patients must allow a new style of doctor–patient relationship to evolve.

Shock wave lithotripsy has generated waves that have reverberated throughout the health care industry, far beyond the few hundred thousand stone-forming patients. It has irrevocably altered the way patients look at doctors, the way urologists relate to their patients, and the government's view of both.

Patients behave differently these days. They now receive a great deal of their medical information from the media, and are better informed and more involved than ever. However, the "miracles" of modern technology are often reported simplistically and smugly by the press. These superficial, often theatrical presentations establish a perception in the reader's or viewer's mind against which subsequent advice and qualifications by physicians are evaluated. At first, patients were skeptical and fearful of the "blitz bath." Accompanied by experimental protocols and surrounded by anxious, disbelieving doctors observing lithotripsy for the first time, patients required constant support to select and undergo lithotripsy. Now, patients demand the "dunk in the bathtub" for all symptomatic stones. ESWL has gone from an experiment to a right. Patients will not accept their urologist's advice that their stone may be too big or the risk of urosepsis too high. They refuse surgery, they refuse percutaneous tubes, they refuse inpatient therapy. Their expectations exceed even the reported, reproducible success rates, and yet, luckily, few are disappointed.

However, patients not educated as to the realistic expectations for their individual stone are bound to be disappointed. In specific, patients who have never before passed a stone, or who have never had stone surgery, lack a prior experience for comparison, and seem to have more trouble coping with episodes of colic or secondary procedures.

The arrival of ESWL thus represents one more wrinkle in the map of the practice of urologic surgery as taught during the residencies of the 1950s through the 1970s. Heaped upon the recent realization that modern practice involves second opinions, managed care, DRGs, and HMOs is the rediscovery of a basic medical school tenet—that the practice of medicine "rests upon a constantly shifting foundation of knowledge." Currently, our foundation has become more and more closely linked with technologic advances.[1] Surgeons must adapt to this evolution to maintain credibility with their patients.

What does the patient think of his newly evolved physician? It is only recently and in industrialized countries that people regularly reach adulthood without being severely ill or experiencing the death of a family member in their own home. Because of recent advances known to all, serious illness in the first 30 years of life has become less and less frequent. As a result, previously healthy people are unfamiliar with distressing illnesses and the "sick role," and when they fall ill they are plunged unprepared into the full rigors of modern technologic medicine.[2] They are annoyed by their situation and want fast and easy remedies. There is also mounting evidence that younger patients will accept rapidly available high-tech medical care without a continuing relationship with a personal physician. They are, in short, seeking and accepting efficient service rather than the individualized care of a personal physician.[3]

Patients undergoing ESWL are sometimes not quite sure how and what the urologist does. After traditional surgery, daily visits to the surgeon for incision and drain care established a professional yet personal doctor–patient relationship which assured outpatient follow-up in most cases. Now, patients are referred to lithotripsy centers for treatment by a machine which is operated by technologists, x-ray staff, and the urologist. Outpatient treatment, or referral to an ESWL center with "button pushers" rather than endoscopists, confuses the patient as to the professional versus the technical part of the treatments. The patient often spends more time with the anesthesiologist, the nurses, and the room technician than he does with the urologic surgeon. Patients with complications are quick to get in touch with their urologist, but for most patients, the ease and completeness of particle passage makes them satisfied customers of lithotripsy, but not very motivated to return to their urologist for follow-up. Since a recurrent stone can be treated in the same way, lithotripsy has lessened the fear associated with repeat operations; the interest of patients in prevention and metabolic evaluation has waned somewhat.

Urologists, perhaps more than any other specialists, have had to come to grips with the impact of technology on their field. Wave energy, invisible to the human eye, now disintegrates stones to sand. The advent of these instruments has necessitated retraining of urologists and a closer working relationship between users and developers of medical technology. Both physicians and engineers have benefited from the exchange. The joint objectives of bioengineering and biotechnology were summarized quite effectively recently by Eckhard Polzer of Dornier Medical Systems, Inc.[4]

> In our industry, we are not dealing with numbers where the speed of calculation is the determining performance criteria. We are dealing with people. Their needs, their size, their fears determine our product. Our systems, therefore, are a complex mixture of mechanical and electronic components, all geared toward serving a human being in the most direct way possible, through therapy in the hands of a responsible doctor. These are the considerations that have driven our present design and will determine any future solution as well. Our goal cannot be technical perfection, but, rather, an optimal solution that provides the best compromise between patient safety, handling qualities, and economic feasibility.

Many urologists have become involved in the financing and marketing of ESWL. In response to poorly researched and arbitrary reimbursement policies for new technologies, groups of urologists have joined with financial entrepreneurs to design innovative means to make lithotripsy available to their patients, in either hospital or outpatient settings. Much of this innovation has been done using private capital. Sometimes through coercion, sometimes through foresight, regional urologists have cooperated to use a centrally located machine, thus avoiding needless and unproductive competition. The rapid response of the industry specializing in health service products has produced free-standing lithotripsy clinics, clinics associated with hotels, mobile lithotripters, and a cornucopia of endoscopic stents and supplies designed specifically for ESWL.

In answer to those who say physicians should not be involved in such ventures, it is certainly true that lithotripters would not be distributed throughout the U.S. and serving patients successfully if private capital and the entrepreneurial spirit had not prevailed. Physician collectivism in this setting is an example of the manner in which new technologies may be acquired and implemented in the future.

The ethics of the urologist-operators will come under increasing scrutiny. The doctor's susceptibility to conflict of interest is inversely proportional to his or her own integrity. Ethical physicians will continue to act and to conduct their affairs in the best interest of patients, and they will welcome objective review of their results and methods. Evaluation of performance is particularly challenging in this field where objective quantification of results must take into account changing indications, nomadic patient bases, and cooperative data pools.

Finally, the urologist must interact with the government and third-party payers, especially if he or she is associated with an ambulatory unit. Given the recent interest in resource consumption and utilization patterns for new health care technology, ESWL has been closely studied to assess true costs of the treatments and for cost-benefit analyses. Decisions by hospital and government administrators to adopt new technology is often marked by uncertainty of third-party payment, and by disparity between true costs and reimbursement levels. Physician input can decrease the lag time between the decision of third-party payers to cover ESWL and the determination of appropriate reimbursement levels. This will reduce the financial risk for those acquiring the new technology.[5]

The new urologist-operator of shock wave technology has an increased responsibility to all concerned. Has his or her job become easier or harder? Now he must interpret for his patients the technologies derived from the cooperation of engineers, physicists, and physicians. He must develop and communicate his role in the selection and safe application of these procedures.

As he watches the effect of technology on the disease, the urologic surgeon cannot lose sight of the person with the disease. As he bridges the area of potential economic conflict, he cannot forget that the essential qualities of a clinician are virtue and interest in humanity. The secret of patient care is caring for the patient, and the surgeon must accommodate to and manage change for the patient's benefit. The urologist must continue to seek new organizational forms as an opportunity not only to increase efficiency, but to maximize the responsiveness of the system to the patients. Effective decentralization and personalization of care can occur through the new technologies of information management systems.[6]

Nevertheless, the "old order" in health—the proverbial doctor–patient relationship—must be appreciated—modernized, yes, but preserved. Doctors must seek the means to ensure that the precious is not lost with the new.[7] Physicians must become expert at communicating with professionals, patients, the public, and politicians about the great improvements and the unexpected consequences of the treatment of illness and the maintenance of health. The clear presentation of new technologies and their effect on the American people, as well as greater awareness on the part of patients of the involvement of government and corporate authority, will help prevent erosion of the public image of the physician.

Physicians must continue to manage an art that currently transcends what science can provide. Through technology and knowledge, the surgeon has an "aura of healing power," which through his or her personality can be brought to bear on the patient's illness.

> The personality factors, the practice of the art of medicine, can reflect well on all of us. We are the direct heirs of the great physicians of the past. Their history of medicine is longer than that of any nation, any dynasty, or any culture. Throughout all but the most recent past, the great and historic physicians who gave luster to our pedigree, no matter how learned, how dedicated, or how skillful they were, had to rely primarily on their personality factors to help or cure their patients. They practiced mainly the art of medicine. They, our professional ancestors, accomplished much that was good by the best use of their personal powers in the practice of their art. It is incumbent upon us, without slighting our scientific advantages, to save and use the best of their art and make it ours.[7]

Endourology was not the end of urology, just as the lithotripter has not replaced the urologist. Shock wave lithotripsy has forced a reevaluation of the urologist's role—a case of a scientific advance shading and directing the art of the surgeon—and the urologic surgeon will emerge as a possibly better physician with a more keen *tactus eruditus*.

> Each surgeon worked in his own way, yet each shared a specific—he stumbled over the word as a new energy reflected from his eyes—a special design to harness a bit of science to assure his patients some relief from suffering and a better quality of life!

## REFERENCES


1. Fredrickson D: On the cultivation of virtue. Mayo Clin Proc 58:660, 1983
2. Osmond H: God and the doctor. New Engl J Med 302:555, 1980
3. Brazil P: The changing face of medicine: or, art versus science revisited. Hosp Pract, November 15, 1985
4. Polzer E: Dornier Lithotripter status report. Read at the second Symposium on Extracorporeal Shock Wave Lithotripsy, Indianapolis, Indiana, February, 1986
5. Report, National Health Services and Practice Pattern Survey, Institute for Health Policy Analysis, Georgetown University Medical Center.
6. Richardson E: "The old order Changeth, yielding place to the New." Perspectives on the health "revolutions." N Engl J Med 291:283, 1974
7. Vickery E: Our art, our heritage, JAMA 250:913, 1983

# Index

Page numbers followed by *f* represent figures; those followed by *t* represent tables